# Challenges and Perspectives of Social Networks within Social Computing

# Challenges and Perspectives of Social Networks within Social Computing

Guest Editors

**Maria Chiara Caschera**
**Patrizia Grifoni**
**Fernando Ferri**

MDPI

Basel • Beijing • Wuhan • Barcelona • Belgrade • Novi Sad • Cluj • Manchester

*Guest Editors*
Maria Chiara Caschera
Institute of Research on Population and Social Policies (IRPPS)
National Research Council (CNR)
Rome
Italy

Patrizia Grifoni
Institute of Research on Population and Social Policies (IRPPS)
National Research Council (CNR)
Rome
Italy

Fernando Ferri
Institute of Research on Population and Social Policies (IRPPS)
National Research Council (CNR)
Rome
Italy

*Editorial Office*
MDPI AG
Grosspeteranlage 5
4052 Basel, Switzerland

This is a reprint of the Special Issue, published open access by the journal *Big Data and Cognitive Computing* (ISSN 2504-2289), freely accessible at: https://www.mdpi.com/journal/BDCC/special_issues/Network_Social_Computing.

For citation purposes, cite each article independently as indicated on the article page online and as indicated below:

Lastname, A.A.; Lastname, B.B. Article Title. *Journal Name* **Year**, *Volume Number*, Page Range.

**ISBN 978-3-7258-4769-3 (Hbk)**
**ISBN 978-3-7258-4770-9 (PDF)**
**https://doi.org/10.3390/books978-3-7258-4770-9**

# Contents

# About the Editors

**Maria Chiara Caschera**

Maria Chiara Caschera is a researcher at the National Research Council (CNR), specifically within the Institute for Research on Population and Social Policies (IRPPS) in Rome, Italy. Her research primarily focuses on the field of human–computer interaction (HCI), where she designs relevant models, tools, and applications aimed at improving usability, accessibility, and user experience across various contexts. Her main research interests include multimodal interaction, social networks, sentiment analysis, emotion recognition, fake news detection, machine learning, social computing, and user-centered design techniques, as well as collaborative, co-creative, and participative approaches. Throughout her work, she strives to integrate both theoretical and practical innovative aspects. She is the author of over 70 papers published in international journals, conference proceedings, and book chapters. She is the scientific director of the CNR for European projects. She has participated in the Organizing Committee of various international workshops and conferences, and she is involved in numerous program committees for national and international conferences and workshops. She is an Associate Editor for international journals, she is involved in the editorial board of various international journals, and she has been a Guest Editor of various special issues of international journals.

**Patrizia Grifoni**

Patrizia Grifoni graduated in Electronic Engineering in 1990 at the University of Rome "La Sapienza". She is a director of research, and she has worked at the National Research Council since 1990. From 1993 to 2000, she was an adjunct professor of "Image Processing" at the University of Macerata. She is a professor of "Software Engineering and Object Oriented Programming" at the Uninettuno International Telematic University. She is the author of over 250 articles in journals, books, and conferences. She has been the scientific director of the CNR for numerous European and national projects. She leads and participates in numerous national and international research projects. She has experience in organizing numerous international scientific conferences and seminars; she is a member of some program committees of international seminars and conferences, and a member of the editorial board of some international journals. Her main research areas of interest are social computing, social computing, human–machine interaction, multimodal interaction, sketch-based interfaces, multimedia applications, user modeling, databases and knowledge, ontologies, machine learning, statistical databases, spatial information, geographic information systems, responsible research and innovation, ethics, risk management, digital ecosystems, web applications, web 2.0, Internet of the future, and social networks.

**Fernando Ferri**

Fernando Ferri has been the director of research at the National Research Council of Italy since 2001. He was a researcher from 1990 to 2001. He was a contract professor from 1993 to 2000 of "Sistemi di Elaborazione" at the University of Macerata. He is the author of more than 300 papers in international journals, books, and conferences. He received a degree in electronic engineering in 1990 and a Ph.D. in medical informatics at the University of Rome "La Sapienza" in 1993. He has coordinated and participated in several national and international research projects. He has organized several international events (scientific conferences and workshops). His main research areas of interest are social informatics, social computing, responsible research and innovation acceptance, addiction, social media and social networks, e-learning, digital ecosystems, ontologies, machine learning, statistical databases, geographic information systems, data and knowledge bases, human–machine interactions, user–machine natural interactions, user modeling, visual interactions, sketch-based interfaces, geographic information systems, risk management, and medical informatics.

# Preface

Social networks enable individuals to connect and share their thoughts, opinions, and emotions through various forms of content, including documents, photos, and videos. This widespread use of social networks has led to the generation of vast amounts of data concerning conversations, text, audio, and video—referred to as multimodal data.

The significance of this data lies in its sheer volume, the diversity of topics covered, and the constantly evolving nature of the language used.

The significant amount of data presents both technological and social challenges. Specifically, it can be leveraged to tackle emerging issues such as fake news and security threats. This can be achieved using a range of methodologies, technologies, and tools related to various fields including human–computer interaction, social media and artificial intelligence, social networks and big data, virtual and augmented reality, e-learning environments, the Internet of Things, security, entertainment, video indexing and retrieval, and communication systems for crisis management (such as during pandemics, wars, and natural disasters).

To address these challenges, methods and tools like machine learning, deep learning, emotion recognition, fake news detection, pattern recognition, semantic knowledge discovery, social network mining, text mining, multimedia data mining, and studies in social and educational contexts can be employed.

This Special Issue focuses on the methods, tools, and applications related to analyzing sentiment scope in social networks and the examination of the massive amount of text data generated by social media for evidence-based policy-making. It discusses the use of web data scraping to gather information from TripAdvisor's social pages for smart tourism consultancy. Additionally, the Special Issue explores the implications of Social Intelligence Mining, which enhances our understanding of the dynamics of social phenomena, influences strategies, fosters innovation, and adds value across various sectors. The impact of digital transformation on Korean society is also analyzed by examining social media data to assess the societal and economic effects triggered by advancements in digital technology. Furthermore, this Special Issue investigates the influence of social media on faculty job performance, specifically through the mediating roles of internal communication and teamwork. The Special Issue also presents an end-to-end learning model for selecting the most influential nodes in a social network, an effective multimodal emotion recognition system that enhances emotion recognition efforts focusing on both audio and text modalities, and a comprehensive review of cyber-hate sentiment analysis in multiple languages. Lastly, it introduces a domain-specific strategy within a multi-domain fake news detection framework designed to address class imbalance in fake news detection across various domains.

**Maria Chiara Caschera, Patrizia Grifoni, and Fernando Ferri**
*Guest Editors*

*big data and cognitive computing*

*Article*

# A Space-Time Framework for Sentiment Scope Analysis in Social Media

**Gianluca Bonifazi, Francesco Cauteruccio, Enrico Corradini, Michele Marchetti, Luigi Sciarretta, Domenico Ursino * and Luca Virgili**

Department of Information Engineering (DII), Polytechnic University of Marche, Via Brecce Bianche 12, 60121 Ancona, Italy
* Correspondence: d.ursino@univpm.it

**Abstract:** The concept of scope was introduced in Social Network Analysis to assess the authoritativeness and convincing ability of a user toward other users on one or more social platforms. It has been studied in the past in some specific contexts, for example to assess the ability of a user to spread information on Twitter. In this paper, we propose a new investigation on scope, as we want to assess the scope of the sentiment of a user on a topic. We also propose a multi-dimensional definition of scope. In fact, besides the traditional spatial scope, we introduce the temporal one, which has never been addressed in the literature, and propose a model that allows the concept of scope to be extended to further dimensions in the future. Furthermore, we propose an approach and a related set of parameters for measuring the scope of the sentiment of a user on a topic in a social network. Finally, we illustrate the results of an experimental campaign we conducted to evaluate the proposed framework on a dataset derived from Reddit. The main novelties of this paper are: (i) a multi-dimensional view of scope; (ii) the introduction of the concept of sentiment scope; (iii) the definition of a general framework capable of analyzing the sentiment scope related to any subject on any social network.

**Keywords:** spatial scope; temporal scope; sentiment analysis; social network analysis; Reddit

## 1. Introduction

Suppose we are at the shore of a lake on a becalmed day with a flat lake surface. Suppose now that we throw a stone into it. We can see how, starting from the point where the stone falls, the water begins to ripple, and small waves are created. These waves are higher near the point where the stone fell while they become smaller and smaller as we move away from it, until they disappear. The heavier the stone thrown, the higher the waves and the farther they propagate. As time passes, the height of the waves tends to decrease until, if no more stones are thrown into the lake, they disappear and the lake surface becomes motionless again. In our opinion, this image describes better than any other what is meant by "scope". From a more formal point of view, in the Concise Oxford Dictionary (Concise Oxford Dictionary—https://en.oxforddictionaries.com (accessed on 15 September 2022)), *scope* is defined as "*the extent of the area or subject matter that something deals with or to which it is relevant*".

Certainly, there are several similarities between the concept of scope and some other ones used in sociology. Consider, for example, the concepts of centrality, reliability, power, reputation, influence, trust, diffusion, etc. In fact, scope goes beyond these concepts and, at the same time, embraces all of them. In fact, they can be seen as different aspects of scope, which certainly exert their influence on it.

Scope has already been studied in past literature. For example, Ref. [1] analyzed the scope of users and hashtags in Twitter, while [2] proposed an approach to compute the scope of a smart object in a Multi-IoT context. In Refs. [3–8], the authors presented approaches to analyze some aspects of scope (e.g., reliability, trust, and influence) for users and/or hashtags. In Ref. [9], the authors studied the distribution of the influence of a user across the network, while in Ref. [10], the authors analyzed the attractiveness of users in networks.

In the social network context, which we focus on in this paper, another much analyzed concept is that of user sentiment. Sentiment analysis is one of the most active research strands regarding social networks and, more generally, artificial Intelligence and data analysis [11–14]. In fact, nowadays, millions and millions of people express their sentiments on the most disparate topics through social networks [15–18]. The knowledge of such sentiments and their evolution in space and time is a valuable source of information for various professionals, such as marketers, politicians, journalists, decision makers, and so on. Finally, knowing how a user's sentiment about a topic can propagate to her neighbors, the neighbors of her neighbors, etc., and how that propagation evolves over time, and being able to measure this through appropriate techniques and metrics, represent challenging issues with enormous practical implications.

One way to address these issues might be integrating the concepts of scope and sentiment of one or more users on a certain topic. In fact, to the best of our knowledge, there is not yet an approach that integrates the concepts of scope and sentiment and treats scope from both a spatial and a temporal point of view (with further points of view, or dimensions, likely to emerge in the future). This paper aims to address this issue by proposing a model and a related approach to investigate the spatial and temporal scope of a user's sentiment about a topic in a social platform.

Our proposed model is based on two graphs. The first is a bipartite one that stores all available information about users, their posts, comments, and sentiments on certain topics. It can be employed as an information source from which it is possible to extract all the necessary data for the processing required by the approach proposed in this paper and any future approaches. The second is derived from the first; it is a single-mode graph representing users and their interactions and is employed for the analyses proposed in this paper. Our model also includes several complement functions that can be exploited to obtain specific information from the first graph or to perform certain supporting processing.

Our approach consists of several steps. First, it identifies the topics emerging from the posts and comments published by users. Then, for each topic, it determines the sentiments of the various users who covered it. For these two activities it employs techniques already proposed in the literature [11,13,19–22]. In other words, for these two activities, our approach is independent of the techniques adopted to perform them. Next, it exploits the concepts, measures, and techniques of Social Network Analysis [23] and graph theory [24–26] to define spatial scope. This definition has a dual nature, since the spatial scope is defined as a set of pairs or a rooted graph. Both definitions aim to indicate the users involved in the spread of sentiment and the intensity degree of the latter for each user involved. The two-fold definition of spatial scope allows us to use both set theory and graph theory for its analysis. Using them, our approach provides a set of metrics and measures for assessing the spatial scope of a user's sentiment on a topic.

In a completely similar way, our approach acts to ensure the temporal scope assessment of a user's sentiment on a topic. In this case, the temporal scope is defined as an ordered list of pairs. As mentioned above, spatial and temporal scopes are orthogonal and represent two views or dimensions that could be integrated, and possibly enriched with other views in the future.

To evaluate the potential of the proposed framework, we conducted a series of experiments on a dataset derived from Reddit, one of the most popular social networks. As we will see in the paper, our model proved to be capable of extracting interesting information that may be useful to various professionals interested in the knowledge of scope.

In summary, the gap in the literature that this paper aims to fill concerns the lack of a general framework capable of supporting a multi-dimensional analysis of the scope of the sentiment of users on any topic in any social network. In fact, as will become clearer in the following, some papers in the past literature analyze the scope of a user or a smart object, others investigate user sentiment, but none analyze the scope of the sentiment of one or more users. Moreover, all previous papers that analyze scope take only the spatial dimension into account and none of them consider other dimensions, such as the temporal one. Finally, most of the studies on sentiment refer to either a well-defined social network or a well-defined subject, and were not designed to operate on any social network and any subject. Our paper aims to fill this gap. Specifically, its contributions are as follows: (i) it introduces the concept of scope of a user's sentiment on one or more topics in a social platform; (ii) it proposes a multi-dimensional definition of scope; and (iii) it presents a framework for studying the scope of a user's sentiment on one or more topics and extracting information from the corresponding data; this framework can operate on any social platform and can evaluate the scope of the user sentiment on topics concerning any subject.

The outline of this paper is as follows: In Section 2, we illustrate the Related Literature. In Section 3, we describe our model for scope representation. In Section 4, we present our approach. In Section 5, we illustrate the experimental campaign we conducted. Finally, in Section 6, we draw our conclusions and look at possible future developments.

## 2. Related Literature

### 2.1. Preface

In recent years, social networks have been exploited by groups of users to discuss a variety of topics, from politics to gossip, from health to sport, and so on. The pervasive spread of social networks has prompted researchers to study the behavior of users who join social networks in various reference contexts [27–29]. Alongside the analysis of the structure of networks, which already provides very interesting knowledge patterns about the behavior of users accessing them, researchers have also begun to examine the content posted and exchanged between users [30–33]. Regarding the latter, elements of particular interest to them are the extraction of topics from texts and the assessment of the sentiment that users have about a given topic. Our work aims to integrate these two research streams because it aims to analyze how the scope of the sentiment of a user on a specific topic propagates both spatially and temporally across a social network.

To better perform our analysis, we divide this section into two parts. In the first, we analyze works dealing with scope and related concepts, while in the second we focus on the analysis of sentiment diffusion within a community.

### 2.2. Related Literature on the Concept of Scope

In the past, the scope of users in social networks has been studied in Ref. [1]. In this paper, the authors analyze the scope of an entity on Twitter. Specifically, they define a framework to measure various aspects of scope (e.g., influence, reliability, popularity) simultaneously and for multiple entities (e.g., users, hashtags). In this way, they can measure the overall scope and several aspects of it by comparing the latter with each other and for different entities. Such comparisons make it possible to extract knowledge patterns (indicating, for example, the presence of anomalies and outliers) that can then be used in several application domains (e.g., information dissemination). In Ref. [2], the authors extend the concept of scope from people to smart objects in a multiple IoT context. In particular, they formalize two scope definitions for smart objects and illustrate some real-world applications of the knowledge patterns thus extracted. Returning to the people-to-people social network context, many authors analyze single aspects of the concept of scope, such as reliability, trust, and influence for users and/or hashtags [3,5–7]. Regarding user influence, the authors of Ref. [9] use the PageRank algorithm to analyze the distribution of influence across the network. In Ref. [10], the authors analyze the attractiveness of users in a social network. The approach they propose characterizes a user based on the new users she is able to interrelate with over time. The authors propose to perform influence maximization, but they do not consider the topics covered by users.

Other approaches investigate the evolution of topic trends in Twitter by analyzing the use of hashtags. Indeed, the latter allow a natural division of posts according to their topics [4,8,34,35]. In particular, the authors of Ref. [8] measure the topic-sensitive influence of users on Twitter by means of an approach based on PageRank. They analyze how a social network user can influence other ones on a topic. Moreover, they propose a metric to compute the influence of users on specific topics. Our approach differs from the one proposed in Ref. [8] in that the latter does not aim to describe the features characterizing the influence of one user on another but only gives a quantitative estimate of that influence. Furthermore, the approach proposed in Ref. [8] does not consider the value of sentiment in its analysis. In Ref. [4], the authors propose an approach that, given a hashtag on Twitter, uses the corresponding comments to construct its profile in order to predict its popularity. This approach has some similarities with ours. In fact, it can be seen as a method for predicting the spatial scope of a hashtag, and thus how far the latter will spread. Our approach differs from the one described in Ref. [4] in that, in investigating the spread of a user's sentiment toward a topic, it analyzes the contribution made by other users of her neighborhood. Therefore, our analysis is more user-centric than the one proposed in Ref. [4]. Furthermore, the approach of Ref. [4] does not consider sentiment when analyzing the expansion of a topic. Finally, it is not intended to propose a multi-dimensional analysis of a hashtag's popularity unlike our approach, which proposes both a spatial and a temporal scope and leaves open the possibility of further dimensions of scope in the future.

In Ref. [36], the authors present a study of the spread of negative sentiment in Online Social Media. The main similarity between the approach of Ref. [36] and ours concerns the idea of studying the spread of user sentiment in online platforms. However, the approach proposed in Ref. [36] focuses on hate speech and does not consider topics, unlike our approach, which precisely analyzes the scope of user sentiment on topics. In Ref. [37], the authors study the propagation of negative sentiment in messages exchanged on the Chinese Sina microblog. This approach is therefore specific to a social platform, and, in this feature, it differs from our approach that it has been designed to operate on any social platform. In Ref. [38], the authors investigate how the spread of sentiment can cause a viral spread of fake news in social media. This paper focuses on the analysis of a specific phenomenon (i.e., the relationship between sentiment and fake news). Instead, our paper aims to propose a general framework capable of studying the space-time evolution of the scope of user sentiment on one or more topics.

### 2.3. Related Literature on the Sentiment of Users

The second part of our analysis on related literature concerns the sentiment of a user on one or more topics. Many approaches to face this problem have been proposed in the past literature. Some of them address this issue from a static point of view; in particular, they generally employ opinion mining techniques to understand the sentiment emerging from a given text [39–42]. In contrast, other techniques address this problem from a dynamic point of view; in fact, given the characteristics of a sentiment on a topic, they want to understand how those characteristics affect the spread of that sentiment both among users and over time [43–49]. Specifically, in Ref. [43], the authors propose a model that combines sentiment and opinion propagation to assess the global sentiment on a given topic. Similarly to our approach, the one of Ref. [43] considers the sentiment emerging from a text and wants to understand how it propagates. However, the authors of Ref. [43] do not aim to provide a formalization of such propagation.

In Ref. [44], the authors present the MISNIS framework, which aims to identify the most influential users on specific topics. It also divides user messages into three categories, based on the results obtained after performing a sentiment analysis task. To carry out topic mining, it analyzes all the words in the message and not only the hashtags; thus, it is able to achieve a higher accuracy. MISNIS and our approach share the goal of analyzing how influential a user is with respect to a specific topic. However, the concepts of topic and sentiment are kept separate in MISNIS, while they are integrated into our approach because it aims to assess the sentiment of users on a topic. Topic and sentiment analysis is also the core of the approach described in Ref. [50]. This approach considers topics and sentiments from Reddit posts and comments and aims to analyze them for extracting information without creating a model.

In Ref. [51], the authors analyze and represent the spread of microblogging topics and sentiments through two graphs and propose metrics to measure the influence of stakeholders in that spread. In pursuing this goal, the approach of Ref. [51] has several similarities with ours. For instance, both are graph-based and define metrics to measure sentiment diffusion. Unlike Ref. [51], which considers topics and sentiments separately, our approach integrates these two entities because it analyzes the spread of users' sentiments on topics. The approach of Ref. [51] is based on a global analysis that examines the whole network to identify the stakeholders of interest. In contrast, our approach tends to work on partitions of the network and not on the global one; in fact, it analyzes how the sentiment of a user on a topic propagates to its neighborhood. Finally, it considers two orthogonal types of scope diffusion, namely spatial and temporal ones. This concept is not present in the approach of Ref. [51].

In Refs. [52,53], the authors analyze changes in topics and sentiment during emergencies. For this purpose, they consider different types of emergencies. In addition, they analyze the spread and propagation of topics and sentiment for different user stereotypes, e.g., governments, celebrities, and media. Our paper differs from Refs. [52,53] because it does not analyze sentiment but the scope of sentiment. Moreover, it proposes a multi-dimensional (in particular, spatio-temporal) view of scope. Finally, our framework is general and can be applied to any social network for analyzing the sentiment on topics under any circumstances. Instead, Refs. [52,53] focus on emergencies.

In particular, as far as temporal scope is concerned, we point out that various studies have been proposed in the literature, which aim to assess how several single aspects of scope (e.g., influence, reliability, popularity) evolve over time [54–57]. However, none of these approaches comprehensively consider the concept of scope but only assesses individual aspects of it. Moreover, in the reference contexts, the goals they set and the techniques they use to achieve those goals are very different from the ones adopted in our approach.

## 3. The Proposed Model

### 3.1. A Formal Representation of the Context of Interest

Before presenting our model, it is necessary to provide a formalization of the context in which it operates. This context concerns a social platform whose users can publish posts and comments. We assume that both posts and comments consist mainly of text; if there are other types of content, these are only to accompany the text. A user publishes a comment when she wants to reply to a previously published post or comment.

We employ the symbol $\mathcal{U} = \{u_1, \cdots, u_l\}$ to represent the set of users operating in our context, the symbol $\mathcal{P} = \{p_1, \cdots, p_m\}$ to denote the set of posts published by the users of $\mathcal{U}$ in a time interval $T$, and the symbol $\mathcal{C} = \{c_1, \cdots, c_n\}$ to indicate the sets of comments posted by these users in $T$. Given a user $u_i \in \mathcal{U}$, we adopt the symbol $\mathcal{P}_i$ (resp., $\mathcal{C}_i$) to denote the subset of the posts (resp., comments) of $\mathcal{P}$ (resp., $\mathcal{C}$) published by her.

As specified in the Introduction, one of the most important factors to consider in our analysis is time. Therefore, a way to model it is in order. To this end, given an overall time interval of interest, we can think of modeling it as an ordered sequence of $z$ time slices, $T = T_1, \cdots, T_z$. For example, $T$ could be a certain month, say August 2022, and it could be represented as a succession of 31 time slices, one for each day. It is advisable that our representation of time should allow the indexing of the sequence of time slices. In other words, it should be possible to select only a particular interval of contiguous time slices of $T$ (e.g., the second decade of August 2022). To this end, our time model uses the notation $T[x..y]$, $1 \leq x \leq y \leq z$, to denote the interval of contiguous time slices in $T$ that begins at $T_x$ and ends at $T_y$. If $x = y$, then it means that we want to take a single time slice; in this case, we will use the abbreviated notations $T_x$ or $T[x]$ to represent $T[x..x]$. If $x = 1$ and $y = z$, then it means we are considering the overall interval of interest; in this case, we will use the abbreviated notation $T$, instead of $T[1..z]$, to denote that interval.

The previous notation about time intervals and slices can be extended to the other sets of the model. Specifically, we denote by $\mathcal{P}[x..y] \subseteq \mathcal{P}$ (resp., $\mathcal{C}[x..y] \subseteq \mathcal{C}$) the subset of posts (resp., comments) published in the time interval $T[x..y]$ and by $\mathcal{P}[x]$ (resp., $\mathcal{C}[x]$) the subset of posts (resp., comments) published in the time slice $x$. Finally, we use the abbreviated notation $\mathcal{P}$ (resp., $\mathcal{C}$) to indicate the overall set of posts $\mathcal{P}[1..z]$ (resp., comments $\mathcal{C}[1..z]$) published in the overall time interval $T$.

Two additional concepts that play a key role in our context are the ones of topic and sentiment tag. A topic is an abstract concept discussed in one or more posts and comments. Natural Language Processing (NLP) researchers have long studied the issues of topic modeling and extraction, and proposed a variety of interesting solutions [19,20]. A sentiment tag is a keyword used to summarize the sentiment expressed on a particular topic. Typical sentiment tags are "pos", "neg", and "neu", to indicate a positive, a negative, and a neutral sentiment, respectively. In the following, we denote by $\mathcal{T} = \{t_1, \cdots, t_q\}$ the set of topics extracted from the posts of $\mathcal{P}$ and the comments of $\mathcal{C}$, while we denote by $\mathcal{S} = \{s_1, \cdots, s_r\}$ the set of available sentiments (tags). In the following, to simplify the discussion, we will use the term "sentiment" instead of "sentiment tag". Given a topic $t_j \in \mathcal{T}$ and a sentiment $s_k \in \mathcal{S}$, we use the pair $(t_j, s_k)$ to indicate that $t_j$ has been tagged with $s_k$, i.e., that $s_k$ has been associated with $t_j$.

#### 3.1.1. Identifying Topics from Posts and Comments

Our framework is independent of the technique adopted for constructing the set $\mathcal{T}$ of topics related to posts and comments. Recall that, in our context, the latter consist mainly of text and that any other content is only an accompaniment to the text. Consequently, to construct $\mathcal{T}$, we can use any approach for identifying topics from a given text proposed in the past literature (see Refs. [19–21] for some surveys on it). Therefore, in the following, we assume that, given a post $p \in \mathcal{P}$ (resp., a comment $c \in \mathcal{C}$), our framework can employ a technique capable of deriving topics from $p$ (resp., $c$), adding them to the overall set $\mathcal{T}$ of topics and associating them with the post $p$ (resp., comment $c$) from which they were derived.

#### 3.1.2. Identifying the Sentiments Characterizing Posts and Comments

Our framework is also independent of the technique used for identifying the sentiments associated with a text. This issue has been extensively studied in sentiment analysis research. Here, researchers have proposed several techniques capable of defining, characterizing, and extracting the sentiment expressed in a text (see Refs. [11,13,22] for some surveys about this topic). In this context, the terms "sentiment tag", "sentiment value", or, simply, "sentiment" have been used equivalently [58].

The technique adopted for identifying sentiments from posts and comments must examine each post (resp., comment) $p \in \mathcal{P}$ (resp., $c \in \mathcal{C}$), acquire the sentiments that emerge in it and associate them with the corresponding topics of $\mathcal{T}$ referring to $p$ (resp., $c$). In more detail, it proceeds as follows: let $p$ (resp., $c$) be a post (resp., comment) of $\mathcal{P}$ (resp., $\mathcal{C}$). It could consist of a simple text, expressing a single sentiment, or a complex text, expressing several sentiments that may even conflict with each other. Clearly, the former hypothesis is a special case of the latter; so, in the following, we will consider the latter directly. In this case, we assume that $p$ (resp., $c$) consists of a succession $p_1, p_2, \cdots, p_w$ (resp., $c_1, c_2, \cdots, c_w$) of texts such that each of them expresses a single sentiment. In what follows, we will use the term "fragment" to refer to each textual content $p_k$ (resp., $c_k$), $1 \leq k \leq w$; in addition, we will use the symbol $f_k$ to generalize $p_k$ and $c_k$. Now, the technique described in Section 3.1.1 can be applied on the fragment $f_k$ to obtain the set $\mathcal{T}_{f_k}$ of topics considered in $f_k$. Then, any technique proposed in the literature to derive the sentiment expressed in a simple text (see [11,13,22]) can be applied on $f_k$ to determine the sentiment $s_k$ characterizing it. At this point, for each topic $t_j \in \mathcal{T}_{f_k}$, we have a pair $(t_j, s_k)$ indicating that $s_k$ is the sentiment on $t_j$ expressed in $f_k$. The set of all the sentiments extracted from all the posts of $\mathcal{P}$ and all the comments of $\mathcal{C}$ form the set $\mathcal{S}$ of the sentiments that characterize our reference context.

### 3.2. The Proposed Model

After formalizing the context of interest, in this section we describe our proposed model. The first element of it is a bipartite support graph $\mathcal{B}$, aiming to enable the storage of the key information of the reference context. As we will see below, from this rather rich and complex graph, it is possible to derive more agile ones, which allow us to perform our analyses more effectively and efficiently. $\mathcal{B}$ is defined as follows:

$$\mathcal{B} = \langle N' \cup N'', E' \rangle \quad (1)$$

$N' \cup N''$ represents the set of nodes of $\mathcal{B}$. Specifically, the nodes in $N'$ are associated with users in $\mathcal{U}$. Indeed, each node $n_i \in N'$ corresponds to a user $u_i \in \mathcal{U}$, and vice versa. Since there is a biunivocal correspondence between a node of $N'$ and a user of $\mathcal{U}$, we will use these two terms interchangeably in the following. Each node $n_{jk} \in N''$ corresponds to a pair $(t_j, s_k)$, where $t_j$ is a topic of $\mathcal{T}$ and $s_k$ is a sentiment of $\mathcal{S}$. It indicates that $t_j$ has been tagged with $s_k$ in at least one post of $\mathcal{P}$ or comment of $\mathcal{C}$.

$E'$ represents the set of edges of $\mathcal{B}$; an edge $(n_i, n_{jk}) \in E'$ between a node $n_i \in N'$ and a node $n_{jk} \in N''$ indicates that the user $u_i$ published at least one post or comment in which she expressed the sentiment $s_k$ on the topic $t_j$. Since $u_i$ may have carried out this task more times in the time interval $T$, we associate a label $l_{ijk}$ with $(n_i, n_{jk})$. It indicates the list of timestamps of the posts and/or comments published by $u_i$ in which she expressed the sentiment $s_k$ on the topic $t_j$.

$\mathcal{B}$ contains all potentially useful information to enable us to investigate the context of our interest. However, being a two-mode graph, it is not easy to analyze and manipulate. Graph theory suggests constructing one or more one-mode graphs from it, each focusing on a single aspect of interest and operating on them [23]. Now, the object of our analysis is the spatial and temporal evolution of the scope of the sentiment of users in a social platform. Consequently, the key aspect to focus on consists of users and their interactions; given these premises, it is reasonable to construct a user-centered single-mode graph from $\mathcal{B}$ in such a way as to operate directly on it instead of $\mathcal{B}$. This graph is defined as follows:

$$\mathcal{A} = \langle N, E \rangle \quad (2)$$

$N$ represents the set of nodes of $\mathcal{A}$. A node $n_i \in N$ corresponds to a user $u_i \in \mathcal{U}$, and vice versa. Again, since there is a biunivocal correspondence between a node $n_i \in N$ and a user $u_i \in \mathcal{U}$, we will employ these two terms interchangeably in the following. Clearly, $N$ is equivalent to the set of nodes $N'$ of $\mathcal{B}$. $E$ is the set of edges of $\mathcal{A}$. An edge $e_{ih} = (n_i, n_h)$ belonging to $E$ indicates that the users $u_i$ and $u_h$ published at least one post/comment on the same topic and, at least once, $u_i$ published a comment on a post/comment of $u_h$, or vice versa.

As can be seen from its definition, the graph $\mathcal{A}$ is very agile and streamlined so that the analyses performed on it are effective and efficient. In some of these analyses there may be a need to use data present in $\mathcal{B}$ that we do not deem necessary to report in $\mathcal{A}$ so as not to burden this graph (for example, because such data are only rarely used). In these cases, we define ad hoc functions that retrieve the necessary information from $\mathcal{B}$ and complement our model. For example, if the set of posts on a given topic $t_j$ published by a certain user $u_i$ in the time interval $T[x..y]$ were needed, we could define a function that receives $u_i$, $t_j$ and $T[x..y]$, and returns the desired set of posts. In Section 3.2.1, we list the functions needed for our approach and that complement our model.

Given the graph $\mathcal{A}$ and a topic $t_j$ of $\mathcal{T}$, we define the projection $\overline{\mathcal{A}^j}$ of $\mathcal{A}$ onto $t_j$ as the graph obtained from $\mathcal{A}$ by considering only the nodes corresponding to users who published at least one post or comment having $t_j$ as their topic. More formally:

$$\overline{\mathcal{A}^j} = \langle \overline{N^j}, \overline{E^j} \rangle \tag{3}$$

$\overline{N^j}$ represents the set of nodes of $\overline{\mathcal{A}^j}$. A node $n_i \in \overline{N^j}$ corresponds to a user $u_i \in \mathcal{U}$ who published at least one post or comment on the topic $t_j$. $\overline{E^j}$ represents the set of edges of $\overline{\mathcal{A}^j}$. There exists an edge $(n_i, n_h)$ in $\overline{E^j}$ if there exists a corresponding edge $(n_i, n_h)$ in the graph $\mathcal{A}$.

Given the graph $\mathcal{A}$ and the time interval $T[x..y]$, we denote by $\mathcal{A}[x..y]$ the "projection of $\mathcal{A}$" in $T[x..y]$:

$$\mathcal{A}[x..y] = \langle N[x..y], E[x..y] \rangle \tag{4}$$

A node $n_i$ belongs to $N[x..y]$ if the corresponding user $u_i$ published at least one post/comment in $T[x..y]$. An edge $e_{ih} \in E[x..y]$ indicates that $u_i$ and $u_h$ published at least one post/comment on the same topic in the time interval $T[x..y]$ and that, in the same interval, at least once, $u_i$ published a comment on a post or a comment of $u_h$, or vice versa. Clearly, $\mathcal{A}[x] = A[x..x]$ is the "projection of $\mathcal{A}$" in the time slice $T_x$ and $\mathcal{A}[1..z]$ is equivalent to $\mathcal{A}$.

### 3.2.1. Functions Complementing Our Model

In this section, we present some support functions that complement our model. They will be used to formalize the activities performed by our approach. Before describing them, we feel it is appropriate to introduce some concepts concerning the prevalence or ambivalence of the sentiment of a user or a community on a topic.

Let $u_i$ be a user of $\mathcal{U}$ and let $t_j$ be a topic of $\mathcal{T}$. We define the *positive* (resp., *negative, neutral*) *sentiment degree* of $u_i$ on $t_j$ as the fraction of posts and/or comments on $t_j$ published by $u_i$ with which a positive (resp., negative, neutral) sentiment was associated after the application of the approach described in Section 3.1.2.

Having made these premises, we can now introduce our complement functions. They are:

- $\sigma^+(u_i, t_j)$: It receives a user $u_i$ and a topic $t_j$ and computes the positive sentiment degree $\delta^+_{ij}$ of $u_i$ on $t_j$. $\delta^+_{ij}$ ranges in the real interval $[0, 1]$; the higher its value, the higher the strength of the positive sentiment. $\sigma^+(u_i, t_j)[x..y]$ represents the "projection of $\sigma^+(u_i, t_j)$" in the time interval $T[x..y]$; it performs the same computation as $\sigma^+(u_i, t_j)$ but considers only the posts and comments published in the time interval $T[x..y]$. We indicate by $\delta^+_{ij}[x..y]$ the corresponding result. Clearly, $\delta^+_{ij}[x] = \delta^+_{ij}[x..x]$ is the "projection of $\delta^+_{ij}$" in the time slice $T_x$ and $\delta^+_{ij}[1..z]$ is equivalent to $\delta^+_{ij}$;
- $\sigma^=(u_i, t_j)$: It receives a user $u_i$ and a topic $t_j$ and computes the neutral sentiment degree $\delta^=_{ij}$ of $u_i$ on $t_j$. $\delta^=_{ij}$ ranges in the real interval $[0, 1]$; the higher its value, the higher the strength of the neutral sentiment. $\sigma^=(u_i, t_j)[x..y]$ represents the projection of $\sigma^=(u_i, t_j)$ in the time interval $T[x..y]$; $\delta^=_{ij}[x..y]$ denotes the corresponding result;
- $\sigma^-(u_i, t_j)$: It receives a user $u_i$ and a topic $t_j$ and computes the negative sentiment degree $\delta^-_{ij}$ of $u_i$ on $t_j$. $\delta^-_{ij}$ ranges in the real interval $[0, 1]$; the higher its value, the higher the strength of the negative sentiment. $\sigma^-(u_i, t_j)[x..y]$ represents the projection of $\sigma^-(u_i, t_j)$ in the time interval $T[x..y]$; $\delta^-_{ij}[x..y]$ denotes the corresponding result;
- $\nu(n_i, \lambda, gr)$: It receives a graph $gr = \langle \hat{N}, \hat{E} \rangle$, a node $n_i \in \hat{N}$ and a positive integer $\lambda$ and returns a set of nodes representing the neighborhood of level $\lambda$ of $n_i$ in $gr$. Formally speaking:

  $$\nu(n_i, \lambda, gr) = \{n_h | n_h \in \hat{N}, \langle n_i, n_h \rangle = \lambda\} \tag{5}$$

  Here, $\langle n_i, n_h \rangle$ represents the length of the shortest path from $n_i$ to $n_h$ in $gr$;

- $\overline{\nu}(n_i, gr)$: It receives a graph $gr = \langle \hat{N}, \hat{E} \rangle$ and a node $n_i \in \hat{N}$ and returns the set of nodes directly connected to $n_i$ in $gr$. In other words, $\overline{\nu}(n_i, gr) = \nu(n_i, 1, gr)$;
- $size(gr)$: It receives a graph $gr$ and returns its size, i.e., the number of its nodes;
- $diameter(gr)$: It receives a graph $gr$ and returns its diameter, i.e., the length of the longest shortest path between any pair of nodes in $gr$;
- $sps(u_i, t_j)$: It receives a user $u_i$ and a topic $t_j$ and returns `true` if $u_i$ has a strongly positive sentiment on $t_j$. This happens when $\sigma^+(u_i, t_j) > \sigma^-(u_i, t_j)$ and $\sigma^+(u_i, t_j) \geq \sigma^=(u_i, t_j)$. In all the other cases it returns `false`. $sps(u_i, t_j)[x..y]$ represents the projection of $sps(u_i, t_j)$ in the time interval $T[x..y]$;
- $wps(u_i, t_j)$: It receives a user $u_i$ and a topic $t_j$ and returns `true` if $u_i$ has a weakly positive sentiment on $t_j$. This happens when $\sigma^+(u_i, t_j) > \sigma^-(u_i, t_j)$ and $\sigma^+(u_i, t_j) < \sigma^=(u_i, t_j)$. In all the other cases it returns `false`. $wps(u_i, t_j)[x..y]$ represents the projection of $wps(u_i, t_j)$ in the time interval $T[x..y]$;
- $sns(u_i, t_j)$: It receives a user $u_i$ and a topic $t_j$ and returns `true` if $u_i$ has a strongly negative sentiment on $t_j$. This happens when $\sigma^-(u_i, t_j) > \sigma^+(u_i, t_j)$ and $\sigma^-(u_i, t_j) \geq \sigma^=(u_i, t_j)$. In all the other cases it returns `false`. $sns(u_i, t_j)[x..y]$ represents the projection of $sns(u_i, t_j)$ in the time interval $T[x..y]$;
- $wns(u_i, t_j)$: It receives a user $u_i$ and a topic $t_j$ and returns `true` if $u_i$ has a weakly negative sentiment on $t_j$. This happens when $\sigma^-(u_i, t_j) > \sigma^+(u_i, t_j)$ and $\sigma^-(u_i, t_j) < \sigma^=(u_i, t_j)$. In all the other cases it returns `false`. $wns(u_i, t_j)[x..y]$ represents the projection of $wns(u_i, t_j)$ in the time interval $T[x..y]$.

At the end of this section, we observe that the four functions $sps()$, $wps()$, $sns()$, and $wns()$ are *mutually exclusive*, in the sense that at most one of them must be `true`, and *complete*, in the sense that at least one of them must be `true`. It follows that, in a given time interval $T[x..y]$, given a user $u_i \in \mathcal{U}$ and a topic $t_j \in \mathcal{T}$, exactly one of these functions returns `true` and all the others return `false`. This allows us to determine the concept of *sentiment type* of $u_i$ on $t_j$ in the time interval $T[x..y]$; it represents the sentiment type associated with that function among the four indicated above that returns `true` in the time interval $T[x..y]$. Clearly, the possible sentiment types are: (i) strongly positive (hereafter, *sp*); (ii) weakly positive (hereafter, *wp*); (iii) weakly negative (hereafter, *wn*); (iv) strongly negative (hereafter, *sn*).

## 4. The Proposed Approach

### 4.1. Objective and Research Questions

The main objective of this paper is to introduce a multi-dimensional view of the scope of the sentiment of a user on one or more topics on any social platform. The paper also wants to define a framework for sentiment scope evaluation. Getting more specific, the main research questions the paper wants to answer are the following:

- **RQ1**: Is it possible to introduce the concept of sentiment scope? In fact, in the past, scope was defined for users and smart objects, but never for sentiments.
- **RQ2**: Is it possible to define a temporal view of scope? In fact, in the past literature, the only view of scope considered was the spatial one.
- **RQ3**: Is it possible to define a framework for evaluating the space-time scope of a sentiment of one or more users on any topic in any social platform?
- **RQ4**: Are there any differences between the scope of negative and positive sentiments?
- **RQ5**: Are there any differences between the scope of strong and weak sentiments?
- **RQ6**: How does the scope of the sentiment of a user on one or more topics propagate to her neighbors?
- **RQ7**: What kind of behavior do users generally exhibit with respect to a sentiment on a topic? In other words, is their sentiment stable or swinging?
- **RQ8**: In showing their sentiments on topics, are users posed and balanced or are they biased toward positive sentiments or negative ones?

In the next sections, we aim to answer all these research questions.

### 4.2. Determining the Spatial Scope of the Sentiment of a User on a Topic

In this section, we illustrate our approach to determine the spatial scope of the sentiment of a user $u_i \in \mathcal{U}$ on a topic $t_j \in \mathcal{T}$. In Section 3.2.1, we have seen that there exist four possible sentiment types. Consequently, it is possible to determine four kinds of scope, one for each sentiment type. In this section, we examine all of them starting with the scope associated with a strongly positive sentiment.

First, let us specify how the spatial scope of $u_i$ on $t_j$ can be represented. A first possibility consists of a set $\Sigma^+_{ij}$ of pairs, as shown in Equation (6):

$$\Sigma^+_{ij} = \{(u_1, \delta^+_{1j}), (u_2, \delta^+_{2j}), \cdots, (u_g, \delta^+_{gj})\} \tag{6}$$

Each pair $(u_h, \delta^+_{hj})$, $h \neq i$, belongs to $\Sigma^+_{ij}$ and consists of a user $u_h$, directly or indirectly connected to $u_i$, and the corresponding positive sentiment degree $\delta^+_{hj}$ on $t_j$.

A second representation consists of a subgraph $\overline{\mathcal{A}^{ji}} = \langle \overline{N^{ji}}, \overline{E^{ji}} \rangle$ of $\overline{\mathcal{A}^j}$. A node $n_h$ belongs to $\overline{N^{ji}}$ if the corresponding user $u_h$ is present in $\Sigma^+_{ij}$. Furthermore, $n_i$ belongs to $\overline{N^{ji}}$. An arc $(n_h, n_u)$ belongs to $\overline{E^{ji}}$ if an arc $(n_h, n_u)$ also exists in $\overline{E^j}$. We call *origin of* $\overline{\mathcal{A}^{ji}}$ the node $n_i$ corresponding to the user $u_i$.

At this point we can define our approach for computing the spatial scope associated with a strongly positive sentiment (hereafter referred to as *strongly positive spatial scope*) of $u_i$ on $t_j$. We represent this approach by defining a function $\psi^+()$ shown in Equation (7). It receives $u_i$, $t_j$ and the initially empty set $\Sigma^+_{ij}$ as parameters. It basically performs a depth-first search on $\overline{\mathcal{A}^j}$, starting from $u_i$ and selecting a node only if certain constraints are satisfied for it. It can be formalized as shown in Equation (7):

$$\psi^+(u_i, t_j, \Sigma^+_{ij}) = \begin{cases} \{(u_i, \delta^+_{ij})\} \cup \bigcup_{n_h \in \overline{\nu}(n_i, \overline{\mathcal{A}^j})} \psi^+(u_h, t_j, \Sigma^+_{ij} \cup \{(u_i, \delta^+_{ij})\}) & \textit{if } sps(u_i, t_j) = \texttt{true} \\ & \text{and } (u_i, \delta^+_{ij}) \notin \Sigma^+_{ij} \\ \emptyset & \text{otherwise} \end{cases} \tag{7}$$

In other words, the function $\psi^+()$, when applied on $u_i$ and $t_j$, first checks whether $u_i$ has a strongly positive sentiment on $t_j$. If this is true and the pair $(u_i, \delta^+_{ij})$ is not already present in $\Sigma^+_{ij}$, then $\psi()^+$ adds this pair to $\Sigma^+_{ij}$. Afterwards, it recursively calls itself by passing as input each node directly connected to $n_i$ in $\overline{\mathcal{A}^j}$. In contrast, if $u_i$ has not a strongly positive sentiment on $t_j$ or the pair $(u_i, \delta^+_{ij})$ is already present in $\Sigma^+_{ij}$, then $\psi()$ simply returns $\emptyset$ and the recursion stops.

The *strongly negative spatial scope* can be defined in a similar way (see Equation (8)). Again, we can introduce a function $\psi^-()$ that receives $u_i$, $t_j$ and the initially empty set $\Sigma^-_{ij}$. It has an identical behavior to the function $\psi^+()$, except that $\delta^+_{ij}$ is replaced by $\delta^-_{ij}$ and the function $sps()$ is replaced by the function $sns()$, defined in Section 3.2.1. Its formalization is shown in Equation (8):

$$\psi^-(u_i, t_j, \Sigma^-_{ij}) = \begin{cases} \{u_i, \delta^-_{ij}\} \cup \bigcup_{n_h \in \overline{\nu}(n_i, \overline{\mathcal{A}^j})} \psi^-(u_h, t_j, \Sigma^-_{ij} \cup \{(u_i, \delta^-_{ij})\}) & \textit{if } sns(u_i, t_j) = \texttt{true} \\ & \text{and } (u_i, \delta^-_{ij}) \notin \Sigma^-_{ij} \\ \emptyset & \text{otherwise} \end{cases} \tag{8}$$

The *weakly positive* (resp., *negative*) *spatial scope* is defined similarly to the strongly positive (resp., negative) spatial scope. In this case, we introduce a function $\xi^+()$ (resp., $\xi^-()$) that receives $u_i$, $t_j$ and an initially empty set $\Pi^+_{ij}$ (resp., $\Pi^-_{ij}$). Its behavior is identical to the one of the function $\psi^+()$ (resp., $\psi^-()$) except that the function $sps()$ (resp., $sns()$) is replaced by the function $wps()$ (resp., $wns()$). The formalization of $\xi^+()$ and $\xi^-()$ is shown in Equations (9) and (10):

$$\xi^{+}(u_i, t_j, \Pi_{ij}^{+}) = \begin{cases} \{u_i, \delta_{ij}^{+}\} \cup \bigcup_{n_h \in \overline{\nu}(n_i, \overline{\mathcal{A}^j})} \xi^{+}(u_h, t_j, \Pi_{ij}^{+} \cup \{(u_i, \delta_{ij}^{+})\}) & \textit{if } wps(u_i, t_j) = \texttt{true} \\ & \text{and } (u_i, \delta_{ij}^{+}) \notin \Pi_{ij}^{+} \\ \emptyset & \text{otherwise} \end{cases} \tag{9}$$

$$\xi^{-}(u_i, t_j, \Pi_{ij}^{-}) = \begin{cases} \{u_i, \delta_{ij}^{-}\} \cup \bigcup_{n_h \in \overline{\nu}(n_i, \overline{\mathcal{A}^j})} \xi^{-}(u_h, t_j, \Pi_{ij}^{-} \cup \{(u_i, \delta_{ij}^{-})\}) & \textit{if } wns(u_i, t_j) = \texttt{true} \\ & \text{and } (u_i, \delta_{ij}^{-}) \notin \Pi_{ij}^{-} \\ \emptyset & \text{otherwise} \end{cases} \tag{10}$$

At this point, we have defined the functions for computing the strongly positive (resp., negative) spatial scope $\Sigma_{ij}^{+}$ (resp., $\Sigma_{ij}^{-}$) and the weakly positive (resp., negative) spatial scope $\Pi_{ij}^{+}$ (resp., $\Pi_{ij}^{-}$)). We have also previously seen that it is possible to provide a graph-based representation of such a scope. In the following, in order not to burden the notation, we will use the symbol $\mathcal{SG}^{+}$ (resp., $\mathcal{SG}^{-}$, $\mathcal{WG}^{+}$, $\mathcal{WG}^{-}$) to denote the graph-based representation corresponding to $\Sigma_{ij}^{+}$ (resp., $\Sigma_{ij}^{-}$, $\Pi_{ij}^{+}$, $\Pi_{ij}^{-}$). Its formalization is reported in Equation (11):

$$\begin{aligned} \mathcal{SG}^{+} &= \langle SN^{+}, SE^{+} \rangle \\ \mathcal{SG}^{-} &= \langle SN^{-}, SE^{-} \rangle \\ \mathcal{WG}^{+} &= \langle WN^{+}, WE^{+} \rangle \\ \mathcal{WG}^{-} &= \langle WN^{-}, WE^{-} \rangle \end{aligned} \tag{11}$$

By studying some properties of these graphs, it is possible to define a variety of information regarding the scope of the sentiment of $u_i$ on $t_j$.

In what follows we will perform all our analyses with regard to the graph $\mathcal{SG}^{+}$, although everything we will see can be straightforwardly extended to the other three graphs.

The first two properties of the scope of the sentiment of $u_i$ on $t_j$ that we consider are its breadth and its depth. Regarding the breadth, it is immediate to think that it can be obtained by considering the size of $\mathcal{SG}^{+}$, that is, the number $|SN^{+}|$ of its nodes. As far as the depth is concerned, we recall that $\mathcal{SG}^{+}$ derives from a depth-first search performed on $\overline{\mathcal{A}^j}$ starting from the node $n_i$, which we have also called the origin of $\mathcal{SG}^{+}$. Therefore, the depth of the scope can be determined by computing the diameter of $\mathcal{SG}^{+}$, that is, the maximum length of the minimum paths from $n_i$ to any other node of $\mathcal{SG}^{+}$.

An important investigation consists in determining how the strongly positive sentiment degree varies as we move away from $n_i$ in $\mathcal{SG}^{+}$. To do this, we can consider the neighborhood of level $\lambda, 1 \leq \lambda \leq d, d = diameter(\mathcal{SG}^{+})$, obtained by applying the function $\nu$ on $n_i$, $\lambda$ and $\mathcal{SG}^{+}$. For each neighborhood, it is then possible to compute the average strongly positive sentiment degree of the nodes belonging to it. Generally, if there were no interference, as we move away from $n_i$, the average strongly positive sentiment degree of a neighborhood should decrease because the influence that $n_i$ exerts on nodes tends to decrease. However, it could be the case that, once we move away from $n_i$, there is another node different from it that exerts an influence on the nodes of the neighborhood of $n_i$. If the new "influencer" has a discordant sentiment with $n_i$, we might see a steep decrease in the average strongly positive sentiment degree, or even a reversal of sentiment polarity. By contrast, if the new "influencer" has a concordant sentiment with $n_i$, we may see a slowdown in the decline of the average strongly positive sentiment degree, or even a new growth of it. The correlation that can arise between two scopes is a challenging topic that is, however, beyond the objective of this paper. Here, we simply provide a tool for computing the variation in the average strongly positive sentiment degree as we move away from $n_i$.

Let $\nu(n_i, \lambda, \mathcal{SG}^{+})$ be the neighborhood of level $\lambda, 1 \leq \lambda \leq d = diameter(\mathcal{SG}^{+})$ of $n_i$ in $\mathcal{SG}^{+}$. The average positive sentiment degree $\overline{\delta_{ij_\lambda}^{+}}$ of $\nu(n_i, \lambda, \mathcal{SG}^{+})$ is defined in Equation (12):

$$\overline{\delta^+_{ij\lambda}} = \frac{\sum_{n_h \in \nu(n_i,\lambda,\mathcal{SG}^+)} \delta^+_{hj}}{size(\nu(n_i,\lambda,\mathcal{SG}^+))} \tag{12}$$

In other words, it is obtained by computing the average strongly positive sentiment degree of all the nodes belonging to $\nu(n_i, \lambda, \mathcal{SG}^+)$. $\overline{\delta^+_{ij\lambda}}$ ranges in the real interval $[0,1]$; the higher its value, the higher the strength of the average positive sentiment.

At this point, we have at our disposal a succession of values $\varrho^+_0, \varrho^+_1, \cdots, \varrho^+_d$ such that $\varrho^+_0 = \delta^+_{ij}$, $\varrho^+_h = \overline{\delta^+_{ij_h}}$, $1 \leq h \leq d$, $d = diameter(\mathcal{SG}^+)$. The examination of that succession can give us some interesting insights into how the average strongly positive sentiment degree evolves as we move away from $n_i$. It takes into account the decreasing influence of $n_i$ as we move away from it, as well as the possible presence of any interference from other "influencers".

By plotting the values of $\varrho^+_0, \varrho^+_1, \cdots, \varrho^+_d$, we get a "spectrum" of the trend of the strongly positive sentiment degree in the spatial scope of $u_i$. In fact, several interesting pieces of information can be derived from that spectrum. These include:

- The variation in the average strongly positive sentiment degree in the $h$th section of the spectrum, defined in Equation (13):

$$\Delta^+_h = \varrho^+_h - \varrho^+_{h-1} \tag{13}$$

- The relative variation in the average strongly positive sentiment degree in the $h$th section of the spectrum, defined in Equation (14):

$$\overline{\Delta^+_h} = \frac{\varrho^+_h - \varrho^+_{h-1}}{\varrho^+_{h-1}} \tag{14}$$

- The mean variation in the average strongly positive sentiment degree in the $h$th section of the spectrum, defined in Equation (15):

$$\widehat{\Delta^+_h} = \frac{\varrho^+_h - \varrho^+_0}{h} \tag{15}$$

- The maximum variation in the average strongly positive sentiment degree in the $h$th section of the spectrum, defined in Equation (16):

$$\Delta^{M+} = \max_{h=1..v} |\Delta^+_h| \tag{16}$$

- The minimum variation in the average strongly positive sentiment degree in the $h$th section of the spectrum, defined in Equation (17):

$$\Delta^{m+} = \min_{h=1..v} |\Delta^+_h| \tag{17}$$

Finally, we can analyze the monotonicity of the succession $\varrho^+_0, \varrho^+_1, \cdots, \varrho^+_d$. In particular, we are interested in whether it is monotonically non-increasing. This occurs when $\varrho^+_h \leq \varrho^+_{h-1}$. In fact, if such a condition is not satisfied, we can say that, as we move away from $n_i$, there is at least one further "influencer" with a sentiment concordant with the one of $n_i$ that is acting on the nodes of the neighborhoods of $n_i$. Otherwise, it could be that there is no other "influencer" interfering with $n_i$ or that such an "influencer" is present but with a discordant sentiment with the one of $n_i$.

What we have seen now are just some of the analyses we can perform on spatial scope. They allow us to give an idea of the potential of this concept. Many other analyses could be thought of simply by applying concepts from mathematical analysis to the succession $\varrho^+_0, \varrho^+_1, \cdots, \varrho^+_d$ or concepts from graph theory to the graph $\mathcal{SG}^+$.

Finally, it is worth emphasizing again that all the analyses we have previously done on $\mathcal{SG}^+$ could be straightforwardly extended to $\mathcal{SG}^-$, $\mathcal{WG}^+$ and $\mathcal{WG}^-$.

*4.3. Determining the Temporal Scope of the Sentiment of a User on a Topic*

In this section, we illustrate our approach to determine the temporal scope of the sentiment of the user $u_i \in \mathcal{U}$ on a topic $t_j \in \mathcal{T}$. In Section 3.2.1, we introduced two concepts on sentiment scope, namely *sentiment type* and *sentiment degree* of $u_i$ on $t_j$. These two concepts play a key role in the analysis of temporal scope. Recall that the sentiment type of $u_i$ on $t_j$ can be strongly positive ($sp$), weakly positive ($wp$), weakly negative ($wn$), and strongly negative ($sn$). Instead, the sentiment degree of $u_i$ on $t_j$ is given by the value of the parameter $\delta_{ij}^+$, in case the sentiment type is $sp$ or $wp$, or the value of the parameter $\delta_{ij}^-$, in case it is $wn$ or $sn$.

The temporal scope of $u_i$ on $t_j$ in the time interval $T[x..y]$ can be represented by an ordered list of pairs, as shown in Equations (18)–(20).

$$\Theta_{ij}[x..y] = [(\tau_x, \theta_x), (\tau_{x+1}, \theta_{x+1}), \cdots, (\tau_y, \theta_y)] \tag{18}$$

$$\tau_b = \begin{cases} sp & \textit{if } sps(u_i, t_j)[b] = \texttt{true} \\ wp & \textit{if } wps(u_i, t_j)[b] = \texttt{true} \\ wn & \textit{if } wns(u_i, t_j)[b] = \texttt{true} \\ sn & \textit{if } sns(u_i, t_j)[b] = \texttt{true} \end{cases} \tag{19}$$

$$\theta_b = \begin{cases} \delta_{ij}^+[b] & \textit{if } (sps(u_i, t_j)[b] = \texttt{true}) \text{ or } (wps(u_i, t_j)[b] = \texttt{true}) \\ \delta_{ij}^-[b] & \textit{if } (wns(u_i, t_j)[b] = \texttt{true}) \text{ or } (sns(u_i, t_j)[b] = \texttt{true}) \end{cases} \tag{20}$$

Recall that $sps(u_i, t_j)[b]$ (resp., $wps(u_i, t_j)[b]$, $wns(u_i, t_j)[b]$, $sns(u_i, t_j)[b]$) represents the projection of $sps(u_i, t_j)$ (resp., $wps(u_i, t_j)$, $wns(u_i, t_j)$, $sns(u_i, t_j)$) in the time slice $T_b$ (see Section 3.2.1). Analogously $\delta_{ij}^+[b]$ (resp., $\delta_{ij}^-[b]$) denotes the value returned by the function $\sigma^+(ui, t_j)$ (resp., $\sigma^-(ui, t_j)$) when projected in the time slice $T_b$ (see, again, Section 3.2.1).

Clearly, by moving from a time instant $T_b$ to a time instant $T_{b+1}$ the value of $\tau$ can remain unvaried or change and the value of $\theta$ can increase, decrease, or remain constant. Each combination of the trend of these two parameters at the transition from $T_b$ to $T_{b+1}$ gives us interesting information about the time trend of the sentiment degree of $u_i$ on $t_j$. For example:

- If both $\tau_b$ and $\tau_{b+1}$ are equal to $sp$:
  - if $\theta_{b+1} > \theta_b$, it means that the sentiment degree is strengthening;
  - if $\theta_{b+1} = \theta_b$, it means that the sentiment degree is static;
  - if $\theta_{b+1} < \theta_b$, it means that, although a strongly positive sentiment still characterizes $u_i$, it is weakening.
- If $\tau_b = sp$ and $\tau_{b+1} = wp$, it means that the posts and comments on $t_j$ published by $u_i$ in which she shows a neutral sentiment, are increasing. This increase is such that they exceed the ones in which $u_i$ shows a positive sentiment. The number of posts/comments with a positive sentiment continues to be greater than the number of posts/comments with a negative sentiment. However, at the time slice $T_{b+1}$ we are seeing a weakening of the positivity of the sentiment of $u_i$ on $t_j$, compared to the time slice $T_b$.
- If $\tau_b = sp$ and $\tau_{b+1} = wn$, it means that $u_i$ is changing her sentiment on $t_j$. This change is not yet radical, since there is a prevalence of neutral posts/comments over negative ones.
- If $\tau_b = sp$ and $\tau_{b+1} = sn$, it means that $u_i$ has completely changed her sentiment on $t_j$. The greater the gap between $\theta_b$ and $\theta_{b+1}$ and the greater the change occurred.

Similarly, suitable information can be extracted in case $\tau_b = wp$, $\tau_b = wn$ or, finally, $\tau_b = sp$.

Analogously to what we have seen for spatial scope, several measures can also be defined for temporal scope. They allow us to get a quantitative view of the changes in the sentiment degree of $u_i$ on $t_j$ over a time interval. Some of these measures are the following (in defining these measures, we will refer to the time slices $T_b$ and $T_{b-1}$, instead of the time slices $T_b$ and $T_{b+1}$, to bring their definition in line with that of the metrics for spatial scope, explained in Section 4.2.):

- The variation in the sentiment degree between the time slices $T_{b-1}$ and $T_b$. It can be defined in Equation (21):

$$\Lambda_b = \theta_b - \theta_{b-1} \tag{21}$$

- The relative variation in the sentiment degree between the time slices $T_{b-1}$ and $T_b$. It can be defined in Equation (22):

$$\overline{\Lambda_b} = \frac{\theta_b - \theta_{b-1}}{|\theta_{b-1}|} \tag{22}$$

- The mean variation in the sentiment degree in the time interval $T[x..y]$. It is defined in Equation (23):

$$\widehat{\Lambda} = \frac{\theta_y - \theta_x}{y - x} \tag{23}$$

- The maximum variation in the sentiment degree in the time interval $T[x..y]$. It is defined in Equation (24):

$$\Lambda^{M+} = \max_{b=x..y} |\Lambda_b| \tag{24}$$

- The minimum variation in the sentiment degree in the time interval $T[x..y]$. It is defined in Equation (25):

$$\Lambda^{m+} = \min_{b=x..y} |\Lambda_b| \tag{25}$$

In addition to defining appropriate metrics to measure the change in the sentiment of $u_i$ on $t_j$, we can check whether the succession of the values of the sentiment degree in the interval $T[x..y]$ is monotonic or not. This information must be closely coupled with that related to sentiment type. In particular, if the succession of values $\theta_x, \theta_{x+1}, \cdots, \theta_y$ is monotonically non-increasing, it means that, in the time interval $T[x..y]$, the sentiment of $u_i$ on $t_j$ is not strengthening and, rather, it is presumably decreasing. Such a decrease could cause the sentiment type to go from strongly positive to weakly positive, weakly negative, or even strongly negative. On the other hand, if the previous succession of values is monotonically non-decreasing, it means that, in the time interval $T[x..y]$, the sentiment of $u_i$ on $t_j$ is not weakening, and, rather, it is presumably strengthening. In this case, we might see reverse transitions from the previous case, e.g., from strongly negative to weakly negative, weakly positive, and strongly positive.

The previous succession may also not be monotonic. In this case, the measures on changes in sentiment degree defined above could be extremely useful. It might also be useful to determine how often the change from one type of sentiment to another occurs, or how often the change from an increasing to a decreasing trend occurs, or vice versa.

Analogously to the spatial scope, those seen above are just some of the analyses that can be performed on the temporal scope. Many other analyses could be performed by applying the concepts of mathematical analysis or time series analysis to the succession of values $\theta_x, \theta_{x+1}, \cdots, \theta_y$.

## 5. Experimental Campaign

### *5.1. Dataset Description*

To build a dataset capable of supporting our experiments, we chose Reddit as the reference social platform. We carried out such a choice because: (i) Reddit is very popular (in fact, it currently ranks 11th among the most visited sites according to Visual Capitalist (www.visualcapitalist.com (accessed on 12 September 2022))); (ii) it allows posts and comments on any topic; and (iii) its data are easily accessible through `pushshift.io` [59]; the latter is a data repository that allows people to download data related to Reddit comments and posts through a suitable API.

In building our dataset we focused on the posts and comments of one particular subreddit, namely /r/worldnews. The reasons for this choice lie in the fact that it has already been used as a reference subreddit in previous analyses (see Refs. [60–62]) and in the fact that it is one of the most complete and neutral news-related subreddits.

Specifically, through `pushshift.io`, we retrieved all posts and comments, along with the corresponding metadata, published in this subreddit from 25 February 2022 to 25 March 2022. The number of posts taken into account is equal to 9884 while the number of comments is equal to 633,371.

Once the data of interest were downloaded from `pushshift.io`, we performed ETL (Extraction, Transformation, and Loading) activities on them. Specifically: (i) we removed all posts and comments published by users who had left Reddit; (ii) we removed all posts and comments that did not have textual content or were written in a language other than English; (iii) we selected only those posts and comments related to a specific discussion theme. Regarding the latter, the choice was complex as it was important to select a specific, but sufficiently broad, theme with many facets, and thus many topics. Based on this reasoning, our choice fell on the armed conflict in Ukraine that began on 24 February 2022.

After filtering and other ETL activities, the final number of posts in the dataset is 2703, which is 27.12% of the initial ones. In contrast, the final number of comments is 82,617, which is 13.21% of the initial ones. In Table 1, we report some of the main characteristics of the final dataset. In addition to the information mentioned above, this table reports some further interesting information. In particular, we can see that the number of authors in our dataset is 4219. Among them only 119 published both posts and comments. This number is clearly very low; in particular, it is 26.50% of the authors publishing posts and 3.14% of those publishing comments.

**Table 1.** Some main parameters of the dataset adopted for our experiments.

| *Parameter* | *Value* |
|---|---|
| No. of posts | 2703 |
| No. of comments | 82,617 |
| No. of (distinct) authors | 4219 |
| No. of (distinct) authors publishing posts | 449 |
| No. of (distinct) authors publishing comments | 3787 |
| No. of (distinct) authors publishing both posts and comments | 119 |

In Figure 1, we show the distribution of comments against posts, while in Figure 2 we report the distribution of comments against score. Both figures are in log-log scale. By examining them we can observe that both distributions follow power laws. Table 2 reports the values of the corresponding coefficients $\alpha$ and $\delta$.

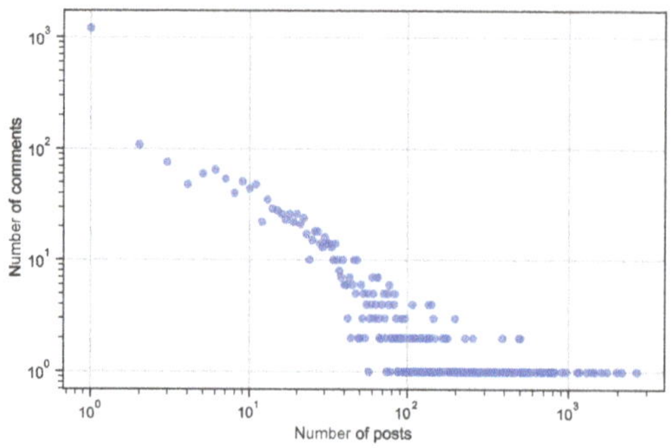


**Figure 1.** Distribution of comments against posts (log-log scale).

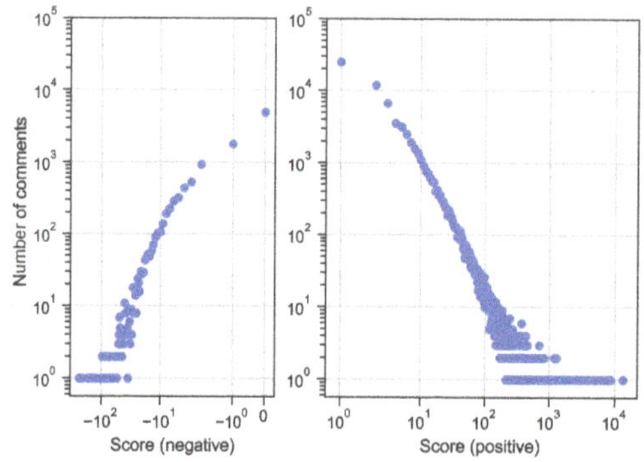


**Figure 2.** Distribution of comments against score (log-log scale).

**Table 2.** Values of $\alpha$ and $\delta$ of the power law distributions for the considered dataset—* These values were computed considering the absolute values of scores.

| *Distribution* | $\alpha$ | $\delta$ |
|---|---|---|
| Figure 1 | 1.8408 | 0.0419 |
| Figure 2 (left) * | 2.9262 | 0.0418 |
| Figure 2 (right) | 2.0383 | 0.0136 |

### 5.2. *Identification of Topics and Sentiments*

In Section 3.1.1, we saw that our framework is independent of the technique used for constructing the set $\mathcal{T}$ of topics. In our experimental campaign, we adopted BERTopic [63] to obtain $\mathcal{T}$. BERTopic is based on BERT (Bidirectional Encoder Representation from Transformers). The latter is a powerful deep learning-based framework for performing NLP tasks on texts. More specifically, it is a topic modeling technique that exploits transformers [64] and c-TF-IDF [65] to create dense clusters from which easily interpretable topics can be derived. BERTopic receives a set of documents as input and returns the list of topics covered in them. It also associates each topic thus obtained with a description and a count. The former consists of the set of words characterizing the topic. The latter indicates the number of documents mentioning it. Given a document, BERTopic is always able to determine a set of topics that characterize it.

We applied BERTopic to the 2703 posts and 82,617 comments in the dataset and obtained a set $\mathcal{T}$ of 101 topics. Table 3 shows some examples of the extracted topics.

**Table 3.** Some examples of the topics and their descriptions extracted by BERTopic.

| *Topic* | *Description* |
|---|---|
| $t_1$ | {invasion, invade, mission} |
| $t_2$ | {nato, defence, member, treaty} |
| $t_3$ | {bunker, underground} |

After constructing the set $\mathcal{T}$ of topics, we turned to consider the sentiments characterizing the posts and comments published by users. In this activity, we used roBERTa-base [66]. This system was trained on approximatively 124 million tweets published from January 2018 to December 2021. Next, it was expressively fine-tuned for sentiment analysis using the TweetEval benchmark [67]. We decided to use roBERTa-base because there is a strongly similarity between the shape of texts characterizing tweets and the one of texts in posts and comments of Reddit. In fact, in both cases, we are in the presence of fast-paced messages employed to express opinions and thoughts in general.

The set of sentiments that can be derived by roBERTa-base are those typically used in sentiment analysis, namely "pos", "neg", and "neu". They are also the sentiments considered in our model, as we have seen in Section 3.1. Therefore, the set $\mathcal{S}$ of sentiments is $\mathcal{S}$ = { "pos", "neg", "neu" }. Table 4 shows some examples of fragments, along with the corresponding sentiments, derived by roBERTa-base. Let $f_k$ be a fragment of a comment or a post (as mentioned in Section 3.1.2, $f_k$ can coincide with a whole comment or a whole post, if these are characterized by a single sentiment.) characterized by a single sentiment. Let $s_k$ be the sentiment that roBERTa-base derived for $f_k$. Finally, let $\mathcal{T}_{f_k}$ be the set of topics of $f_k$ identified by BERTopic. Then, the joint use of BERTopic and roBERTa-base on $f_k$ allows us to extract a pair $(t_j, s_k)$ for each element $t_j$ of $\mathcal{T}_{f_k}$. Such a pair indicates that the sentiment $s_k$ was associated with the topic $t_j$ in $f_k$. As previously pointed out, 101 topics were identified in our dataset. From them, 302 pairs of the type $(t_j, f_k)$ were obtained.

**Table 4.** Some examples of fragments and their sentiments derived by roBERTa-base (swear words are partially masked).

| *Fragment* | *Sentiment* |
|---|---|
| "It makes me hopeful too. We need to find a way to get NATO forces engaged." | pos |
| "But it's a f***ing kid that got killed by that c**t" | neg |
| "Anyone know when this interview took place? NBC has no time stamp on the video" | neu |

### 5.3. Descriptive Analysis of the Graphs $\mathcal{A}$, $\mathcal{SG}^+$, $\mathcal{SG}^-$, $\mathcal{WG}^+$, and $\mathcal{WG}^-$

In this section, we present a descriptive analysis of the graphs $\mathcal{A}$, $\mathcal{SG}^+$, $\mathcal{SG}^-$, $\mathcal{WG}^+$. and $\mathcal{WG}^-$ obtained from our dataset. This analysis allows us to identify some features of these graphs that will be useful in the next experiments. It also allows us to identify the first differences among the four graphs $\mathcal{SG}^+$, $\mathcal{SG}^-$, $\mathcal{WG}^+$, and $\mathcal{WG}^-$, and thus among the trends of the various sentiment types that we defined in this paper.

We begin our analysis from the graph $\mathcal{A}$. Recall that this is a user-centered, single mode graph representing user interactions. Specifically, an edge in $\mathcal{A}$ indicates that the users associated with the corresponding nodes published at least one post or comment on the same topic and, further, that one of the two users commented on at least one post or comment of the other.

In Table 5, we report the values of some features of the graph $\mathcal{A}$. In particular, we consider the number of nodes, the number of arcs, the density, and the clustering coefficient. Clearly, the number of nodes of $\mathcal{A}$ is equal to the number of distinct authors in the dataset, and thus to 4219. The number of arcs of $\mathcal{A}$ is 32,648 and, consequently, the density is 0.0018, which is a very low value. This can be explained both by taking into account the average number of comments posted by each user, which is 19.58, and by considering that the condition of existence of an arc in $\mathcal{A}$ is very stringent. In fact, an arc exists in $\mathcal{A}$ if at least one of the comments of one of its nodes refers to a post or comment of the other node. The clustering coefficient is equal to 0.0349. This value is quite high if we consider the low density of $\mathcal{A}$. It implies that this graph consists of several components strongly internally connected and weakly coupled together. Some of these components may also be disconnected from all the others. This is already an interesting result found through our analysis. Indeed, it tells us that in the r/worldnews subreddit, users tend to organize themselves into high cohesive and weakly coupled communities.

**Table 5.** Some basic properties of the graph $\mathcal{A}$.

| *Property* | *Value* |
|---|---|
| Number of nodes | 4219 |
| Number of edges | 32,648 |
| Density | 0.0018 |
| Clustering coefficient | 0.0349 |

Having analyzed the graph $\mathcal{A}$, we now turn to the analysis of the graphs $\mathcal{SG}^+$, $\mathcal{SG}^-$, $\mathcal{WG}^+$, and $\mathcal{WG}^-$. As we saw in Section 4.2, each of these graphs is related to a pair $(u_i, t_j)$, where $u_i$ is a user and $t_j$ is a topic. The four graphs are associated with the four possible sentiment types; in particular, the graph $\mathcal{SG}^+$ (resp., $\mathcal{SG}^-$, $\mathcal{WG}^+$ and $\mathcal{WG}^-$) is associated with the strongly positive (resp., strongly negative, weakly positive, weakly negative) spatial scope. Essentially, these graphs represent the spatial spread of the scope related to a user $u_i$ discussing a topic $t_j$. Recall that, given the pair $(u_i, t_j)$, only one of the four graphs can exist in a given time interval, depending on the sentiment type that $u_i$ had shown for $t_j$ in that time interval. Since each graph is associated with a pair $(u_i, t_j)$, in our analysis we considered all possible pairs of users $(u_i, t_j)$ in the various time slices of the dataset and, for each of them, we calculated its breadth (which coincides with the number of its nodes) and its depth (which coincides with its diameter). Finally, we aggregated the results based on sentiment type, obtaining average values for each graph types. These are shown in Table 6.

**Table 6.** Average values of breadth and depth for the graphs $\mathcal{SG}^+$, $\mathcal{SG}^-$, $\mathcal{WG}^+$, and $\mathcal{WG}^-$.

| *Property* | $\mathcal{SG}^+$ | $\mathcal{SG}^-$ | $\mathcal{WG}^+$ | $\mathcal{WG}^-$ |
|---|---|---|---|---|
| Average breadth | 143 | 187 | 89 | 124 |
| Average depth | 7.8 | 8.4 | 6.9 | 7.3 |

The examination of this table reveals additional interesting insights. First of all, the differences among the four graphs under examination mainly concern the average breadth, while the values of the average depth are more similar. In addition, we can observe that for both the average breadth and the average depth the graphs associated with negative sentiments have higher values than the corresponding graphs associated with positive sentiments. This is in line with several researches proposed in the past literature whose authors found that negative sentiments tend to spread more easily than positive ones [68–72]. Finally, we can observe that, for both average breadth and average depth, the graphs associated with weak sentiments have lower values than the corresponding graphs associated with strong ones. This is in line with other studies proposed in the past literature where it has been shown that the stronger a sentiment is, the more people resonate with it, and the likelier it is they will spread it to others [71,73,74].

### 5.4. Experiments on Spatial Scope

#### 5.4.1. Variation of the Spatial Scope against the Neighborhood Level

We began our experiments on spatial scope by analyzing how it varies against the neighborhood level and whether this variation differs for the different sentiment types. To conduct this analysis we proceeded as follows.

Let us first consider the case in which the sentiment type is strongly positive. In Section 4.2, we have seen that, in this case, the graph representing the scope is $\mathcal{SG}^+$ and the average positive sentiment degree of the neighbors of level $\lambda$ of the user $u_i$ on the topic $t_j$ is $\overline{\delta^+_{ij_\lambda}}$, as shown in Equation (12). We have also seen that the trend of this degree against $\nu$ is given by a succession of values $\varrho^+_0, \varrho^+_1, \cdots, \varrho^+_d$ such that $\varrho^+_0 = \delta^+_{ij}$, $\varrho^+_h = \overline{\delta^+_{ij_h}}$, $1 \leq h \leq d$, $d = diameter(\mathcal{SG}^+)$. In other words, this succession measures the variation of the $u_i$'s capability of influencing the sentiment on $t_j$ as we move away from her in the social platform, also taking into account the possible interference of other users.

In Section 5.3, we have seen that the average depth (which coincides with the average diameter) of $\mathcal{SG}^+$ is 7.8. Therefore, in the current analysis, we consider a value of $h$ ranging from 0 to 7.

Consider, now, all possible pairs of users $(u_i, t_j)$ such that $u_i$ showed a strongly positive sentiment on $t_j$. For each of these pairs, we performed all the computations specified above and constructed the succession $\varrho^+_0, \varrho^+_1, \cdots, \varrho^+_d$. The latter tells us how the average value of the sentiment degree on $t_j$ of the neighbors of level $h$, $0 \leq h \leq 7$, of the users showing a strongly positive sentiment on $t_j$ varies against the increase of $h$. In other words, it shows how the influence of the users having a strongly positive sentiment on $t_j$ varies as we move away from them in the social platform, also taking into account the possible interference of other users. Finally, we computed the mean of all the values of $\varrho^+_0, \varrho^+_1, \cdots, \varrho^+_d$ over the possible pairs of users $(u_i, t_j)$. These mean values are graphically reported in Figure 3.

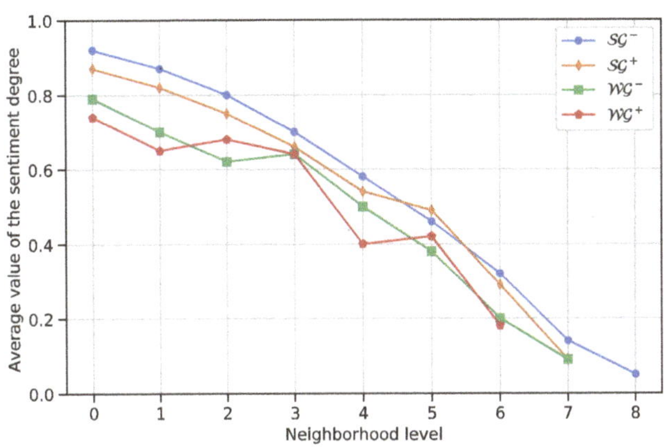


**Figure 3.** Variation of the mean value of the sentiment degree on $t_j$ of users against the neighborhood level they belong to.

Similarly, we computed the succession of the mean values of the average sentiment degree on $t_j$ of the neighbors of level $h$ of the users showing a strongly negative (resp., weakly positive, weakly negative) sentiment on $t_j$. It indicates the variation of the influence of the users having a strongly negative (resp., weakly positive, weakly negative) sentiment on $t_j$ as we move away from them in the social platform, also taking into account the possible interference of other users. In this case, based on Table 6, $h$ should range from 0 to 8 (resp., 6, 7). The values of this succession are graphically reported in Figure 3.

From the analysis of this figure, we can deduce several useful information about the trend of the spatial scope for the different sentiment types. As we might expect, whatever the sentiment type, as the neighborhood level increases, the average sentiment degree (and, thus, the influence of the corresponding users regarding the sentiment on a topic) decreases. As for the different types of scope, we can observe that the users with negative sentiment have a greater influence than the ones with positive sentiment, and the users with strong sentiment have a greater influence than the ones with a weak sentiment. This is in line with the results described in Section 5.3 and those found in the past literature [68–74].

The analysis of Figure 3 shows that the influence of users with strongly negative sentiment degree, besides being generally strong, decreases smoothly. This suggests that it is not affected by any interference from other users. When we turn to users with strongly positive sentiment degree, we can see that there is always a decrease of their influence, but this is somewhat more irregular. This indicates that the influence of this type of users may be affected, although not decisively, by the interference from other users. At some time slices, this interference can accelerate the influence decrease, while, at other time slices, it is able to slow it down. However, it is not able to reverse the trend. As for users with weak sentiment degree, we can observe that the trend is more irregular. Overall, the values are lower than the corresponding ones of the users with strongly negative sentiment degree. In addition, the interference from other users is stronger. In fact, it does not only make the decrease irregular, but is also able to reverse the trend at some points, although only for short stretches. All the peculiarities characterizing the influence of users with weakly negative sentiment degree occur even more strongly for the influence of users with weakly positive sentiment degree. In this case, the trend is even more irregular and its inversions are more frequent and pronounced.

As we have seen in Section 4.2, starting from the successions shown in Figure 3, we can derive several other interesting information. In particular, as for the succession corresponding to the average strongly positive sentiment degree, we have that:

- $\Delta_1^+ = -0.05$; $\Delta_2^+ = -0.07$; $\Delta_3^+ = -0.10$; $\Delta_4^+ = -0.12$; $\Delta_5^+ = -0.12$; $\Delta_6^+ = -0.14$; $\Delta_7^+ = -0.18$.
- $\overline{\Delta_1^+} = -\frac{0.05}{0.92} = -0.05$; $\overline{\Delta_2^+} = -\frac{0.07}{0.87} = -0.08$; $\overline{\Delta_3^+} = -0.13$; $\overline{\Delta_4^+} = -0.17$; $\overline{\Delta_5^+} = -0.21$; $\overline{\Delta_6^+} = -0.30$; $\overline{\Delta_7^+} = -0.56$.
- $\widehat{\Delta_1^+} = -\frac{0.87-0.92}{1} = -0.05$; $\widehat{\Delta_2^+} = -\frac{0.80-0.92}{2} = -0.06$; $\widehat{\Delta_3^+} = -0.07$; $\widehat{\Delta_4^+} = -0.09$; $\widehat{\Delta_5^+} = -0.09$; $\widehat{\Delta_6^+} = -0.10$; $\widehat{\Delta_7^+} = -0.11$.
- $\Delta^{M+} = max(0.05, 0.07, 0.10, 0.12, 0.12, 0.14, 0.18) = 0.18$.
- $\Delta^{m+} = min(0.05, 0.07, 0.10, 0.12, 0.12, 0.14, 0.18) = 0.05$.

Similarly, we can compute the corresponding parameter values for the other successions we examined above.

Furthermore, we can say that the successions related to $\mathcal{SG}^-$ and $\mathcal{SG}^+$ are monotonically non-increasing. In contrast, the successions related to $\mathcal{WG}^-$ and $\mathcal{WG}^+$ are non-monotone.

Finally, as specified in Section 4.2, many other analyses can be conducted on the successions and on the graphs $\mathcal{SG}^+, \mathcal{SG}^-, \mathcal{WG}^+$ and $\mathcal{WG}^-$ for extracting more information. For example, we can observe that the succession corresponding to $\mathcal{WG}^-$ presents only one trend reversal while the succession corresponding to $\mathcal{WG}^+$ shows two trend reversals, which also have a larger magnitude. This also allows us to say numerically and objectively that the latter succession is more irregular than the former.

### 5.4.2. Relationship between Density and Clustering Coefficient and Spatial Scope

In the previous sections, we have seen some analyses allowing us to derive information on a spatial scope from its representation through a graph. In particular, we have illustrated what information on a spatial scope can be derived from the breadth and depth of the corresponding graph, as well as from the analysis of the variation of the sentiment against the neighborhood levels. In this section, we want to continue in this direction by considering some Social Network Analysis and graph theory parameters and seeing if and how they can support us in gaining a deeper understanding of scope. In particular, we will focus on density and average clustering coefficient.

In Section 5.3, we have computed the values of these two parameters for the graph $\mathcal{A}$, and we have seen that they are low; then, we have provided an explanation for this behavior. In this section, we want to see what happens for the graphs $\mathcal{SG}^+$, $\mathcal{SG}^-$, $\mathcal{WG}^+$, and $\mathcal{WG}^-$ associated with the various sentiment types.

To answer that question, we computed the density and the average clustering coefficient for all the graphs of types $\mathcal{SG}^+$ (resp., $\mathcal{SG}^-$, $\mathcal{WG}^+$, and $\mathcal{WG}^-$) associated with the pairs $(u_i, t_j)$ such that $u_i$ showed a strongly positive (resp., strongly negative, weakly positive, weakly negative) sentiment on $t_j$. Then, we averaged the values obtained for each graph type. In Table 7, we report the corresponding results.

**Table 7.** Average values of density and average clustering coefficient for the graphs of type $\mathcal{SG}^+$, $\mathcal{SG}^-$, $\mathcal{WG}^+$ and $\mathcal{WG}^-$.

| *Property* | $\mathcal{SG}^+$ | $\mathcal{SG}^-$ | $\mathcal{WG}^+$ | $\mathcal{WG}^-$ |
|---|---|---|---|---|
| Average Density | 0.0242 | 0.0288 | 0.162 | 0.0184 |
| Average Clustering Coefficient | 0.2215 | 0.2417 | 0.1918 | 0.2012 |

From the analysis of this table, we can see that the values of the density and the average clustering coefficient of the graph $\mathcal{SG}^+$ (resp., $\mathcal{SG}^-$, $\mathcal{WG}^+$ and $\mathcal{WG}^-$) are much higher than those of the graph $\mathcal{A}$. This can be explained by considering how the graph $\mathcal{SG}^+$ (resp., $\mathcal{SG}^-$, $\mathcal{WG}^+$ and $\mathcal{WG}^-$) is constructed. In fact, such a construction starts from a node serving as the root and gradually adds nodes belonging to the various neighborhoods of the root, along with the corresponding arcs, as long as the conditions expressed in the function $\psi^+()$ (respectively, $\psi^-()$, $\xi^+()$, $\xi^-()$) in Equation (7) (resp., (8)–(10)) are satisfied. This way of proceeding tends to favor the construction of dense and compact graphs obtained as subgraphs of the connected component of $\mathcal{A}$ on which their root node is located. When the boundary of the connected component is reached, the construction of $\mathcal{SG}^+$ (resp., $\mathcal{SG}^-$, $\mathcal{WG}^+$ and $\mathcal{WG}^-$) stops. Such a construction also tends to stop when arriving at sparse areas of the graph $\mathcal{A}$.

Another important information we can derive from examining Table 7 concerns the fact that the density of the graphs associated with strong sentiments is greater than that of the graphs associated with weak sentiments. This difference becomes much less marked if we consider the average clustering coefficient instead of the density. In contrast, there is no great difference between the parameters of the graphs associated with positive sentiments and those of the graphs associated with negative sentiments. This result, coupled with the ones obtained in the previous sections, suggests to us that the negativity of a sentiment is able to increase the intensity of its transmission but it is not able to increase, except marginally, the number of connections activated by users for its transmission.

### 5.5. Experiments on Temporal Scope

#### 5.5.1. Variation of the Scope over Time for Each Sentiment Type

This test is dual to the one we conducted for the spatial scope in Section 5.4.1. In fact, it aims to evaluate the trend of the sentiment degree over time and how it differs for different sentiment types. The time interval we considered is the reference interval for our dataset, which is the interval from 25 February 2022 to 25 March 2022.

In Section 4.3, we have seen that, given a user $u_i$ and a topic $t_j$, the temporal scope of $u_i$ on $t_j$ in the time interval $T[x..y]$ is represented by an ordered list of pairs (see Equation (18)), one for each time slice in the interval. The generic pair $(\tau_b, \theta_b)$ denotes the sentiment type $(\tau_b)$ and the sentiment degree $(\theta_b)$. Recall that our model associates only one sentiment type with a user $u_i$ and a topic $t_j$ in a time slice $T_b$. Both values can vary when passing from one time slice to another.

We carried out this experiment as follows: given a time slice $T_b$ (which coincided in practice with a day of the time interval relative to our dataset), we identified all possible pairs $(u_i, t_j)$ such that, in the time interval $T_b$, the user $u_i$ expressed a sentiment on the topic $t_j$. Then, for each of these pairs, we determined the sentiment type $\tau_{ij_b}$ expressed by $u_i$ on $t_j$ in $T_b$ and the corresponding sentiment degree $\theta_{ij_b}$.

At this point, we partitioned the pairs $(u_i, t_j)$ based on the corresponding sentiment types in $T_b$ and, for each partition, we computed the average value of the sentiment degree. In this way, we obtained four average values of sentiment degree, i.e., $\overline{\theta_b^{sn}}$, $\overline{\theta_b^{sp}}$, $\overline{\theta_b^{wn}}$ and $\overline{\theta_b^{wp}}$, one for each sentiment type. Finally, we repeated these tasks for each time slice of the considered interval. The results obtained are shown in Figure 4, while in Table 8 we report the values of some statistical measures computed over the whole time period of interest for the four cases under consideration.

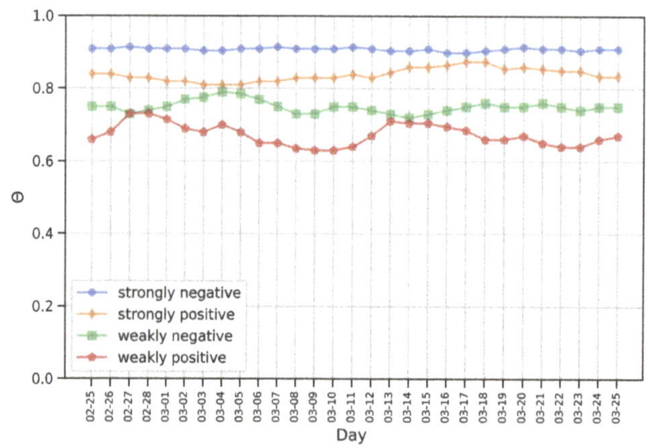


**Figure 4.** Variation over time of the average value of the sentiment degree associated with each sentiment type.

**Table 8.** Values of some statistic measures computed over the whole time period for *sn*, *sp*, *wn*, and *wp*.

| *Parameter* | *sn* | *sp* | *wn* | *wp* |
|---|---|---|---|---|
| Max | 0.92 | 0.88 | 0.79 | 0.73 |
| Min | 0.90 | 0.81 | 0.72 | 0.63 |
| Mean | 0.91 | 0.84 | 0.75 | 0.67 |
| Standard deviation | 0.003 | 0.018 | 0.017 | 0.029 |

From the analysis of Figure 4 and Table 8 we can derive some interesting knowledge patterns on temporal scope. First, we observe that the values of the average sentiment degree are generally very high, since they range from a maximum of 0.92 to a minimum of 0.63.

In addition, we can observe that the trend of the average sentiment degree for strong sentiments is generally higher than that for the corresponding weak sentiments. In fact, Table 8 shows that the average sentiment degree is equal to 0.91 and 0.84 for strong sentiments, while it is equal to 0.75 and 0.67 for weak ones. This confirms what we had already found in Section 5.4.1 for spatial scope. In addition to this, we can observe that the trend over time for strong sentiments is more constant than for weak ones. In fact, in Table 8, we can see that the standard deviation of the sentiment degree is equal to 0.003 and 0.018 for strong sentiments, while it is equal to 0.017 and 0.029 for weak ones. This is a new knowledge pattern about the trend of sentiment degree that we were able to obtain thanks to the introduction of temporal scope in this paper. It is in line with the previous results in the literature regarding strong and weak sentiments [71,73,74]. It can be explained by considering that strong sentiments correspond to very marked polarizations, and thus are unlikely to change over time, contrary to what happens for weak sentiments.

A second interesting result that can be observed from Figure 4 and Table 8 concerns the trends of negative versus positive sentiments. In fact, we can observe that the values of negative sentiment degrees are on average higher than those of positive sentiment degrees. As evidence of this, Table 8 shows that the average sentiment degree is equal to 0.91 and 0.75 for negative sentiments, while it is equal to 0.84 and 0.67 for positive ones. This represents a confirmation of the results already found for the spatial scope in Section 5.4.1. In addition to this, we can observe that the time trend for negative sentiments is more constant than that for the corresponding positive sentiments. In fact, in Table 8 the standard deviation of the sentiment degree is equal to 0.003 and 0.017 for negative sentiments, while it is equal to 0.018 and 0.029 for positive ones. The latter knowledge pattern is new to the literature and could only be extracted due to the introduction of temporal scope. It is in line with the previous results found in the literature regarding positive and negative sentiments [68–72].

By integrating all the derived information, it follows that the strongest and most stable sentiment is the strongly negative one; it is unlikely to be changed over time. In contrast, the ficklest sentiment is the weakly positive one. Indeed, it can be modified over time by acting appropriately on users. As for the modification possibility, the strongly positive and the weakly negative sentiments lie somewhere between the two extremes.

#### 5.5.2. Analysis of User Stereotypes

In the previous section, we focused on the temporal variation of the average values of sentiment degree. Instead, in this section, we want to analyze the temporal variation of the sentiment degree of single users on specific topics. In particular, we want to define some user stereotypes and check whether and to what extent they are present in our dataset. More specifically, the stereotypes we define are reported in Table 9. It is worth pointing out that these are stereotypes defined by us taking into consideration the semantics of the various sentiment types and the potential usefulness of them. However, new stereotypes may be defined in the future, should the need arise.

**Table 9.** Some possible user stereotypes.

| *User Stereotype* | **Definition** |
|---|---|
| sp-user (strongly positive user) on $t_j$ | This is a user who always showed a sentiment of type $sp$ on $t_j$ when she expressed her opinions during the time interval $T[x..y]$. |
| sn-user (strongly negative user) on $t_j$ | Similar to the sp-user but with $sn$ instead of $sp$. |
| wp-user (weakly positive user) on $t_j$ | Similar to the sp-user but with $wp$ instead of $sp$. |
| wn-user (weakly negative user) on $t_j$ | Similar to the sp-user but with $wn$ instead of $sp$. |
| nn-user (non-negative user) on $t_j$ | Similar to the sp-user but with $sp$ or $wp$ instead of $sp$ (Recall that a user can show only one sentiment type on a topic $t_j$ in a time slice; however, she can show different sentiments on $t_j$ in different time slices of $T[x..y]$). |
| np-user (non-positive user) on $t_j$ | Similar to the sp-user but with $sn$ or $wn$ instead of $sp$. |
| w-user (weak user) on $t_j$ | Similar to the sp-user but with $wp$ or $wn$ instead of $sp$. |

**Table 9.** *Cont.*

| *User Stereotype* | **Definition** |
|---|---|
| s-user (strong user) on $t_j$ | Similar to the sp-user but with $sp$ or $sn$ instead of $sp$. |
| super-sp-user (super strongly positive user) | This is a user who always showed a sentiment of type $sp$ on all the topics she discussed during $T[x..y]$. |
| super-sn-user (super strongly negative user) | Similar to the super-sp-user but with $sn$ instead of $sp$. |
| super-wp-user (super weakly positive user) | Similar to the super-sp-user but with $wp$ instead of $sp$. |
| super-wn-user (super weakly negative user) | Similar to the super-sp-user but with $wn$ instead of $sp$. |
| super-nn-user (super non-negative user) | Similar to the super-sp-user but with $sp$ or $wp$ instead of $sp$. |
| super-np-user (super non-positive user) | Similar to the super-sp-user but with $sn$ or $wn$ instead of $sp$. |
| super-w-user (super weak user) | Similar to the super-sp-user but with $wp$ or $wn$ instead of $sp$. |
| super-s-user (super strong user) | Similar to the super-sp-user but with $sp$ or $sn$ instead of $sp$. |
| sw-user (swinging user) on $t_j$ | This is a user who showed all the four sentiment types on $t_j$ during $T[x..y]$. |
| ss-user (super swinging user) | This is a user who behaved as a sw-user on every topic she discussed during $T[x..y]$. |
| p-user (posed user) | This is a user who was sp-user for at least one topic, sn-user for at least a second topic, wp-user for at least a third topic, and wn-user for at least a fourth topic. In other words, she demonstrated the ability to express the full range of sentiments depending on the topic. |

After defining stereotypes, we computed how many users in our dataset could be associated with each of them. Recall that our dataset includes 4219 users and 101 topics. The number of potential pairs $(u_i, t_j)$, such that $u_i$ is a user and $t_j$ is a topic, is 426,119, while the number of actual pairs in the dataset is 130,794. The number of users associated with the various stereotypes that we defined is reported in Table 10.

From the analysis of this table, we can deduce some interesting insights. First, we observe that: *(i)* the number of sn-users (resp., super-sn-users) is greater than the one of sp-users (resp., super-sp-users); *(ii)* the number of wn-users (resp., super-wn-users) is greater than the one of wp-users (resp., super-wp-users); *(iii)* the number of np-users (resp., super-np-users) is greater than the one of nn-users (resp., super-nn-users). This is in line with what we have seen in the previous sections and in the literature regarding the trend of positive and negative user sentiments. Similarly, we can observe that: *(i)* the number of sn-users (resp., super-sn-users) is greater than the one of wn-users (resp., super-wn-users); *(ii)* the number of sp-users (resp., super-sp-users) is greater than the one of wp-users (resp., super-wp-users). This result is also in line with what we have seen in the previous sections and in the literature regarding strong and weak sentiments.

On the other hand, it is interesting to note that the number of w-users (resp., super-w-users) is greater than the one of s-users (resp., super-s-users). This might seem a contradiction to the previous results in this section and to the ones in the previous sections. In fact, this is not the case; this phenomenon can be explained by taking into account that the sentiments $wp$ and $wn$ are somewhat "contiguous". Therefore, it is easier for a user to switch from one to the other without ever reaching strong sentiments. In contrast, the sentiments $sp$ and $sn$ are extreme; to be s-user or super-s-user, one could have to oscillate between these two extreme sentiments without ever going through the weak sentiments that lie in between. This is much more difficult than a context in which a user oscillates between two weak sentiments.

A further observation concerns the very low number of swinging users. This is explained by the fact that it is really difficult for a user to have four different sentiment types on the same topic. Even the number of super-s-users is so low that we can assume that their presence is more of a bias than anything else.

Finally, it is worth pointing out that more than half of the users in our dataset are p-users. In our opinion, this is an extremely positive result because it tells us that the users in our dataset were really able to express the full range of possible sentiments depending on the reference topic and time slice.

**Table 10.** Number of users associated with each stereotype.

| *Stereotype* | Number of Users |
|---|---|
| sp-user (strongly positive user) on $t_j$ | 1211 |
| sn-user (strongly negative user) on $t_j$ | 1274 |
| wp-user (weakly positive user) on $t_j$ | 1058 |
| wn-user (weakly negative user) on $t_j$ | 1142 |
| nn-user (non-negative user) on $t_j$ | 2119 |
| np-user (non-positive user) on $t_j$ | 2497 |
| w-user (weak user) on $t_j$ | 1714 |
| s-user (strong user) on $t_j$ | 1134 |
| super-sp-user (super strongly positive user) | 72 |
| super-sn-user (super strongly negative user) | 88 |
| super-wp-user (super weakly positive user) | 48 |
| super-wn-user (super weakly negative user) | 53 |
| super-nn-user (super non-negative user) | 221 |
| super-np-user (super non-positive user) | 244 |
| super-w-user (super weak user) | 174 |
| super-s-user (super strong user) | 118 |
| sw-user (swinging user) on $t_j$ | 42 |
| ss-user (super swinging user) | 2 |
| p-user (posed user) | 2284 |

#### 5.5.3. Discussion

In this section, we present a discussion regarding the framework proposed in this paper. In particular, we present a brief overview of its main strengths, limitations, and practical applications.

Regarding the first point, the main strengths of our framework are the following: (i) it defines a multi-dimensional view of the concept of scope; (ii) it can operate on any social platform as long as the messages exchanged in the latter are predominantly text-based; and (iii) it can assess the scope of user sentiment on any topic.

Our framework also has some limitations. Specifically: (i) it operates on only one social platform at a time, whereas the various platforms are currently interconnected because many users join simultaneously on multiple platforms, acting as bridges among them; (ii) it can currently handle only two possible dimensions, namely space and time; (iii) it is unable to evaluate and analyze the possible interference that different users may exert on a given user in defining her sentiment on a topic; (iv) it is based on text analysis and, consequently, works on social platforms where the messages exchanged are predominantly textual.

A first possible practical application of our framework involves supporting information diffusion. In fact, the knowledge of scope and its dynamics can facilitate in identifying new strategies to spread certain messages as widely as possible. A second application, dual to the previous one, consists in countering fake news. The latter, in fact, often arouse a strongly negative sentiment. Exploiting this characteristic and the concept of scope makes it possible to define an approach for identifying fake news and countering their spread.

Last but not least, one could use our framework in order to identify certain user stereotypes (think, for instance, of the swinging and posed users introduced in Section 5.5.2).

## 6. Conclusions

In this paper, we have proposed a framework to determine the spatial and temporal scope of the sentiment of a user on a topic in a social platform. First, we have presented the concept of scope and we have seen that it summarizes several concepts, such as centrality, reputation, and diffusion, introduced in the past Social Network Analysis literature. In fact, all these concepts represent different aspects of the concept of scope. Then, we have introduced the concept of scope of the sentiment of a user on a topic and we have defined a model capable of representing and handling a multi-dimensional view of scope. Afterwards, we have proposed a set of parameters and an approach for evaluating the spatial and the temporal scope of the sentiment of a user on a topic in a social platform. Finally, we have performed a set of tests to evaluate the proposed framework on a real dataset obtained from Reddit.

The main novelties of this paper are the following: (i) it proposes a multi-dimensional view of scope, particularizing it to space and time dimensions; (ii) it introduces the concept of scope of the sentiment of a user on one or more topics; and (iii) it presents a general framework for extracting information about the scope of the sentiment of a user on topics of any subject; this framework is capable of operating on any social platform.

The main results and findings obtained by applying our framework on our Reddit dataset can be summarized as follows: (i) negative sentiments tend to spread more easily than positive ones; (ii) strong sentiments tend to spread more easily than weak ones; (iii) the influence of a user on the sentiment felt by her neighbors tends to decrease when the neighbors' distance from her increases; (iv) users with negative sentiments influence their neighbors more than users with positive sentiments; (v) users with strong sentiments influence their neighbors more than users with weak sentiments; (vi) the influence of users with strongly negative sentiment is not affected by any interference from other users; (vii) the negativity of a sentiment can increase the intensity of its transmission but cannot increase the number of connections activated by users for its transmission; (viii) the temporal trend for strong sentiments is more constant than the one for weak sentiments; (ix) the average degree of strong sentiments over time is generally higher than the average degree of weak sentiments; (x) the average degree of negative sentiments over time is generally higher than the average degree of positive sentiments; (xi) the time trend of negative sentiments is more constant than the one of positive sentiments; (xii) the number of users with negative sentiments is greater than the number of users with positive sentiments; (xiii) the number of swinging users (i.e., users who felt all the four possible sentiment types for the same topic) is negligible; and (xiv) the number of posed users (i.e., users who were capable of feeling the full range of sentiments depending on the topic) is very high.

The ideas proposed in this paper have several possible future developments. First, we plan to extend the concepts proposed here from a single network to a Social Internetworking System, that is, a set of interrelated networks in which each user may join one or more of them. Second, we would like to study further dimensions of the scope of the sentiment of a user on a topic in a social network, in addition to the spatial and temporal ones considered in this paper. Third, we would like to further study the interference of multiple users on the sentiment of a user $u_i$ on a topic $t_j$. In particular, we would like to analyze the case in which the interfering users are very close, as well as the case in which they have high scope and, therefore, the interference caused by each of them may be significant or, even, decisive. Last but not least, we plan to investigate the possible use of our framework to health economics applications [75] and computational linguistics [76].

**Author Contributions:** Conceptualization, G.B. and L.V.; methodology, F.C. and D.U.; software, L.S.; validation, E.C. and M.M.; formal analysis, G.B. and F.C.; investigation, E.C. and D.U.; resources, M.M. and L.V.; data curation, G.B. and E.C.; writing—original draft preparation, F.C. and L.V.;

writing—review and editing, M.M. and D.U.; visualization, L.S.; supervision, G.B., E.C. and M.M.; project administration, F.C., D.U. and L.V. All authors have read and agreed to the published version of the manuscript.

**Funding:** This research received no external funding.

**Institutional Review Board Statement:** Not applicable.

**Informed Consent Statement:** Not applicable.

**Data Availability Statement:** We used data accessible through `pushshift.io`.

**Conflicts of Interest:** The authors declare no conflict of interest.

## References


1. Leggio, D.; Marra, G.; Ursino, D. Defining and investigating the scope of users and hashtags in Twitter. In Proceedings of the International Conference on Ontologies, DataBases, and Applications of Semantics (ODBASE 2014), Amantea, Italy, 27–31 October 2014; pp. 674–681.
2. Cauteruccio, F.; Cinelli, L.; Fortino, G.; Savaglio, C.; Terracina, G.; Ursino, D.; Virgili, L. An Approach to Compute the Scope of a Social Object in a Multi-IoT Scenario. *Pervasive Mob. Comput.* **2020**, *67*, 101223. [CrossRef]
3. Kempe, D.; Kleinberg, J.; Tardos, É. Maximizing the spread of influence through a social network. In Proceedings of the International ACM SIGKDD Conference on Knowledge Discovery and Data Mining (KDD 2003), Washington, DC, USA, 24–27 August 2003; pp. 137–146.
4. Ma, Z.; Sun, A.; Cong, G. Will this #Hashtag be Popular Tomorrow? In Proceedings of the ACM SIGIR International Conference on Research and Development in Information Retrieval (SIGIR 2012), Portland, OR, USA, 12–16 August 2012; pp. 1173–1174.
5. Ma, Z.; Sun, A.; Cong, G. On Predicting the Popularity of Newly Emerging Hashtags in Twitter. *J. Am. Soc. Inf. Sci. Technol.* **2013**, *64*, 1399–1410. [CrossRef]
6. Miller, Z.; Dickinson, B.; Deitrick, W.; Hu, W.; Wang, A.H. Twitter Spammer Detection Using Data Stream Clustering. *Inf. Sci.* **2014**, *260*, 64–73. [CrossRef]
7. Romero, D.; Galuba, W.; Asur, S.; Huberman, B. Influence and passivity in social media. In Proceedings of the International Conference on World Wide Web (WWW'11), Hyderabad, India, 28 March–1 April 2011; pp. 113–114.
8. Weng, J.; Lim, E.; Jiang, J.; He, Q. TwitterRank: Finding Topic-sensitive Influential Twitterers. In Proceedings of the ACM International Conference on Web Search and Data Mining (WSDM 2010), New York, NY, USA, 3–6 February 2010; pp. 261–270.
9. Cataldi, M.; Caro, L.D.; Schifanella, C. Emerging Topic Detection on Twitter Based on Temporal and Social Terms Evaluation. In Proceedings of the International Workshop on Multimedia Data Mining (MDMKDD 2010), Washington, DC, USA, 25 July 2010; pp. 4–13.
10. Qasem, Z.; Jansen, M.; Hecking, T.; Hoppe, H. On the detection of influential actors in social media. In Proceedings of the International Conference on Signal-Image Technology & Internet-Based Systems (SITIS'15), Sorrento, Italy, 26–29 November 2015; pp. 421–427.
11. Yue, L.; Chen, W.; Li, X.; Zuo, W.; Yin, M. A survey of sentiment analysis in social media. *Knowl. Inf. Syst.* **2019**, *60*, 617–663. [CrossRef]
12. Pozzi, F.A.; Fersini, E.; Messina, E.; Liu, B. Challenges of sentiment analysis in social networks: An overview. *Sentim. Anal. Soc. Netw.* **2017**, 1–11. [CrossRef]
13. Yadav, A.; Vishwakarma, D. Sentiment analysis using deep learning architectures: A review. *Artif. Intell. Rev.* **2020**, *53*, 4335–4385. [CrossRef]
14. Birjali, M.; Kasri, M.; Beni-Hssane, A. A comprehensive survey on sentiment analysis: Approaches, challenges and trends. *Knowl.-Based Syst.* **2021**, *226*, 107134. [CrossRef]
15. Cortis, K.; Davis, B. Over a decade of social opinion mining: A systematic review. *Artif. Intell. Rev.* **2021**, *54*, 4873–4965. [CrossRef]
16. Basile, V.; Cauteruccio, F.; Terracina, G. How dramatic events can affect emotionality in social posting: The impact of COVID-19 on reddit. *Future Internet* **2021**, *13*, 29. [CrossRef]
17. Lai, M.; Tambuscio, M.; Patti, V.; Ruffo, G.; Rosso, P. Stance polarity in political debates: A diachronic perspective of network homophily and conversations on Twitter. *Data Knowl. Eng.* **2019**, *124*, 101738. [CrossRef]
18. Ramachandran, D.; Parvathi, R. A novel domain and event adaptive tweet augmentation approach for enhancing the classification of crisis related tweets. *Data Knowl. Eng.* **2021**, *135*, 101913. [CrossRef]
19. Jelodar, H.; Wang, Y.; Yuan, C.; Feng, X.; Jian, X.; Li, Y.; Zhao, L. Latent Dirichlet Allocation (LDA) and topic modeling: Models, applications, a survey. *Multimed. Tools Appl.* **2019**, *78*, 15169–15211. [CrossRef]
20. Vayansky, I.; Kumar, S. A review of topic modeling methods. *Inf. Syst.* **2020**, *94*, 101582. [CrossRef]
21. Qiang, J.; Qian, Z.; Li, Y.; Yuan, Y.; Wu, X. Short Text Topic Modeling Techniques, Applications, and Performance: A Survey. *IEEE Trans. Knowl. Data Eng.* **2022**, *34*, 1427–1445. [CrossRef]
22. Ravi, K.; Ravi, V. A survey on opinion mining and sentiment analysis: Tasks, approaches and applications. *Knowl.-Based Syst.* **2015**, *89*, 14–46. [CrossRef]

23. Tsvetovat, M.; Kouznetsov, A. *Social Network Analysis for Startups: Finding Connections on the Social Web*; O'Reilly Media, Inc.: Newton, MA, USA, 2011.
24. Moazzami, D. Toughness of the Networks with Maximum Connectivity. *J. Algorithms Comput.* **2015**, *46*, 51–71.
25. Khoshnood, A.; Moazzami, D. A Survey on Tenacity Parameter—Part I. *J. Algorithms Comput.* **2021**, *53*, 181–196.
26. Moazzami, D.; Khoshnood, A. A Survey on Tenacity Parameter—Part II. *J. Algorithms Comput.* **2022**, *54*, 47–72.
27. Bonchi, F.; Castillo, C.; Gionis, A.; Jaimes, A. Social network analysis and mining for business applications. *ACM Trans. Intell. Syst. Technol. (TIST)* **2011**, *2*, 1–37. [CrossRef]
28. Scott, J. Social network analysis: Developments, advances, and prospects. *Soc. Netw. Anal. Min.* **2011**, *1*, 21–26. [CrossRef]
29. Cantini, R.; Marozzo, F.; Talia, D.; Trunfio, P. Analyzing political polarization on social media by deleting bot spamming. *Big Data Cogn. Comput.* **2022**, *6*, 3. [CrossRef]
30. Bayrakdar, S.; Yucedag, I.; Simsek, M.; Dogru, I.A. Semantic analysis on social networks: A survey. *Int. J. Commun. Syst.* **2020**, *33*, e4424. [CrossRef]
31. Pankong, N.; Prakancharoen, S.; Buranarach, M. A combined semantic social network analysis framework to integrate social media data. In Proceedings of the International Conference on Knowledge and Smart Technology (KST'12), Chonburi, Thailand, 7–8 July 2012; pp. 37–42.
32. Xia, Z.; Bu, Z. Community detection based on a semantic network. *Knowl.-Based Syst.* **2012**, *26*, 30–39. [CrossRef]
33. Ismail, H.; Khalil, A.; Hussein, N.; Elabyad, R. Triggers and Tweets: Implicit Aspect-Based Sentiment and Emotion Analysis of Community Chatter Relevant to Education Post-COVID-19. *Big Data Cogn. Comput.* **2022**, *6*, 99. [CrossRef]
34. Yeasmin, N.; Mahbub, N.I.; Baowaly, M.K.; Singh, B.C.; Alom, Z.; Aung, Z.; Azim, M.A. Analysis and Prediction of User Sentiment on COVID-19 Pandemic Using Tweets. *Big Data Cogn. Comput.* **2022**, *6*, 65. [CrossRef]
35. Poulopoulos, V.; Wallace, M. Social Media Analytics as a Tool for Cultural Spaces—The Case of Twitter Trending Topics. *Big Data Cogn. Comput.* **2022**, *6*, 63. [CrossRef]
36. Gallacher, J.; Bright, J. Hate Contagion: Measuring the spread and trajectory of hate on social media. *PsyArXiv* **2021**. [CrossRef]
37. Yin, F.; Xia, X.; Pan, Y.; She, Y.; Feng, X.; Wu, J. Sentiment mutation and negative emotion contagion dynamics in social media: A case study on the Chinese Sina Microblog. *Inf. Sci.* **2022**, *594*, 118–135. [CrossRef]
38. Pröllochs, N.; Bär, D.; Feuerriegel, S. Emotions explain differences in the diffusion of true vs. false social media rumors. *Sci. Rep.* **2021**, *11*, 22721. [CrossRef]
39. Almars, A.; Li, X.; Zhao, X. Modelling user attitudes using hierarchical sentiment-topic model. *Data Knowl. Eng.* **2019**, *119*, 139–149. [CrossRef]
40. Yang, Z.; Kotov, A.; Mohan, A.; Lu, S. Parametric and non-parametric user-aware sentiment topic models. In Proceedings of the International ACM SIGIR Conference on Research and Development in Information Retrieval (SIGIR'15), Santiago, Chile, 9–13 August 2015; pp. 413–422.
41. Naskar, D.; Mokaddem, S.; Rebollo, M.; Onaindia, E. Sentiment analysis in social networks through topic modeling. In Proceedings of the International Conference on Language Resources and Evaluation (LREC'16), Portorož, Slovenia, 23–28 May 2016; pp. 46–53.
42. Liu, B.; Zhang, L. A survey of opinion mining and sentiment analysis. In *Mining Text Data*; Springer: Boston, MA, USA, 2012; pp. 415–463.
43. Neves-Silva, R.; Gamito, M.; Pina, P.; Campos, A.R. Modelling influence and reach in sentiment analysis. *Procedia CIRP* **2016**, *47*, 48–53. [CrossRef]
44. Carvalho, J.; Rosa, H.; Brogueira, G.; Batista, F. MISNIS: An intelligent platform for twitter topic mining. *Expert Syst. Appl.* **2017**, *89*, 374–388. [CrossRef]
45. Ferrara, E.; Yang, Z. Quantifying the effect of sentiment on information diffusion in social media. *PeerJ Comput. Sci.* **2015**, *1*, e26. [CrossRef]
46. Zhao, K.; Yen, J.; Greer, G.; Qiu, B.; Mitra, P.; Portier, K. Finding influential users of online health communities: A new metric based on sentiment influence. *J. Am. Med. Inform. Assoc.* **2014**, *21*, e212–e218. [CrossRef]
47. Cao, N.; Lu, L.; Lin, Y.; Wang, F.; Wen, Z. Socialhelix: Visual analysis of sentiment divergence in social media. *J. Vis.* **2015**, *18*, 221–235. [CrossRef]
48. Kušen, E.; Strembeck, M.; Cascavilla, G.; Conti, M. On the influence of emotional valence shifts on the spread of information in social networks. In Proceedings of the IEEE/ACM International Conference on Advances in Social Networks Analysis and Mining (ASONAM '17), Sydney Australia, 31 July–3 August 2017; pp. 321–324.
49. Zafarani, R.; Cole, W.D.; Liu, H. Sentiment propagation in social networks: A case study in livejournal. In Proceedings of the International Conference on Social Computing, Behavioral Modeling, and Prediction (SBP'10), Bethesda, MD, USA, 29 March–1 April 2010; pp. 413–420.
50. Melton, C.A.; Olusanya, O.A.; Ammar, N.; Shaban-Nejad, A. Public sentiment analysis and topic modeling regarding COVID-19 vaccines on the Reddit social media platform: A call to action for strengthening vaccine confidence. *J. Infect. Public Health* **2021**, *14*, 1505–1512. [CrossRef]
51. An, L.; Zhou, W.; Ou, M.; Li, G.; Yu, C.; Wang, X. Measuring and profiling the topical influence and sentiment contagion of public event stakeholders. *Int. J. Inf. Manag.* **2021**, *58*, 102327. [CrossRef]

52. Cai, M.; Luo, H.; Meng, X.; Cui, Y. Topic-emotion propagation mechanism of public emergencies in social networks. *Sensors* **2021**, *21*, 4516. [CrossRef]
53. Cai, M.; Luo, H.; Meng, X.; Cui, Y. A Study on the Topic-Sentiment Evolution and Diffusion in Time Series of Public Opinion Derived from Emergencies. *Complexity* **2021**, *2021*, 2069010. [CrossRef]
54. Xu, Y.; Li, Y.; Liang, Y.; Cai, L. Topic-sentiment evolution over time: A manifold learning-based model for online news. *J. Intell. Inf. Syst.* **2020**, *55*, 27–49. [CrossRef]
55. Wang, X.; Jin, D.; Musial, K.; Dang, J. Topic enhanced sentiment spreading model in social networks considering user interest. In Proceedings of the AAAI Conference on Artificial Intelligence (AAAI'20), New York, NY, USA, 7–12 February 2020; Volume 34, pp. 989–996.
56. Tsugawa, S.; Ohsaki, H. Negative messages spread rapidly and widely on social media. In Proceedings of the International Conference on Online Social Networks (COSN'15), Palo Alto, CA, USA, 2–3 November 2015; pp. 151–160.
57. Heimbach, I.; Hinz, O. The impact of content sentiment and emotionality on content virality. *Int. J. Res. Mark.* **2016**, *33*, 695–701. [CrossRef]
58. Majumder, N.; Poria, S.; Peng, H.; Chhaya, N.; Cambria, E.; Gelbukh, A. Sentiment and Sarcasm Classification With Multitask Learning. *IEEE Intell. Syst.* **2019**, *34*, 38–43. [CrossRef]
59. Baumgartner, J.; Zannettou, S.; Keegan, B.; Squire, M.; Blackburn, J. The pushshift Reddit dataset. In Proceedings of the International AAAI Conference on Web and Social Media (ICWSM'20), Atlanta, Georgia, USA, 8–11 June 2020; Volume 14, pp. 830–839.
60. Mills, R. Reddit.Com: A Census of Subreddits. In Proceedings of the International Web Science Conference (WebSci'15), Oxford, UK, 28 June–1 July 2015; pp. 49:1–49:2.
61. Guimaraes, A.; Balalau, O.; Terolli, E.; Weikum, G. Analyzing the Traits and Anomalies of Political Discussions on Reddit. In Proceedings of the International Conference on Web and Social Media (ICWSM 2019), Münich, Germany, 11–14 June 2019; pp. 205–213.
62. Horne, B.; Adali, S. The impact of crowds on news engagement: A reddit case study. In Proceedings of the International AAAI Conference on Web and Social Media (ICWSM'17), Montreal, QC, Canada, 15–18 May 2017; p. 11.
63. Grootendorst, M. BERTopic: Neural topic modeling with a class-based TF-IDF procedure. *arXiv* **2022**, arXiv:2203.05794.
64. Vaswani, A.; Shazeer, N.; Parmar, N.; Uszkoreit, J.; Jones, L.; Gomez, A.N.; Kaiser, Ł.; Polosukhin, I. Attention is All you Need. *Adv. Neural Inf. Process. Syst.* **2017**, *30*. [CrossRef]
65. Buitinck, L.; Louppe, G.; Blondel, M.; Pedregosa, F.; Mueller, A.; Grisel, O.; Niculae, V.; Prettenhofer, P.; Gramfort, A.; Grobler, J.; et al. API design for machine learning software experiences from the scikit-learn project. In Proceedings of the European Conference on Machine Learning and Principles and Practices of Knowledge Discovery in Databases (ECMP/PKDD 2013), Prague, Czech Republic, 23–27 September 2013; pp. 108–122.
66. Loureiro, D.; Barbieri, F.; Neves, L.; Anke, L.; Camacho-collados, J. TimeLMs: Diachronic Language Models from Twitter. In Proceedings of the Annual Meeting of the Association for Computational Linguistics: System Demonstrations (ACL'22), Dublin, Ireland, 22–27 May 2022; pp. 251–260.
67. Barbieri, F.; Camacho-Collados, J.; Anke, L.E.; Neves, L. TweetEval: Unified Benchmark and Comparative Evaluation for Tweet Classification. In Proceedings of the Findings of the Association for Computational Linguistics (EMNLP'20), Online, 16–20 November 2020; pp. 1644–1650.
68. Yu, H.; Yang, C.; Yu, P.; Liu, K. Emotion diffusion effect: Negative sentiment COVID-19 tweets of public organizations attract more responses from followers. *PLoS ONE* **2022**, *17*, e0264794. [CrossRef]
69. Schöne, J.P.; Parkinson, B.; Goldenberg, A. Negativity spreads more than positivity on Twitter after both positive and negative political situations. *Affect. Sci.* **2021**, *2*, 379–390. [CrossRef]
70. Cinelli, M.; Pelicon, A.; Mozetič, I.; Quattrociocchi, W.; Novak, P.K.; Zollo, F. Dynamics of online hate and misinformation. *Sci. Rep.* **2021**, *11*, 1–12. [CrossRef]
71. Stieglitz, S.; Dang-Xuan, L. Emotions and information diffusion in social media—sentiment of microblogs and sharing behavior. *J. Manag. Inf. Syst.* **2013**, *29*, 217–248. [CrossRef]
72. Suh, B.; Hong, L.; Pirolli, P.; Chi, E.H. Want to be retweeted? large scale analytics on factors impacting retweet in twitter network. In Proceedings of the International Conference on Social Computing (SOCIALCOM '10), Minneapolis, MN, USA, 20–22 August 2010; pp. 177–184.
73. Rimé, B. The social sharing of emotion as an interface between individual and collective processes in the construction of emotional climates. *J. Soc. Issues* **2007**, *63*, 307–322. [CrossRef]
74. Rimé, B.; Finkenauer, C.; Luminet, O.; Zech, E.; Philippot, P. Social sharing of emotion: New evidence and new questions. *Eur. Rev. Soc. Psychol.* **1998**, *9*, 145–189. [CrossRef]
75. Shin, E. Physician Connectedness and Referral Choice. *Oxford Bulletin of Economics and Statistics*; Wiley Online Library: New York, NY, USA, 2022.
76. Ott, M.; Choi, Y.; Cardie, C.; Hancock, J. Finding deceptive opinion spam by any stretch of the imagination. *arXiv* **2011**, arXiv:1107.4557.

*big data and cognitive computing*

MDPI

*Article*

# Using an Evidence-Based Approach for Policy-Making Based on Big Data Analysis and Applying Detection Techniques on Twitter

**Somayeh Labafi [1,*], Sanee Ebrahimzadeh [2], Mohamad Mahdi Kavousi [3], Habib Abdolhossein Maregani [4] and Samad Sepasgozar [5]**

1 Iranian Research Institute for Information Science and Technology (IranDoc), Tehran 1314156545, Iran
2 Department of Media Management, University of Tehran, Tehran 1411713114, Iran
3 Institute of Higher Education of Ershad Damavand, Tehran 1349933851, Iran
4 Department of Business Management, University of Tehran, Tehran 1738953355, Iran
5 Faculty of Arts, Design and Architecture, The University of New South Wales, Sydney, NSW 2052, Australia
* Correspondence: labafi@irandoc.ac.ir

**Abstract:** Evidence-based policy seeks to use evidence in public policy in a systematic way in a bid to improve decision-making quality. Evidence-based policy cannot work properly and achieve the expected results without accurate, appropriate, and sufficient evidence. Given the prevalence of social media and intense user engagement, the question to ask is whether the data on social media can be used as evidence in the policy-making process. The question gives rise to the debate on what characteristics of data should be considered as evidence. Despite the numerous research studies carried out on social media analysis or policy-making, this domain has not been dealt with through an "evidence detection" lens. Thus, this study addresses the gap in the literature on how to analyze the big text data produced by social media and how to use it for policy-making based on evidence detection. The present paper seeks to fill the gap by developing and offering a model that can help policy-makers to distinguish "evidence" from "non-evidence". To do so, in the first phase of the study, the researchers elicited the characteristics of the "evidence" by conducting a thematic analysis of semi-structured interviews with experts and policy-makers. In the second phase, the developed model was tested against 6-month data elicited from Twitter accounts. The experimental results show that the evidence detection model performed better with decision tree (DT) than the other algorithms. Decision tree (DT) outperformed the other algorithms by an 85.9% accuracy score. This study shows how the model managed to fulfill the aim of the present study, which was detecting Twitter posts that can be used as evidence. This study contributes to the body of knowledge by exploring novel models of text processing and offering an efficient method for analyzing big text data. The practical implication of the study also lies in its efficiency and ease of use, which offers the required evidence for policy-makers.

**Keywords:** evidence-based policy; evidence; social media; evidence detection; twitter

## 1. Introduction

Social media play a special role in our daily lives. These platforms, which are based on the ideology and technology of Web 2 [1], have transformed the way people communicate and interact [2]. They work based on user participation in content generation and have led to the emergence of active and dynamic users instead of passive citizens, so that users are more engaged in diverse social developments [3]. In the meantime, such media are increasingly used to provide feedback on various social issues. In fact, with the emergence of participatory social media, a new ecosystem has come to light that facilitates citizen participation in social events [4]. Sharing content by citizens can give rise to social values, such as building a social discourse or raising awareness on political and economic issues [5].

The use of values created in this environment by the governments can lead to the formation of good governance in all countries, especially developing countries [6].

In the wake of the platform revolution, the role of policy-makers has changed. They have taken on new roles, including actively analyzing and extracting knowledge from the opinions of citizens using digital platforms [7]. One of the best ways to identify and understand citizens' views in a bid to take them into account in the policy-making process is using social media platforms [8,9]. Policy-makers need to be aware of citizens' opinions expressed in the form of social media posts, as they are published through minimum gatekeeping [10]. This creates an exceptional chance for policy-makers to interact more effectively with the citizens, learn about their needs and opinions, and take them into account in the policy-making process [11]. In fact, social media can serve as a channel between users and policy-makers and can be used as a new source for engaging citizens in formulating and implementing policies [12]. This also creates a great opportunity for governments to learn about their citizens' views and communicate effectively with them. Social media analysis has drawn more attention in recent years, with governments seeking to take advantage of this opportunity to boost user participation on social media [13].

Social media data analysis is not new to the literature, and researchers and experts have already analyzed various datasets in different countries through diverse methods and techniques over the past few years [14,15]. However, the application of these data and their analysis in the field of policy-making is a novel concept and there are many gaps and challenges that should be addressed [12–16]. Policy-makers have started using social media data in various sectors, including education [17], health [18], and communications [10]. Analyzing these data can contribute to improving the performance of governments, boosting the quality of services, creating new and developed forms of interaction with citizens, and promoting the welfare of citizens [3]. Doing so can not only help governments improve their decision-making and governance, but also revolutionize the creation and provision of such services [19,20].

Based on the evidence-based policy approach, the policy-maker is obliged to use various types of evidence. Traditionally, the basis of "evidence" is the knowledge elicited from applied research, statistics, surveys and focus groups, and so on. However, now, policy-makers cannot ignore the evidence on various issues that is being widely produced by citizens on social media [21,22]. Since the question of how data retrieved from social media can be used in the policy-making process has not been answered yet [12], finding a response is of great significance. Even though various studies have been conducted on social media analysis and policy-making in recent years [20,22,23], social media have not been viewed as a source of much-needed evidence in the policy-making process. The significance of social media data used as evidence in the policy-making process is undisputed [24,25], but the unanswered problem is how to organize a large body of scattered data and use them as evidence in the policy-making process. Therefore, policy-makers need to distinguish "evidence" from "non-evidence" among a plethora of social media posts. There is an indication that Iranian policy-makers welcome the use of Twitter analytic tools, including monitoring dashboards, to obtain insights into diverse areas, including public health, education, technology, etc. To date, however, the Iranian technology policy-makers have not systematically used social media data in the policy-making process, despite the incontrovertible significance of such data, which can reflect the opinions and feedback of the users. This is set against the backdrop of the fact that the absence of social media users' views who are already sensitive to policy issues can challenge the formulation, implementation, and evaluation of policies in the future. To this end, the present study sought to provide a model for detecting evidence in Twitter posts. After being designed and tested, a model was proposed to be used during the process of technology policy-making in Iran.

In the first phase of this study, the authors drafted the characteristics of the tweets to be considered as evidence in the field of technology by reviewing the related literature and also by conducting semi-structured interviews with technology policy experts in Iran.

By evidence, the authors mean tweets that are relevant, critical, and representative of the facts on the ground, which are capable of creating a criterion for the policy-makers to base their decisions on an evidence-based, rather than an intuition-based, approach. By doing so, the evidence-based approach can help reduce uncertainties and bridge the gap between speculations and facts. In the second phase, the researchers collected the 6-month data from Twitter accounts that pertained to the field of technology and labeled them as evidence and non-evidence. Next, they developed a model, which was based on the six-month Twitter data, to distinguish evidence from non-evidence.

In the following sections, first, the related background studies will be reviewed, then the steps of conducting the research study will be explained, and finally, its applications will be discussed.

## 2. Background

Technology policy was chosen as a controversial area, given the challenges facing its implementation and the public outcry it occasionally triggers. As a developing country, Iran has been facing complexities in terms of emerging technology policy. The multiplicity of formal and informal policy-making institutions, high conflict of interest among policy stakeholders, lack of transparency in policy formulation and implementation, insufficient knowledge of the technologies, the uncertainty of social, economic, and cultural effects of these technologies, and the complications of the economic environment are among such complexities [26,27]. Needless to say, the policies related to technology in Iran, as in many countries, are limited and too general, lagging behind the emerging technologies and preventing the policy-makers from responding appropriately to the rising problems. To reduce the potential inefficiencies, the policy-makers must be aware of the needs, interests, and desires of all stakeholders in the field of technology.

By January 2020, about 70 percent of Iranians had access to the Internet, a sharp increase of 11 percent from 2019. It is also estimated that more than 33 million Iranians (40%) use social networks. Many Iranians take to Twitter to express their critical and expert-level views on diverse social and political issues. By focusing on content rather than on the users' profiles, Twitter has created an environment where users can easily debate popular topics, with features such as hashtags and algorithmic timelines [28]. This thematic arrangement of content, as well as the openness and real-timeness of Twitter [29], has enabled experts and policy-makers in various fields to use it as a source of ideas to set their policy priorities [30].

Twitter is ironically expanding its users and is estimated to have over 3.2 million users in the selected country of this study, with nearly 790 million tweets (including 200 million retweets) posted between 21 March 2021 and 20 March 2022. [31]. Twitter users in Iran were estimated to be around 2 million a year ago, with 500 million tweets (including 200 million retweets) published every year. It is evident that the users are mostly from the educated strata of Iranian society, and most of the content is created by the users themselves, and unlike other platforms (such as Instagram), there is less copy and paste. Twitter has become highly politicized in Iran, and most policy-makers, public opinion leaders, and experts in various fields have official Twitter accounts. Via Twitter, Iranians from different walks of life, from ordinary citizens to the members of parliament and cabinet members, can express their views on diverse social, political, and economic issues in society. Iranians have found that launching a Twitter storm is an effective way to spread an idea, belief, demand, or protest. They already use hashtags and share them to make certain issues that have been underreported by the mainstream media trending on Twitter, as part of a strategy to grant themselves a voice [32]. It can be argued that Twitter data, as a representation of public opinions, can influence the policy-making agenda.

Nowadays, policy-makers actively take advantage of social media networks to reach a wider audience, raise awareness of the issues that matter to them, promote their views, mobilize supporters, and receive timely feedback [33]. Given the prevalence of Twitter among Iranian intellectual figures, policy-makers have already kept an eye on the content produced and spread on the platform. If detected and analyzed, such data can be translated into relevant evidence that is much needed by the policy-makers in emerging areas, including technology policy.

## 3. Literature Review

### 3.1. Evidence-Based Policy

Evidence-based policy is defined as the systematic use of evidence in formulating public policies. It is an approach that has its roots in evidence-based medicine and treatment [34,35]. Evidence-based policy became pervasive once the modernization of governments rose to the top of the agenda of countries and the tendency for policy-making based on social science analysis increased [35,36]. Policy-makers and government officials sought to respond to the needs and pressures of citizens for whom social services were inadequate or unsuitable. Frustrated with the low efficiency of policies used in the social programs, the policy-makers and government managers were prompted to seek new policy approaches [37]. Policies based on old ideologies no longer worked, and the policy-makers welcomed more modern approaches [27]. These new orientations were coupled with the tendencies of major NGOs, guilds and other stakeholders in the public space to become involved in solving the existent problems [38]. The awakening of policy-makers opened up an opportunity for the public policy field to offer solutions in the form of new approaches to gain more control over the ambiguous and confusing realities of the policy-making environment. Policy-makers adopted an evidence-based policy approach as part of an effort to solve their problems and increase policy efficiency. This new approach is the result of the emergence of the inefficiencies of government policy-making and its implementation in various fields [8,39] and aims to develop policy alternatives to boost the quality of decisions made by policy-makers.

The evidence-based policy approach is based on two contexts without which proper enforcement of this policy-making approach is not possible. The first context is a favorable political culture that allows the inclusion of transparency and rationality in the policy-making process. The second includes a research culture committed to analytical studies using rigorous scientific methods that generate a wide range of evidence for policy-making [40]. Sufficient information is one of the prerequisites of a good policy [29–41]. Policy-making requires a variety of evidence in complex, variable, and high-risk policy areas. Therefore, generating up-to-date, multidimensional, and multilevel evidence can further help policy-makers in this field [42]. There are several reasons why the traditional evidence collected through applied research, which is available to the policy-makers, is not sufficient for effective policy-making. First, it is not possible to obtain accurate findings on key issues in many areas through research. Second, policy-makers and politicians are often influenced and motivated by many factors other than research evidence [41–43]. Therefore, it can be concluded that the availability of reliable research results alone does not guarantee its effectiveness. Valid, extensive, and multidimensional evidence is largely missing in evidence-based policy. Producing evidence with these characteristics demands consistent and long-term efforts by policy-makers and researchers.

### 3.2. What Is the Evidence?

An evidence-based policy cannot work properly and yield expected results without accurate, sufficient, and appropriate evidence. Searching for accurate and reliable evidence and efficiently using it in the policy-making process is viewed as one of the underlying principles of the evidence-based policy approach [39–44]. The following three types of evidence are important in policy-making: political evidence, analytical and technical evidence, and evidence collected from the field and through professional experience [24]. In fact, these types of evidence offer three kinds of lenses for policy-making. Each is produced with respective knowledge, and professional and political protocols and is subject to policy-makers' interpretations, limitations, and constraints [45].

In evidence-based policy, the scope of evidence must be expanded to include all types of evidence [40–46]. Traditionally, the basis of "evidence" as the foundation of evidence-based policy is the knowledge produced via applied research on comprehensive trends and the explanation of social and organizational phenomena [46]. Based on the type of policy issue and the ability and capabilities of policy-makers in data collection [34–47], governments need to have a certain level of evidence analysis capabilities for policy-making. In practice, however, such a capability does not always exist, and governments often fail to systematically analyze a large body of formal and informal evidence and incorporate it into the policy-making process [48,49].

From an evidence-based policy perspective, the question is as follows: what kind of data/information is needed to produce evidence? Some policy researchers and analysts have started asking the question of whether the persistence of complex social problems is due to the lack of data available to policy-makers [50,51]. However, now, the evidence shows that obtaining more data to fill the gaps will not necessarily lead us to good policy solutions because the most important step to take in policy-making is to reconcile different values and political views with scientific evidence within a policy-making system that is capable of approximating the evidence elicited from the opinions of stakeholders.

Hence, there is disagreement about what kind and quality of evidence can help improve policies. In a broader sense, it can be argued that there is no single basis to define what evidence is [27]. All types of different evidence must be used in a more inclusive context [37], that is, formal and informal evidence needs to be integrated to meet policy needs.

### 3.2.1. Challenges of Using Evidence in Policy-Making

Quantitative data and empirical methods have been used as a tool to provide accurate and reliable evidence for policy-makers for many years. Over the years, the focus has been on the advantages, such as the high accuracy of these methods in producing quantitative and analytical evidence [43–48]. However, there have always been three main challenges confronting the use of this evidence in policy-making. The first stems from the inherently political and value-oriented nature of policy and decision-making [45], which does not allow the use of this type of evidence in policy-making. The second is related to the fact that evidence is produced in different ways by different actors who look at the phenomena through different lenses. The third challenge is the complicated policy network that has made evidence difficult to use. Policy network actors interpret, understand, and prioritize evidence in different ways based on different experiences and values [45]. In the real world, policies do not follow analytical and empirical evidence but are rather based on judgments, values, and so on. Policy-making is a vague, politicized process, and is sometimes contradictory to the original paradigms [52], and requires compromise, adjustment, and agreement that make it difficult to use evidence.

A crisis or emergency, political priorities, and changing social values in public opinion are among the cases where governments face difficulty in using evidence [27,49,53,54]. These are some of the complicated issues faced by policy-makers who will be entangled with problems in producing appropriate and multidimensional evidence, while pursuing the evidence-based policy approach. Instances of such policies can be found in areas related to moral and ethical issues. These are some of the major problems threatening evidence producers, which have validated the use of different types of evidence-based methods on different foundations in the policy-making process to meet the aforementioned challenges.

### 3.2.2. Evidence from Public Engagement

In public environment in which government policies are implemented in the public sphere, or more specifically the public opinion sphere, all the policy stakeholders are present. This environment reacts to different policies and the success or failure of policies is determined in this sphere. The policy-makers must respond to this complex environment in which stakeholders are present and encourage them to implement the policies. There is an emphasis on the need to collect and produce evidence based on the engagement of the public and all policy stakeholders, which is viewed as the basis for producing multidimensional and comprehensive evidence [54]. Engaging all stakeholders in a policy is often considered as a part of the policy-makers' long-term relationship with them, as well as the development of public engagement capabilities [55,56].

Involving citizens' digital participation in policies and employing it in the policy-making process in the form of evidence is already underway. Policy-makers need to use citizens' digital interaction data to learn about public views and the way public discourses are formed, who the main stakeholders are, and what their expert groups and communications are like, which are growing on a daily basis [55–57]. We are now witnessing an increase in the use of new technologies to collect and analyze citizens' online participation data [58,59]. These tools operate as a complement to the conventional data collection and analysis methods to generate evidence. This adds to the questions about how these digital audiences can provide useful evidence for policy-makers.

### 3.3. Social Media Data as Evidence in the Policy-Making Process

Social media data offer a good representation of the big data of any country, opening up many opportunities to raise awareness about the demands of citizens. Social media data enable accurate predictions, create knowledge, bring about new services, and can lead to providing citizens with better facilities [10,16,60]. Using social media data, policy-makers can reduce policy costs and achieve sustainable development in a variety of areas. In addition, social media data can enhance policy transparency, increase the credibility of governments and policy-makers, boost government oversight performance, and narrow the gap between government oversight and the realities in society [61]. Such characteristics of social media data can enable policy-makers to boost the oversight of citizens' digital interactions [12]. In addition, the ideal combination of social media data with policy-makers' background knowledge will result in more relevant information. Moreover, social media can generate large amounts of data, in contrast to the traditional sources of evidence, and can garner novel insights into how stakeholders think about policy issues. The value of social media data as part of the policy-making cycle and using evidence collected from social media is highly practical; therefore, it is not possible to ignore its significance [62]. Using social media data in policy-making is not a new issue, as various tools have already been developed for collecting, analyzing, and visualizing social media content almost on a daily basis [60–63].

Recent research has focused on combining social media data with the policy-making process and how these resources can enable policy-makers to obtain fresh insights. Fernandez et al. [64] employed the citizen participation (CP) framework and linked it to the digital participation created through social media. Panayiotopoulos, Bowen, and Brooker [13] used the crowd capabilities conceptual framework to underline the value of social media data for policy-makers in the policy-making process. They question how policy-makers understand the value of social media data and which of the collective capabilities of social media data can meet the needs of policy-makers for filtering data input in the policy-making process. Edom, Hwang, and Kim [5] examined the roles that communities formed on social media can assume in promoting the social responsibility of governments. They presented a typology that features government communication on social media in which social media data become a new resource for policy-makers and increases communication between policy-makers and citizens. Gintova [63] believes that few studies have shed light on the behaviors and views of users on state social media so far. In her study, she analyzed the experiences of social media users and the way they interact on the Twitter and Facebook pages run by a Canadian federal government agency. Driss, Mellouni, and Trabelsi [12] provide a conceptual framework for employing data generated by citizens on Facebook in policy-making. According to their research results, social media data can be used in two phases of the policy-making cycle, i.e., the definition of the problem and policy evaluation. Napoli [64] argues that social media, based on the framework of "public resources", are a public resource that can and should be used to promote public interests. He believes that there is a positive correlation between extracting and analyzing social media data and improving public life [53] investigated the impact of social media on citizens' participation in events that needed policy intervention. They concluded that more access and activity on social media result in better participation in such affairs. Lee, Lee, and Choi [29] examined the impact of politicians' Twitter communications on advancing their policies, assuming that politicians around the world are increasingly using social media as a channel of direct communication with the public. The study results indicated that communication breakdowns on social media by politicians would have the greatest impact on policy implementation during the process of policy-making, causing the users to distrust the enforcement of the policies. Simonowski, [16] present a framework for analyzing social media data and how to use them in policy-making. They integrate data collected from users' digital participation on electronic platforms and social media, employing them at various stages of the policy-making process.

Previous research works into social media analysis and policy-making already corroborate the importance of using social media data in policy-making. However, some studies have questioned the use of such data as evidence in policy-making [12,41,58,63–65].

## 4. Methods

### 4.1. Proposed Model for Evidence Detection

One of the major objectives that the present research seeks to realize is the identification of the characteristics that the Twitter posts need to possess to be considered as policy evidence. Since labeled data (evidence/non-evidence) were required to train the evidence detection model, the researchers carried out in-depth, semi-structured interviews with the Iranian technology policy practitioners to elicit the characteristics. The experts were selected based on their expertise and several years of practical experience related to technology policy-making. The interviews began with general questions about the effectiveness of Twitter in policy-making and proceeded based on the statements made by the interviewees. Prior to the interview, an interview guide, which contained a series of open-ended questions aimed at further preparing the interviewees, was emailed to them. The interviews were transcribed and analyzed via the thematic analysis technique to extract the characteristics of tweets that were deemed as tech-related policy evidence by the policy-makers, based on which the tweets were labeled as evidence or non-evidence in the next phase of the study. Some challenges arose while conducting the interviews with the experts and policy-makers about identifying the characteristics of the so-called evidence tweets. The definition of what was referred to as "evidence" was a subjective concept and lacked an objective criterion. In addition, some of the definitions had a broader scope and included professional experience, political knowledge, ideas of stakeholders, etc., while some others had a narrower definition of evidence based on statistical comparisons. In aggregate, 96 themes, 32 sub-themes, and 15 concepts (characteristics) were extracted from 480 comments and meaningful sentences. Table 1 lists the characteristics of the tweets considered as evidence by the interviewees.

To detect the evidence tweets, all tweets had to be labeled as evidence or non-evidence, based on the characteristics extracted from interviews with technology policy-making experts, which were conducted earlier. The tweets were divided into two classes, evidence and non-evidence-based, using the above-mentioned characteristics during the labeling process.

### 4.2. Data Set and Feature Engineering

In order to develop an evidence detection model, as one of the other major objectives of this research, the following steps were carried out.

#### 4.2.1. Data Collection

All of the data used in the present study were collected from Persian tweets posted over six months using the Twitter API tool, a data extraction tool employed by the developers.

First, 39 keywords related to "technology policy in Iran" were selected by reviewing the literature in Persian [59–66]. The tweets were searched and collected based on the selected keywords. As there were few tweets for some keywords, the authors decided to remove them from the list and place a greater focus on other more frequently used keywords. There were also some other relevant keywords commonly used by the users that did not exist in the technology policy-making jargon. The authors decided to include these keywords with the aim of collecting more tweets related to technology policy-making. They searched and collected tweets that contained the hashtags of these keywords. Based on the selected keywords, 28,277 tweets were initially collected. Nearly half of the tweets, which included duplicate tweets, retweets, spam, advertisements, or irrelevant tweets, were removed from the dataset during the pre-processing phase, leaving 14,029 tweets.

**Table 1.** The evidence evaluation criteria in the field of technology policy from the perspective of policy-makers.

| No | Evaluation Criteria |
|---|---|
| 1 | Relevance to technology policies |
| 2 | Distinguish individual comments from retweets |
| 3 | Contain a specific need relevant to technology |
| 4 | Contain statistics relevant to technology |
| 5 | Relevance to modern technologies |
| 6 | Provide political knowledge relevant to technology |
| 7 | Provide practical and professional experience relevant to technology |
| 8 | Posted by a technology expert or someone with relevant experience |
| 9 | Contain a critical issue |
| 10 | Indicative of social values in technology |
| 11 | Capable of creating a network effect |
| 12 | The topic has political priorities for the policy-maker |
| 13 | The urgency of the topic mentioned in a tweet |
| 14 | Reveal corruption in technology |
| 15 | Provide analytic and technical knowledge relevant to technology |

#### 4.2.2. Feature Extraction

After the pre-processing step, the appropriate features were extracted. In the previous research on social network user behaviors, account-based and text-based features were employed to analyze user data, implying the successful use of these features in analyzing Twitter posts. Many studies on social media analysis [56–67] have already employed account-based features to evaluate user profiles and text-based features to identify the behavioral patterns used in the text. As far as the authors are concerned, these features have not been used in any of the studies on policy-making to distinguish evidence from non-evidence. Therefore, given the successful use of these features in various studies, the authors decided to employ the selected features in order to distinguish evidence from non-evidence items. The research study also aimed to extract new features that can contribute to detecting evidence posts. Accordingly, both text-based and account-based features were used to distinguish evidence posts from non-evidence posts. Table 2 shows the text-based features used in the study.

Table 3 lists account-based features that showcase the characteristics of the accounts.

#### 4.2.3. Feature Selection

In this phase, to select the best subset of the features, the information gain metric was used to determine the value of each feature. Since the most effective feature for classification is the one that decreases entropy, the information gain metric is used for measuring the amount of entropy decline. The information gain is calculated via the following formula:

$$\text{Gain}(S.A) = Entropy(S) - \sum_{j \in A} \frac{|S_j|}{|S|} Entropy(S_j)$$

**Table 2.** Text-based features used in the study.

| No | Feature Name | Description |
|---|---|---|
| 1 | Swear Word | The tweet contains swear words |
| 2 | Tweet Time | The time a tweet was sent |
| 3 | No_Sentences | The number of sentences in a tweet |
| 4 | No_Lines | The number of lines that a tweet has |
| 5 | No_Mentions | The number of mentions included in a tweet |
| 6 | No_Urls | The number of URLs included in a tweet |
| 7 | No_Hashtags | The number of hashtags included in a tweet |
| 8 | No_Digits | The total number of digits in a tweet |
| 9 | No_Emojis | The number of emojis included in a tweet |
| 10 | No_Spaces | The number of spaces included in a tweet |
| 11 | Length of Tweet | The length of a tweet |
| 12 | Max Length of Words | The maximum length of words that a tweet has |
| 13 | Mean Length of Words | The mean length of words that a tweet has |
| 14 | No_Exclamation Marks | The number of exclamation marks included in a tweet |
| 15 | No_Question Marks | The number of question marks included in a tweet |
| 16 | No_Punctuations | The number of punctuations marks included in a tweet, except for question and exclamation marks |
| 17 | No_Words | Total number of words that a tweet has |
| 18 | No_Characters | The total number of characters that have been used in a tweet |
| 19 | Digits To Chars Ratio | The number of digits to the number of characters ratio in a tweet |
| 20 | Lines To Sentences Ratio | The number of lines to the number of sentences ratio in a tweet |
| 21 | Words To Sentences Ratio | The number of words to the number of sentences ratio in a tweet |
| 22 | Hashtags More Than 2 | The tweet has more than 2 hashtags |
| 23 | No_Words Less Than 3 Chars | Total number of words with less than 3 characters that a tweet has |
| 24 | No_Words More Than 5 Chars | Total number of words with more than 5 characters that a tweet has |
| 25 | Video | The tweet contains a video |
| 26 | Image | The tweet contains an image |

**Table 3.** Account-based features used in the study.

| No | Feature Name | Description |
|---|---|---|
| 1 | No. of_Followers | The number of followers of this Twitter user |
| 2 | No. of_Following | The number of accounts this Twitter user follows |
| 3 | FF_Ratio | The number of followers to the number of followings ratio |
| 4 | Description | Contains a description in the profile |
| 5 | No. of_Likes | The number of user favorites by this Twitter user |
| 6 | URL In Description | Contains a URL in the description of the Twitter user |
| 7 | No. of Lists | The number of lists that this Twitter user added |
| 8 | No. of_Tweets | The number of tweets this Twitter user sent |
| 9 | Profile Image | Contains a profile image in the Twitter profile account |
| 10 | Background Image | Contains a background image in the Twitter profile account |
| 11 | Profile Background Image | Contains a profile background image in the Twitter profile account |

The Table 4 shows the gain values of each feature. Some features extracted via the evidence detection model are more important for the model, while others are less important.

In calculating the information gain of each feature, the value of some features was zero, which may be because the values of those features were the same for both evidence and non-evidence labels. For example, all the user accounts in the dataset had descriptions and profile pictures in their accounts. It was also made clear that the time of posting tweets (around the clock) was almost equally distributed between the evidence and non-evidence categories, so features such as the time of posting the tweets were not considered an important feature in training the algorithm.

Taking into account the values of the information gains of the features, different subsets of data were examined to implement the classification model. Accordingly, features 1 to 25 were chosen as the best subset of features in this study, which can train the model to detect evidence tweets with a higher degree of accuracy. This subset included 20 text-based features and 5 account-based features. The Figure 1 shows the information gained from each feature.

**Table 4.** Information gain of each feature.

| No | Feature Name | IG | No | Feature Name | IG |
|---|---|---|---|---|---|
| 1 | No_Hashtags | 0.07309 | 20 | Max Length of Words | 0.02279 |
| 2 | No_Emojis | 0.07305 | 21 | Mean Length of Words | 0.02258 |
| 3 | No_Digits | 0.07054 | 22 | No_Followings | 0.01878 |
| 4 | No_Lines | 0.06164 | 23 | No_Followers | 0.01825 |
| 5 | Length of Tweet | 0.06003 | 24 | No_Sentences | 0.01737 |
| 6 | Digits to Chars Ratio | 0.05954 | 25 | No_Lists | 0.01227 |
| 7 | Swear Words | 0.05654 | 26 | Words to Sentences Ratio | 0.00597 |
| 8 | Hashtags More Than 2 | 0.04358 | 27 | Ff_Ratio | 0.00544 |
| 9 | No_ Characters | 0.04293 | 28 | Background Image | 0.00313 |
| 10 | No_Spaces | 0.04287 | 29 | Profile Background Image | 0.00313 |
| 11 | No_ Punctuations | 0.04163 | 30 | No_Question Marks | 0.00258 |
| 12 | No_Words | 0.03741 | 31 | URL In Description | 0.00199 |
| 13 | No_Word More Than 5 Chars | 0.03626 | 32 | Tweet Time | 0 |
| 14 | No_Mentions | 0.03428 | 33 | No_Exclamation Marks | 0 |
| 15 | Lines To Sentences Ratio | 0.03036 | 34 | Profile Image | 0 |
| 16 | No_Tweets | 0.02697 | 35 | Description | 0 |
| 17 | No_Words Less Than 3 Chars | 0.02653 | 36 | Video | 0 |
| 18 | No_Urls | 0.02469 | 37 | Image | 0 |
| 19 | No_Likes | 0.02287 | | | |

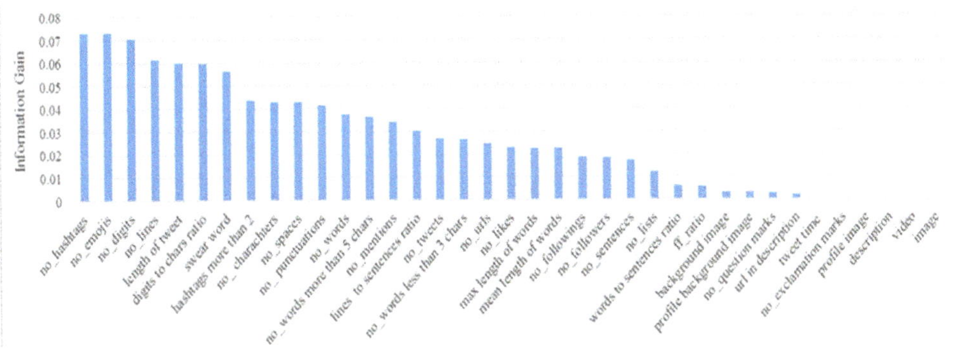


**Figure 1.** Values of information gain of each feature.

## 4.3. Proposed Classification Approach

Machine learning algorithms have been used to distinguish evidence tweets from non-evidence tweets. Since we aim to achieve specific outputs for samples via evidence detection, our approach is based on supervised learning problems and classification techniques. The steps for implementing our proposed model are presented in the framework (Figure 2). There are two main processes in the supervised learning approach, which include training the algorithm and testing the trained algorithm. The data that are used for this purpose have to be prepared based on the extracted feature set for each sample, for example, S is shown as S = {f1, f2, f3, … , fn} and labeled as L. Therefore, each instance in the dataset is shown as Data = {(S1, L1), (S2, L2), (S3, L3), … , (Sn, Ln)}. Then, to train the machine learning (ML) algorithms, the dataset is split into two parts, including train and test sets. In order to train the ML algorithm, the whole samples in the larger part of the data (train set), along with their labels, are given to the ML algorithm (classifier) created in the format of the data above, and then the second part is used for testing it.

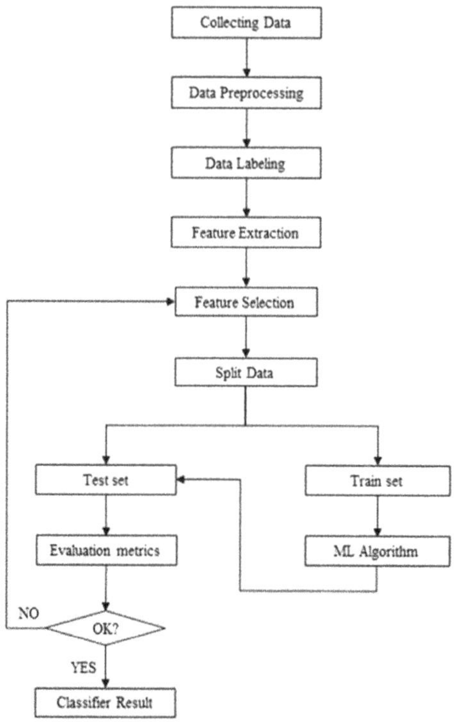


**Figure 2.** Proposed model for detecting evidence tweets.

The test set consists of samples that the ML algorithm has never encountered before. The whole samples in the test set without their labels are given to the trained classifier in the format of Test = {S1, S2, ... , Sn}. Then, the trained classifier predicts and assigns the label of each sample to it. Finally, the predicted class of each sample is compared with its original label. A decision tree is already commonly used as an inductive inference algorithm to solve classification problems and develop prediction models. Decision trees are known as the most popular class of function, as Chapelle and Chang (2011) reported in their study. The decision tree is known as one of the most widely used classification algorithms in supervised learning problems. Due to its simplicity in interpretation and high power in classifying classes, this algorithm is also widely employed as one of the most powerful algorithms in classification problems. The structure of a decision tree is similar to a tree and consists of roots, nodes, and leaves. The best feature is located at the root and each node is compared by the feature values. The leaves of the tree represent the final class results for the samples in question. In Section 5.3 of this study, we tested several algorithms, with the decision tree showing the best performance.

The performance of each classifier can be assessed by evaluation metrics.

### 4.4. Evaluation Metrics

To examine the performance of the evidence detection model, we used the metrics widely used by researchers in classification problems. After comparing the labels predicted by the classifiers with the actual label of the samples, the results were grouped into the following four categories:

- True positive (TP): tweets that belong to class Evidence (E) and are correctly predicted as class E.
- False positive (FP): tweets that do not belong to class E and are incorrectly predicted as class E.
- True negative (TN): tweets that do not belong to class E and are correctly predicted as class non-E.
- False negative (FN): tweets that belong to class E and are incorrectly predicted as class non-E.

The above classifier definitions are displayed via the confusion matrix in the Table 5.

**Table 5.** Confusion matrix.

| | | **Prediction** | |
|---|---|---|---|
| | | **Evidence** | **Non-Evidence** |
| Label | Evidence | True-positive (TP) | False-negative (FN) |
| | Non-evidence | False-positive (FP) | True-negative (TN) |

The performance of a classifier can be evaluated by accuracy, precision, recall, and F-measure metrics. Precision, recall, and F-measure metrics are calculated using confusion matrix values according to the following formulas:

$$\text{Precision} = \text{TP}/(\text{TP} + \text{FP})$$

$$\text{Recall} = \text{TP}/(\text{TP} + \text{FN})$$

$$\text{F-measure} = (2 * \text{Precision} * \text{Recall})/(\text{Precision} + \text{Recall})$$

Recall is defined as the number of correct results divided by the number of results that should have been returned, while precision is how closely the measurements are gathered around a point estimation. The F score is the harmonic mean (average) of the precision and recall. Accuracy is one of the most common metrics in evaluating classifiers' performance. This metric is calculated using the following formula:

$$\text{Accuracy} = (\text{TP} + \text{TN})/(\text{TP} + \text{TN} + \text{FP} + \text{FN})$$

## 5. Experiments and Results

In this section, first, the significance of the topics from the users' viewpoints based on the number of tweets that contain the keywords is reviewed, and then the users' behaviors are analyzed, and finally, the proposed evidence detection model is assessed.

### *5.1. Statistical Report*

The Figure 3 shows the number of extracted tweets. Among the topics related to technology, "filtering" with 8817 tweets and "#filtering" with 1248 tweets were the most frequently posted tweets. This means that they were considered the most important topics from the perspective of users. Tweets that contained the keywords of "Personal Information", "Information Disclosure", "Access to Information" and "Free Access" were viewed respectively as the other most important topics from the users' point of view.

Figure 4 illustrates the most frequently used topics by the users in a word cloud. The word cloud is sorted out according to the number of tweets that contained the related keywords, indicating the significance of various topics in technology policy. They are indicative of the topics that attracted the attention of the users the most in the above-mentioned period.

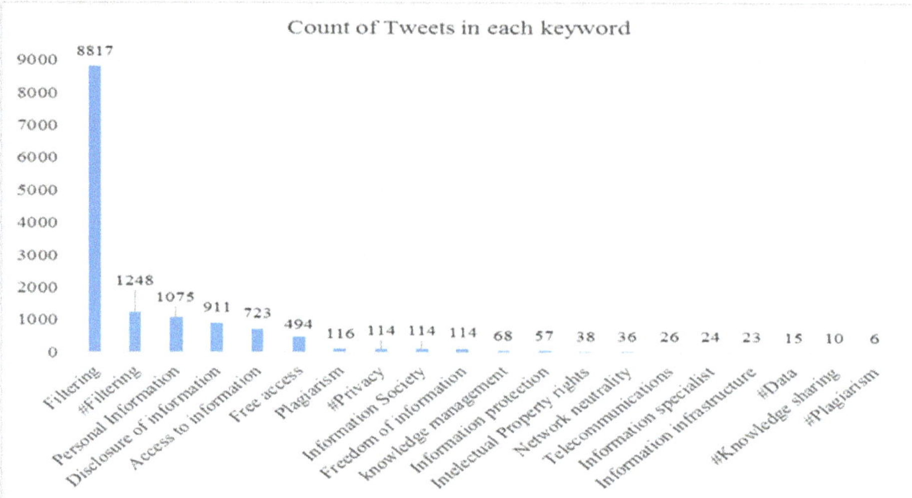


**Figure 3.** The number of tweets related to each keyword.

**Figure 4.** Word cloud of keywords.

### 5.2. Analyzing the Behavior of the Users Posting Evidence Tweets

This section examines the behaviors of the users that posted the evidence tweets. To implement the proposed evidence detection model, first, all of the extracted tweets, regardless of the keywords, were categorized into an integrated data set from which 4560 tweets were randomly selected. As the figure below shows, the evidence generated by Twitter users has its own characteristics. These features are shown in Figure 5. According to Figure 5a, most of the evidence tweets did not contain a hashtag, and very few had only one hashtag. This indicates that the texts that can be considered as evidence are often not hashtagged by the users. Figure 5b shows that most of the evidence tweets lacked emojis. This means that users whose tweets can be considered as evidence mostly do not use emojis in their tweets. Not using emojis in most of the evidence tweets is indicative of the formal setting in which these tweets were posted.

Figure 5c also shows that most evidence-generating users are reluctant to use more than one mention in their tweets, which seems to indicate that the users do not intend to post personal comments or reply to another specific tweet. The users do not also intend to address other users or attract their attention. Figure 5d indicates the absence of URL links in more than 87% of the evidence tweets, with the remaining 13% of tweets containing only one URL link. This indicates that giving references to sources outside Twitter is not very common as far as generating evidence is concerned, as most of the evidence is generated and published within the platform. Figure 5e also shows that evidence-generating users often tend to express their opinions within 1–3 sentences. This also shows the brevity of the evidence tweets.

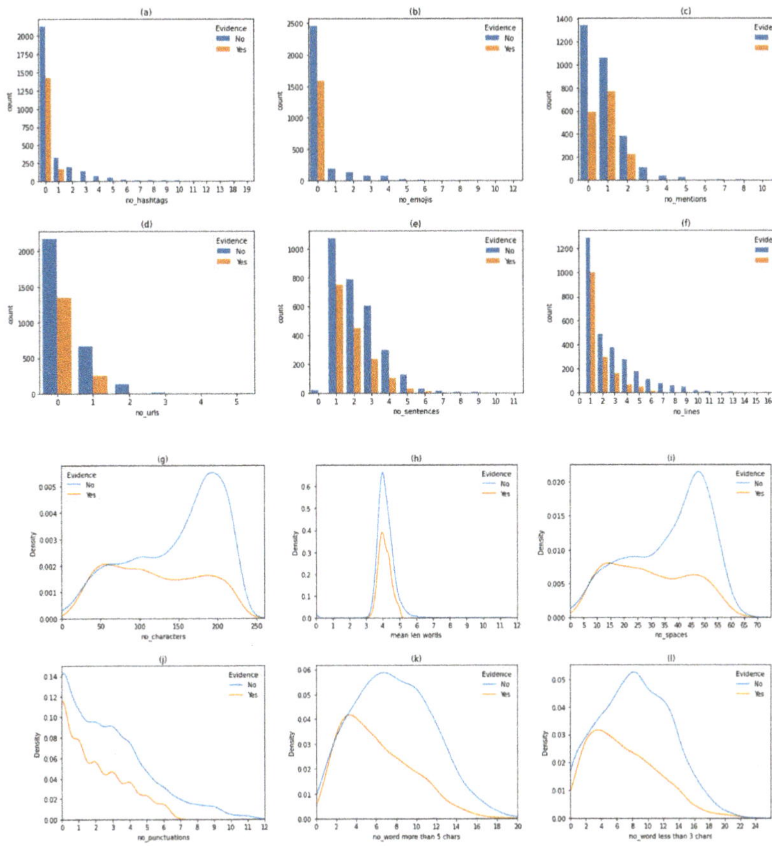


**Figure 5.** Comparative charts related to the features.

In Figure 5f, it can be observed that a larger number of characters (mostly over 200 characters) have been used in non-evidence tweets compared to evidence tweets. Similarly, according to Figure 5g, the users are more inclined to use single-line texts, showing no tendency to post long tweets or threads. Figure 5h shows that the average number of words used in both groups is almost the same (around 4). According to Figure 5i, nearly 70% of the users used up to 40 spaces in the text of the evidence tweets to express their views, with 30% of those generating evidence tweets using spaces beyond that range. Moreover, the patterns extracted from user behavior analysis suggest a less frequent use of punctuation marks, less frequent use of words exceeding five letters, and limited use of words exceeding three letters in evidence tweets in the last three charts.

The figures below also show the analysis of the behavior patterns extracted from account-related features (followers, followings, and lists). It was found that 75% of the users who posted evidence tweets published less than 17,000 tweets, as 25% of those generating evidence had posted more than 17,000 tweets. However, 35% of the users who posted non-evidence tweets had less than 17,000 tweets and 65% of these people had more than 17,000 tweets.

### *5.3. Evaluating Proposed Model*

To evaluate the proposed model, the authors used 4560 tweets, which were divided into 2 groups of evidence and non-evidence tweets. In this dataset, 2978 samples were labeled as evidence and 1583 samples as non-evidence. Thus, the data set was divided into two segments to train and evaluate the proposed model, the train set (including 4104 samples) and the test set (including 254 non-evidence and 202 evidence samples). In this step to detect evidence tweets, six classifiers, including support vector machine (SVM), decision tree (DT), K-nearest neighbor (KNN), linear discriminant analysis (LDA), X-GBoost, and logistic regression (LR) were used [68] and their performances were assessed. First, the classifier models were implemented in the train set, and after the learning process, the trained classifiers were implemented in the test set. Table 6 shows the performance results of each classifier based on the evaluation metrics for each algorithm.

**Table 6.** Performance evaluation comparison based on precision, recall, and F-measure (percent).

| Algorithm | Precision | Recall | F_Measure |
|---|---|---|---|
| Decision tree (DT) | 79.13 | 90.1 | 84.26 |
| XGBoost | 80 | 83.17 | 81.55 |
| K-nearest neighbor (KNN) | 75.54 | 68.81 | 72.02 |
| Logistic regression (LR) | 72.96 | 70.79 | 71.86 |
| Linear discriminant analysis (LDA) | 73.85 | 47.52 | 57.83 |
| Support vector machine (SVM) | 62.8 | 50.99 | 56.28 |

Given that the F-measure is calculated via precision and recall, we decided to choose the F-measure instead of the other two. As can be observed, the decision tree classifier was able to achieve the highest value of F-measure with 84.26 percent, followed by XGBoost, K-nearest neighbor (KNN), logistic regression (LR), linear discriminant analysis (LDA), and support vector machine (SVM).

In terms of accuracy, the results listed in the Table 7 indicate that DT could achieve the highest rate with an accuracy score of 85.09, compared with other classifiers.

**Table 7.** Accuracy of classifiers.

| Algorithm | Accuracy (%) |
|---|---|
| Decision tree (DT) | 85.09 |
| XGBoost | 83.33 |
| K-nearest neighbor (KNN) | 76.32 |
| Logistic regression (LR) | 75.44 |
| Linear discriminate analysis (LDA) | 69.3 |
| Support vector machine (SVM) | 64.91 |

These results indicate that the decision tree surpassed other classifiers both in terms of accuracy and F-measure. The experiments suggest that the performance of the proposed model was sufficient to achieve this study's desired goal, which was as follows: detecting evidence tweets from non-evidence tweets with acceptable accuracy. Since detecting evidence tweets via the proposed model is a new idea and the features used in this study have not been used in previous works in this field, it is taken for granted that the accuracy obtained for this model is desirable. Further analysis of the proposed model reveals that text-based features perform better than account-based features. The reason for this may be the fact that users who post evidence-type tweets can do so at one time and post non-evidence tweets at another time. This technically means that account-based features for a user that posts both evidence and non-evidence tweets will be the same. This is also the case for users who post non-evidence tweets. It is for this reason that the text-based features were found to be more reliable than the account-based tweets in detecting evidence tweets. So, it can be claimed that combining these two kinds of features can increase the accuracy of the model.

The best classifier selected for the proposed model enables the researchers to use data collected from one of the most significant social networks used by Persians in the process of technology policy-making, through an evidence-based policy approach. Labeling the data is one of the challenging stages of preparing the data set, since it is a time-consuming process.

By applying the proposed model to distinguish evidence from non-evidence tweets, the policy-makers can have access to useful tweets that can be considered as evidence and learn about the users' opinions, interests, needs, and capacities. The evidence and learning materials can be considered in making appropriate policy decisions worldwide, especially in the selected case country. Furthermore, the proposed model can be used with larger datasets to improve the training of the model and increase its accuracy.

## 6. Discussion

This paper contributes to the current body of knowledge by offering an evidence detection model to facilitate the use of data collected from Twitter, an influential social media platform used by policy-makers. This technology supports policy-makers by offering an evidence-based policy-making approach. The policy-makers can use the proposed evidence detection model to learn about the users' views, interests, needs, and capacities and make proper policy decisions accordingly. This model can be designed as a dashboard to be used by technology policy-makers.

The development of policy evidence from formal evidence to informal evidence elicited from user-generated data on Twitter is the first theoretical contribution of this research. However, this should be treated with caution, as tweets cannot be deemed as scientific evidence. Formalizing the use of social media data as evidence in the evidence-based policy approach is the second contribution of the study. To this end, features that are capable of turning user-generated social media data into evidence were detected. Some extracted features were general enough to be applied to other policy-making areas. The present study formalized user-generated social media data in the policy-making process by presenting a model based on several scientific fields, including policy-making (evidence-based policy-making), social media, and data science. The theoretical approach adopted in this article was initially derived from evidence-based policy. It also originated from citizens' digital participation in public policy. In this study, to the best of the authors' knowledge, for the first time, using social media data as evidence in policy-making was proposed and accordingly, a model was designed to detect evidence. The proposed model can rid the policy-makers of the daunting task of detecting evidence among a plethora of irrelevant data collected simply based on selected keywords. The model, if applied, can facilitate using such data in the evidence-based policy-making process. The authors decided to use posts published on Twitter, a social media platform in Iran with extensive user-generated content on public policies. In addition, the scope of the research was limited to technology policy, which, due to the emerging problems it deals with, needs to include citizens' views. Such policies can be applied to different communities. Moreover, this research study was conducted in a policy area in Iran where the policy-makers are reluctant to use users' opinions in the policy-making process. The model proposed in this study can facilitate this process and possibly encourage policy-makers to do so. Thus, implementing the model in similar developing countries can prepare the ground for the participation of citizens in shaping public policies.

This study addressed key challenges and the current gap in the literature by focusing on policy-making and addressing some key issues. Some of the issues raised by the researchers regarding using social media data in policy-making can be summarized as follows:

- Lack of text-based analytical tools for detecting policy evidence from the posts on social media [12,69];
- Analytical tools developed so far are more suitable for improving the image of brands or obtaining feedback from customers about the products or services offered by companies and may fail to meet policy-making needs [13,63–70];
- Comments by non-expert users about specialized policy issues, in the form of social media posts, may not be included in the evidence category. Moreover, there is no specific tool to sort out such data [21,71,72];

- The big data shared on social media may have been tainted with bias, further complicating the analysis [70–75];
- The intents of the users that participate in producing posts on social media usually differ from those of the analysts [36,45,71,76–79].

The research also offers a great practical tool for policy-makers to use in various contexts. Due to its acceptable level of accuracy for detecting evidence among large datasets obtained from Twitter, the model verified in the present study can be easily employed by policy-makers. It is also possible to apply it to other social media platforms in other developing countries. The authors suggest that researchers should use the evidence detection model in other communities for detecting evidence on diverse policy issues about which the citizens are sensitive to producing content on social media to provide the policy-makers with evidence elicited from the public opinions on social media. This model enables the policy-makers to add a different, but very important, channel to their evidence-gathering channels.

## 7. Conclusions

The large amount of data shared on social media is considered as one of the obstacles to using this information in the policy-making process. The main reason for this is the complexity and challenging process of analyzing big data created by various users over an extensive time period. This paper facilitated the acceptance of a novel approach to analyzing the data by reducing the size of the data and applying deep learning models. Using evidence in policy-making can facilitate the process, obviating the tremendous task of detecting evidence among a large number of posts on social media. The present study contributes to the efforts to develop a model that can help policy-makers to distinguish evidence from non-evidence.

This study offers innovative contributions at several levels. First, the investigation identified features of user-generated content by using a machine learning approach that can be converted into policy evidence. Such features had not been identified in the previous and relevant literature in the field. Despite being limited to a specific policy area (technology), many of the features are general enough to be used in detecting evidence posts in other areas as well. These features can also be developed based on the selected social media platform and policy area. Second, this study integrated data analysis techniques and applied them to develop a model. Social media data analysis methods have not been used for detecting evidence. This proposed model is limited to tested data and a specific policy area, which is technology in Iran. Given the significance of the first-hand evidence elicited from the end users' feedback in technology policy-making, this research can serve as a role model for the implementation of similar studies in other communities and policy areas. Moreover, the features based on which the model was developed are of a global nature and can be used in similar cases across the world. The present research, for the first time, proposed a model that distinguishes evidence posts from non-evidence posts on Twitter. It can underpin the decisions made by the policy-makers by providing social media users' views on different issues. Given the large amount of data on social media, including users' views and comments, it is not possible to use all the data in the policy-making process. The proposed model is capable of detecting evidence posts and supplying the policy-makers with suitable data and encouraging them to use the data in the policy-making process.

However, distinguishing evidence from non-evidence in social media data cannot be considered the only way to improve evidence-based policy. This is especially the case with complex problems, which can only be solved through different types of evidence. The present study simply aimed to develop different types of evidence and did not intend to criticize other types of evidence. Sole reliance on insights from social media analytics can even result in bipolar views and non-constructive arguments for policy-makers [72] and even mislead them, since the social media data analysis methods are not developed enough to offer comprehensive views on policy-making. However, the development of different concepts, methods, and techniques in this area can be helpful and may increase the efficiency of the policy-making process.

For future work, the authors suggest that other researchers should test the proposed model on other social media platforms to answer the question of whether the features capable of turning social media data into policy evidence on Twitter can gain similar results on other social media platforms. This study was limited to technology policy. The question that future research should address is whether diverse policy areas need different evidence detection models.

**Author Contributions:** Conceptualization, S.L.; methodology, S.L. and M.M.K.; software, M.M.K. and S.E.; validation, S.L. and S.S. and S.E.; formal analysis, S.S. and H.A.M.; investigation, M.M.K.; resources, S.E.; data curation, M.M.K.; writing—original draft preparation, S.L. and H.A.M.; writing—review and editing, H.A.M.; visualization, S.S.; supervision, S.L. and S.S.; project administration, S.L.; All authors have read and agreed to the published version of the manuscript.

**Funding:** This research received no external funding.

**Data Availability Statement:** Not applicable.

**Conflicts of Interest:** The authors declare no conflict of interest.

## References

1. Kaplan, A.M.; Haenlein, M. Users of the world, unite! The challenges and opportunities of social media. *Bus. Horiz.* **2010**, *53*, 59–68. [CrossRef]
2. Nooren, P.; Van Gorp, N.; Van Eijk, N.; Fathaigh, R. Should We Regulate Digital Platforms? A New Framework for Evaluating Policy Options. *Policy Internet* **2018**, *10*, 264–301. [CrossRef]
3. Chan-Olmsted, S.M.; Wolter, L. Perceptions and practices of media engagement: A global perspective. *Int. J. Media Manag.* **2018**, *20*, 1–24. [CrossRef]
4. Fuchs, C. *Social Media: A Critical Introduction*; SAGE: Thousand Oaks, CA, USA, 2017.
5. Edom, S.J.; Hwang, H.; Kim, J.H. Can social media increase government responsiveness? A case study of Seoul, Korea. *Gov. Inf. Q.* **2018**, *35*, 109–122.
6. Evens, T.; Donders, K. *Platform Power and Policy in Transforming Television Markets*; Springer: Berlin/Heidelberg, Germany, 2018.
7. Janssen, M.; Helbig, N. Innovating and changing the policy cycle: Policy-makers be prepared! *Gov. Inf. Q.* **2016**, *12*, 120–132. [CrossRef]
8. Fernando, S.; Díaz López, J.A.; Şerban, O.; Gómez-Romero, J.; Molina-Solana, M.; Guo, Y. Towards a large-scale twitter observatory for political events. *Future Gener. Comput. Syst.* **2019**, *14*, 976–983. [CrossRef]
9. Suzor, N.; Dragiewicz, M.; Harris, B.; Gillett, R.; Burgess, J.; Van Geelen, T. Human Rights by Design: The Responsibilities of Social Media Platforms to Address Gender-Based Violence Online. *Policy Internet* **2019**, *11*, 84–103. [CrossRef]
10. Dekker, R.; van den Brink, P.; Meijer, A. Social media adoption in the police: Barriers and strategies. *Gov. Inf. Q.* **2020**, *37*, 101441. [CrossRef]
11. DePaula, N.; Dincelli, E.; Harrison, M. Toward a typology of government social media communication: Democratic goals, symbolic acts, and self-presentation. *Gov. Inf. Q.* **2018**, *35*, 98–108. [CrossRef]
12. Driss, O.B.; Mellouli, S.; Trabelsi, Z. From citizens to government policy-makers: Social media data analysis. *Gov. Inf. Q.* **2019**, *36*, 560–570. [CrossRef]
13. Panayiotopoulos, P.; Bowen, F.; Brooker, P. The value of social media data: Integrating crowd capabilities in evidence-based policy. *Gov. Inf. Q.* **2017**, *34*, 601–612. [CrossRef]
14. Chen, C.; Wang, Y.; Zhang, J.; Xiang, Y.; Zhou, W.; Min, G. Statistical features-based real-time detection of drifted twitter spam. *IEEE Trans. Inf. Forensics Secur.* **2016**, *12*, 914–925. [CrossRef]
15. Matheus, R.; Janssen, M.; Maheshwari, D. Data science empowering the public: Data-driven dashboards for transparent and accountable decision-making in smart cities. *Gov. Inf. Q.* **2020**, *37*, 101284. [CrossRef]
16. Simonofski, A.; Fink, J.; Burnay, C. Supporting policy-making with social media and e-participation platforms data: A policy analytics framework. *Gov. Inf. Q.* **2021**, *38*, 101590. [CrossRef]
17. Sun, T.Q.; Medaglia, R. Mapping the Challenges of Artificial Intelligence in the Public Sector: Evidence from Public Healthcare. *Gov. Inf. Q.* **2019**, *36*, 368–383. [CrossRef]
18. Clarke, A.; Margetts, H. Governments and citizens getting to know each other? Open, closed, and big data in public management reform. *Policy Internet* **2014**, *6*, 393–417. [CrossRef]
19. Enroth, H. Governance: The art of governing after governmentality. *Eur. J. Soc. Theory* **2014**, *17*, 60–76. [CrossRef]
20. Davies, H.; Nutley, S.; Smith, P. *What Works? Evidence-Based Policy and Practice in Public Services*; Policy Press: Bristol, UK, 2001.
21. Nutley, S.M.; Walter, I.; Davies, H.T. *Using Evidence: How Research Can Inform Public Services*; Policy Press: Bristol, UK, 2007.

22. Picazo-Vela, S.; Gutiérrez-Martínez, I.; Luna-Reyes, L.F. Understanding risks, benefits, and strategic alternatives of social media applications in the public sector. *Gov. Inf. Q.* **2012**, *29*, 504–511. [CrossRef]
23. Prpić, J.; Taeihagh, A.; Melton, J. The fundamentals of policy crowdsourcing. *Policy Internet* **2015**, *7*, 340–361. [CrossRef]
24. Park, C.S.; Kaye, B.K. The tweet goes on: Interconnection of Twitter opinion leadership, network size, and civic engagement. *Comput. Hum. Behav.* **2017**, *69*, 174–180. [CrossRef]
25. Stamatelatos, G.; Gyftopoulos, S.; Drosatos, G.; Efraimidis, P.S. Revealing the political affinity of online entities through their Twitter followers. *Inf. Process. Manag.* **2020**, *57*, 102–172. [CrossRef]
26. Parkhurst, J. *The Politics of Evidence*; from Evidence-Based Policy to the Good Governance of Evidence; Routledge: London, UK, 2017.
27. Bucher, T.; Helmond, A. The affordances of social media platforms. In *The SAGE Handbook of Social Media*; Burgess, J., Marwick, A., Poell, T., Eds.; SAGE Publications Ltd.: London, UK, 2017; pp. 233–253.
28. Styles, K. Twitter is 10 and It's Still Not a Social Network. 2016. Available online: http://thenextweb.com/opinion/2016/03/21/twitter-10-still-not-social-network/ (accessed on 24 June 2020).
29. Lee, E.J.; Lee, H.Y.; Choi, S. Is the message the medium? How politicians' Twitter blunders affect perceived authenticity of Twitter communication. *Comput. Hum. Behav.* **2020**, *104*, 106–188. [CrossRef]
30. Beta Research Center. Report on Social Media Networks in Iran. Available online: http://betaco.ir/%da%af%d8%b2%d8%a7%d8%b1%d8%b4%d8%b4%d8%a8%da%a9%d9%87%d9%87%d8%a7%db%8c-%d8%a7%d8%ac%d8%aa%d9%85%d8%a7%d8%b9%db%8c%db%b1%db%b4%db%b0%db%b0-%d9%85%d8%b1%da%a9%d8%b2%d8%a8%d8%aa%d8%a7/ (accessed on 13 April 2022).
31. Mahdavi, S. Twitter, power and activism in the public sphere. *Q. Mod. Media Stud.* **2019**, *4*, 147–188.
32. Lee, J.; Xu, W. The more attacks, the more retweets: Trump's and Clinton's agenda setting on Twitter. *Public Relat. Rev.* **2018**, *44*, 201–213. [CrossRef]
33. Howlett, M. Policy analytical capacity and evidence-based policy-making: Lessons from Canada. *Can. Public Adm.* **2009**, *52*, 153–175. [CrossRef]
34. Sanderson, I. Evaluation, policy learning and evidence-based policy-making. *Public Adm.* **2002**, *80*, 1–22. [CrossRef]
35. Parsons, W. From muddling through to muddling up—Evidence-based policy-making and the modernization of British government. *Public Policy Adm.* **2002**, *17*, 43–60.
36. Cairney, P. *The Politics of Evidence-Based Policy-Making*; Springer: Berlin/Heidelberg, Germany, 2016.
37. Head, B. Reconsidering evidence-based policy: Key issues and challenges. *Policy Soc.* **2010**, *29*, 77–94. [CrossRef]
38. Shaxson, L. Is your evidence robust enough? Questions for policy-makers and practitioners. *Evid. Policy A J. Res. Debate Pract.* **2005**, *1*, 101–112. [CrossRef]
39. Cartwright, N.; Hardie, J. *Evidence-Based Policy: A Practical Guide to Doing It Better*; Oxford University Press: Oxford, UK, 2012.
40. Young, K.; Ashby, D.; Boaz, A.; Grayson, L. Social Science and the Evidence-based Policy Movement. *Soc. Policy Soc.* **2002**, *1*, 215–224. [CrossRef]
41. Lodge, M.; Wegrich, K. Crowdsourcing and regulatory reviews: A new way of challenging red tape in British government? *Regul. Gov.* **2014**, *9*, 30–46. [CrossRef]
42. Misuraca, G.; Codagnone, C.; Rossel, P. From practice to theory and back to practice: Reflexivity in measurement and evaluation for evidence-based policy making in the information society. *Gov. Inf. Q.* **2013**, *30*, S68–S82. [CrossRef]
43. Koziarski, J.; Lee, J.R. Connecting evidence-based policing and cybercrime. *Polic. Int. J.* **2020**, *43*, 198–211. [CrossRef]
44. Yang, Q. A New Approach to Evidence-Based Practice Evaluation of Mental Health in Psychological Platform under the Background of Internet + Technology. In Proceedings of the 2019 International Conference on Electronic Engineering and Informatics (EEI), Nanjing, China, 8–10 November 2019; IEEE: Piscataway, NJ, USA, 2019; pp. 321–323.
45. Shiroishi, Y.; Uchiyama, K.; Suzuki, N. Better actions for society 5.0: Using AI for evidence-based policy-making that keeps humans in the loop. *Computer* **2019**, *52*, 73–78. [CrossRef]
46. Setiadarma, E.G. Understanding the Evidence-Based Policy Making (EBPM) Discourse in the Making of the Master Plan of National Research (RIRN) Indonesia 2017-2045. *STI Policy Rev.* **2018**, *9*, 30–54.
47. Newman, J.; Cherney, A.; Head, B.W. Policy capacity and evidence-based policy in the public service. *Public Manag. Rev.* **2017**, *19*, 157–174. [CrossRef]
48. Head, B.W. Three lenses of evidence-based policy. *Aust. J. Public Adm.* **2008**, *67*, 1–11. [CrossRef]
49. Freedman, D. *The Politics of Media Policy*; Polity: Cambridge, UK, 2008.
50. Cairney, P.; Oliver, K. Evidence-based policy-making is not like evidence-based medicine, so how far should you go to bridge the divide between evidence and policy? *Health Res. Policy Syst.* **2017**, *15*, 1–11. [CrossRef] [PubMed]
51. Pang, M.S.; Lee, G.; DeLone, W.H. IT resources, organizational capabilities, and value creation in public-sector organizations: A public-value management perspective. *J. Inf. Technol.* **2014**, *29*, 187–205. [CrossRef]
52. Castelló, I.; Morsing, M.; Schultz, F. Communicative dynamics and the polyphony of corporate social responsibility in the network society. *J. Bus. Ethics* **2013**, *118*, 683–694. [CrossRef]
53. Kasabov, E. The challenge of devising public policy for high-tech, science-based, and knowledge-based communities: Evidence from a life science and biotechnology community. *Environ. Plan. C Gov. Policy* **2008**, *26*, 210–228. [CrossRef]
54. Zahra, S.A.; George, G. Absorptive capacity: A review, reconceptualization, and extension. *Acad. Manag. Rev.* **2002**, *27*, 185–203. [CrossRef]

55. McCay-Peet, L.; Quan-Haase, A. A model of social media engagement: User profiles, gratifications, and experiences. In *Why Engagement Matters: Cross-Disciplinary Perspectives and Innovations on User Engagement with Digital Media*; O'Brien, H., Lalmas, M., Eds.; Springer: Berlin/Heidelberg, Germany, 2016.
56. Abayomi-Alli, O.; Misra, S.; Abayomi-Alli, A.; Odusami, M. A review of soft techniques for SMS spam classification: Methods, approaches, and applications. *Eng. Appl. Artif. Intell.* **2019**, *86*, 197–212. [CrossRef]
57. Yan, Z. Big Data and Government Governance. In Proceedings of the International Conference on Information Management and Processing, London, UK, 12–14 January USA; IEEE: Piscataway, NJ, USA; 2018; pp. 110–114.
58. Lundberg, J.; Laitinen, M. Twitter trolls: A linguistic profile of anti-democratic discourse. *Lang. Sci.* **2020**, *8*, 101268. [CrossRef]
59. Bekkers, V.; Edwards, A.; de Kool, D. Social media monitoring: Responsive governance in the shadow of surveillance? *Gov. Inf. Q.* **2013**, *30*, 335–342. [CrossRef]
60. Panagiotopoulos, P.; Shan, L.C.; Barnett, J.; Regan, Á.; McConnon, Á. A framework of social media engagement: Case studies with food and consumer organisations in the UK and Ireland. *Int. J. Inf. Manag.* **2015**, *35*, 394–402. [CrossRef]
61. Williams, M.L.; Edwards, A.; Housley, W.; Burnap, P.; Rana, O.; Avis, N.; Morgan, J.; Sloan, L. Policing cyber neighbourhoods: Tension monitoring and social media networks. *Polic. Soc.* **2013**, *23*, 461–481. [CrossRef]
62. Fernandez, M.; Wandhoefer, T.; Allen, B.; Cano Basave, A.; Alani, H. Using social media to inform policy-making: To whom are we listening? In Proceedings of the European Conference on Social Media (ECSM 2014), Brighton, UK, 10–11 July 2014.
63. Gintova, M. Understanding government social media users: An analysis of interactions on Immigration, Refugees and Citizenship Canada Twitter and Facebook. *Gov. Inf. Q.* **2019**, *36*, 101388. [CrossRef]
64. Napoli, P.M. User data as public resource: Implications for social media regulation. *Policy Internet* **2019**, *11*, 439–459. [CrossRef]
65. Benthaus, J.; Risius, M.; Beck, R. Social media management strategies for organizational impression management and their effect on public perception. *J. Strateg. Inf. Syst.* **2016**, *25*, 127–139. [CrossRef]
66. Fan, W.; Gordon, M.D. The power of social media analytics. *Commun. ACM* **2014**, *57*, 74–81. [CrossRef]
67. Hoffman, D.L.; Fodor, M. Can you measure the ROI of your social media marketing? *Sloan Manag. Rev.* **2010**, *52*, 41–49.
68. Zhang, C.; Liu, C.; Zhang, X.; Almpanidis, G. An up-to-date comparison of state-of-the-art classification algorithms. *Expert Syst. Appl.* **2017**, *82*, 128–150. [CrossRef]
69. Schniederjans, D.; Cao, E.S.; Schniederjans, M. Enhancing financial performance with social media: An impression management perspective. *Decis. Support Syst.* **2013**, *55*, 911–918. [CrossRef]
70. Boyd, D.; Crawford, K. Critical questions for big data. *Inf. Commun. Soc.* **2012**, *15*, 662–679. [CrossRef]
71. Schintler, L.A.; Kulkarni, R. Big data for policy analysis: The good, the bad, and the ugly. *Rev. Policy Res.* **2014**, *31*, 343–348. [CrossRef]
72. Ghanei Raad, M.; Mohammadi, A.; Beigdeloo, N. Reviewing interactive patterns of institutions collaborating with science and technology supreme councils of policy. *Rahyaft* **2011**, *49*, 5–17.
73. Ahmadian, M.; Aqajani, H.; Shirkhodaei, M. Tehranchian A. Designing science & technology policy model based on economic complexity approach. *Public Policy* **2018**, *4*, 27–29.
74. Kalantari, E.; Montazer, G.; Qazinoori, S. Drafting passe scenarios of enhanced science and technology policy structure in Iran. *Strateg. Manag. Res.* **2019**, *74*, 75–102.
75. Sedhai, S.; Sun, A. Semi-supervised spam detection in the Twitter stream. *IEEE Trans. Comput. Soc. Syst.* **2017**, *5*, 169–175. [CrossRef]
76. Mostafa, S.A.; Mustapha, A.; Mohammed, M.A.; Hamed, R.I.; Arunkumar, N.; Abd Ghani, M.K.; Jaber, M.M.; Khaleefah, S.H. Examining multiple feature evaluation and classification methods for improving the diagnosis of Parkinson's disease. *Cogn. Syst. Res.* **2019**, *54*, 90–99. [CrossRef]
77. Ghasemaghaei, M. Are firms ready to use big data analytics to create value? The role of structural and psychological readiness. *Enterp. Inf. Syst.* **2019**, *13*, 650–674. [CrossRef]
78. De Paula, N.O.B.; de Araújo Costa, I.P.; Drumond, P.; Moreira, M.Â.L.; Gomes, C.F.S.; Dos Santos, M.; do Nascimento Maêda, S.M. Strategic support for the distribution of vaccines against Covid-19 to Brazilian remote areas: A multicriteria approach in the light of the ELECTRE-MOr method. *Procedia Comput. Sci.* **2022**, *199*, 40–47. [CrossRef] [PubMed]
79. Moreira, M.Â.L.; Gomes, C.F.S.; Dos Santos, M.; da Silva Júnior, A.C.; de Araújo Costa, I.P. Sensitivity Analysis by the PROMETHEE-GAIA method: Algorithms evaluation for COVID-19 prediction. *Procedia Comput. Sci.* **2022**, *199*, 431–438. [CrossRef] [PubMed]

*big data and cognitive computing*

MDPI

*Article*

# Intelligent Multi-Lingual Cyber-Hate Detection in Online Social Networks: Taxonomy, Approaches, Datasets, and Open Challenges

**Donia Gamal [1,*], Marco Alfonse [1,2], Salud María Jiménez-Zafra [3] and Mostafa Aref [1]**

[1] Computer Science Department, Faculty of Computer and Information Sciences, Ain Shams University, Cairo 11566, Egypt; marco_alfonse@cis.asu.edu.eg (M.A.); mostafa.aref@cis.asu.edu.eg (M.A.)

[2] Laboratoie Interdisciplinaire de l'Université Française d'Égypte (UFEID LAB), Université Française d'Égypte, Cairo 11566, Egypt

[3] Computer Science Department, SINAI, CEATIC, Universidad de Jaén, 23071 Jaén, Spain; sjzafra@ujaen.es

* Correspondence: donia.gamaleldin@cis.asu.edu.eg

**Abstract:** Sentiment Analysis, also known as opinion mining, is the area of Natural Language Processing that aims to extract human perceptions, thoughts, and beliefs from unstructured textual content. It has become a useful, attractive, and challenging research area concerning the emergence and rise of social media and the mass volume of individuals' reviews, comments, and feedback. One of the major problems, apparent and evident in social media, is the toxic online textual content. People from diverse cultural backgrounds and beliefs access Internet sites, concealing and disguising their identity under a cloud of anonymity. Due to users' freedom and anonymity, as well as a lack of regulation governed by social media, cyber toxicity and bullying speech are major issues that need an automated system to be detected and prevented. There is diverse research in different languages and approaches in this area, but the lack of a comprehensive study to investigate them from all aspects is tangible. In this manuscript, a comprehensive multi-lingual and systematic review of cyber-hate sentiment analysis is presented. It states the definition, properties, and taxonomy of cyberbullying and how often each type occurs. In addition, it presents the most recent popular cyberbullying benchmark datasets in different languages, showing their number of classes (Binary/Multiple), discussing the applied algorithms, and how they were evaluated. It also provides the challenges, solutions, as well as future directions.

**Keywords:** cyber-hate; cyberbullying; sentiment analysis; online social networks; machine learning

## 1. Introduction

Sentiment Analysis (SA) is the area of Natural Language Processing (NLP) that focuses on analyzing and studying individuals' sentiments, appraisals, evaluations, emotions, and attitudes writing in texts [1]. The utilization of social media platforms, for example, Twitter, Facebook, and Instagram, have immensely increased the quantity of online social interactions and communications by connecting billions of people who prefer the exchange of opinions. The penetration of social media into the life of internet users is increasing. According to the most recent data, there will be 5.85 billion social media users globally in 2027, a 1.26 percent rise over the previous year [2,3], as shown in Figure 1.

Moreover, social media platforms offer visibility to ideas and thoughts that would somehow be neglected and unspoken by traditional media [4]. The textual content of interactions and communications that signify upsetting, disturbing, and negative phenomena such as online cyber-hate, harassment, cyberbullying, stalking, and cyber threats is increasing [5]. Therefore, this has strongly led to an expansion of attacks against certain users based on different categorizations such as religion, ethnicity, social status, age, etc. Individuals frequently struggle and battle to deal with the results and consequences of such offenses. By employing NLP, several attempts have been put in action to deal with the issue of online cyber-hate and cyberbullying speech detection. This is because the computational analysis of language could be utilized to rapidly identify and distinguish offenses to facilitate and ease the process of dealing with and removing harsh messages [6].

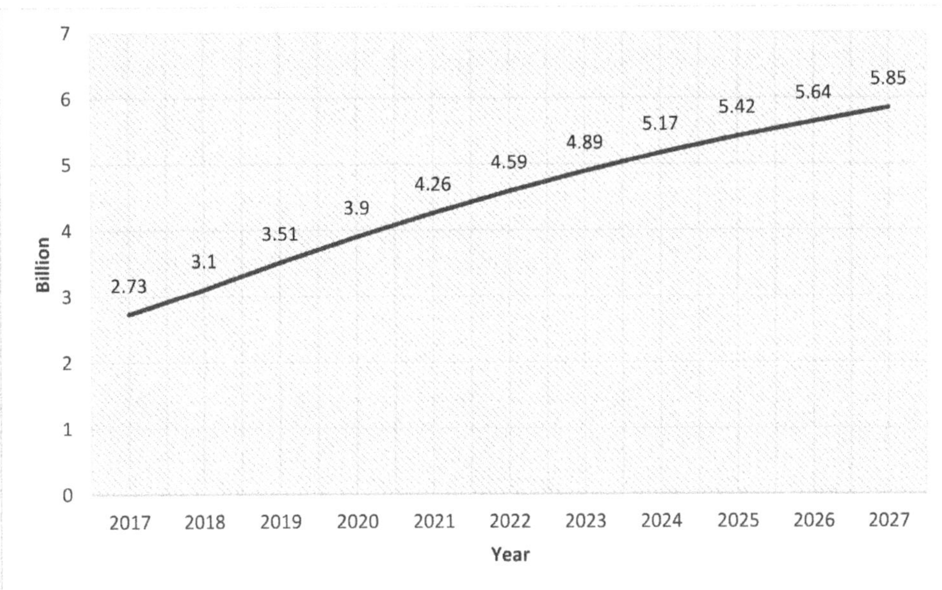


**Figure 1.** Number of Global Social Media Users Within 2017–2027.

The instance of hate speech and brutal communication shown over the Internet is named cyber-hate [7]. Cyber-hate, also known as cyberbullying, is defined as any utilization of electronic communications technologies to spread supremacist, racist, religious, extremist, or terrorist messages. It can target not only individuals but also entire communities [8]. Cyberbullying occurs when someone utilizes the internet to hurt or disturb a child or young person. It can occur on a social media platform, a game, an app, or any other online or electronic service or platform. Examples include posts, comments, texts, messages, chats, livestreams, memes, photos, videos, and emails. Some examples of how the internet can be used to violate someone's self-confidence is:

- Sending derogatory messages about them.
- Sharing humiliating images or videos of them.
- Spreading slanderous web rumors against them.
- Making fake accounts in their name.
- Making the public believe they are someone else.

Various types of violations are committed for many reasons in the online cyber realm through cyber innovation and technology. This insecure environment of online social networks requires consideration to prevent the harm and damage brought by these crimes to society. Several researchers are working in multiple directions to achieve the best results for automated cyberbullying detection using machine-learning techniques. In this manuscript, a taxonomy of multiple techniques being utilized in cyber-hate detection and prediction through different languages will be presented.

The rest of the manuscript is organized as follows: Section 2 provides a summary of the main properties of cyber-hate speech, and a detailed taxonomy of cyber-hate speech is given. The available datasets of cyber-hate in different languages are discussed in Section 3.

Section 4 previews the main approaches of cyber-hate classification. Section 5 introduces the comparative study of binary and multiple classifications over different datasets and various languages. Open Challenges in cyber-hate detection are discussed in Section 6. Finally, Section 7 concludes this manuscript and presents future work.

## 2. Cyber-Hate Speech Properties

### 2.1. Definition

Cyber-hate is the act of threatening, intimidating, harassing, irritating, or bullying any individual or group (for example, non-white people) through communication technology such as social media [9]. For textual content to be considered bullying, the intent of harm, such as physical, emotional, social, etc., should be obvious, as shown in Figure 2.The following situations are examples of cyber-hate speech:

- Posting threats such as physical harm, brutality, or violence.
- Any discussion intended to offend an individual's feelings, including routinely inappropriately teasing, prodding, or making somebody the brunt of pranks, tricks, or practical jokes.
- Any textual content meant to destroy the social standing or reputation of any individual on online social networks or offline communities.
- Circulating inappropriate, humiliating, or embarrassing images or videos on social networks.
- Persistent, grievous, or egregious utilization of abusive, annoying, insulting, hostile, or offensive language.

**Figure 2.** Examples of Online Social Media Cyberbullying.

### 2.2. Taxonomy

There is much more to cyber-hate than meets the eye. For instance, many people once believed that cyber-hate only consisted of physical bullying and name-calling. However, there are ten types of cyber-hate, which range from excluding and gossiping about people to making fun of their race or religion, as shown in Figure 3.

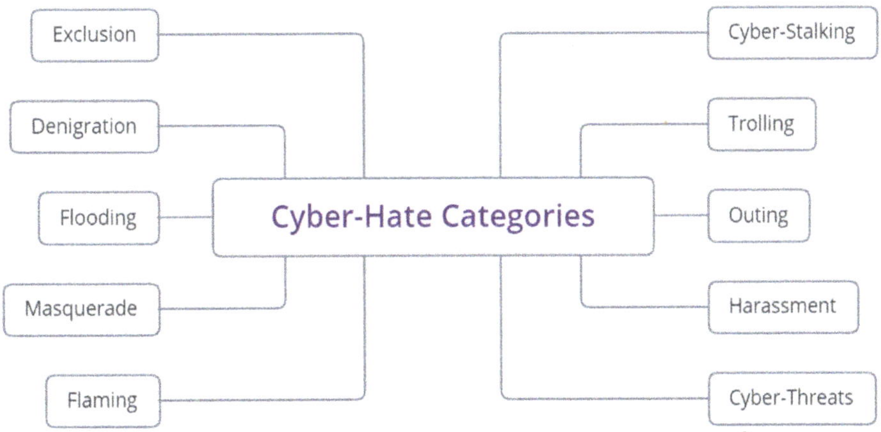


**Figure 3.** Cyber-hate Categories.

The categories that comprise the taxonomy of the term cyber-hate are presented and defined below:

i. Exclusion is defined as ignoring or neglecting the victim in a conversation [10]. Cyber-Exclusion is an intentional and deliberate action to make it clear to people that they do not belong to the group and that their involvement is not needed. On social networking sites, individuals can defriend or block others, which implies their inability to view their profiles, write comments, and so forth.

ii. Denigration involves the practice of demeaning, gossiping, dissing, or disrespecting another individual on social networks [11]. Writing rude, vulgar, mean, hurtful, or untrue messages or rumors about someone to another person or posting them in a public community or chat room falls under denigration. The purpose is to hurt the victim in the eyes of his or her community, as the insults are seen not only by the victim but also others.

iii. Flooding is the posting of a countless number of online social networking messages so the victim cannot post a message [12]. It consists of the bully or harasser repeatedly writing the same comment, posting nonsense comments, or holding down the enter key to not allow the victim to contribute to the chat or conversation.

iv. Masquerade is defined as the process of impersonating another person to send messages that seem to be originated by that person and cause damage or harm [13]. One of the ways to do this is by, for example, hacking into a victim's e-mail account and instantly sending these messages. Moreover, friends sharing passwords can also regularly accomplish this type of access; however, the sophisticated hacker may discover other ways, for example, by systematically testing probable passwords. This strategy is inherently hard and difficult to be recognized or detected.

v. Flaming, blazing, or bashing involves at least two users attacking and assaulting each other on a personal level. In this category of cyber-hate, flaming refers to a conversation full of hostile, unfriendly, irate, angry, and insulting communications and interactions that are regularly unkind personal attacks [14]. Flaming can occur in a diversity of environments, such as online social networking and discussion boards, group chat rooms, e-mails, and Twitter. Anger is frequently expressed by utilizing capital letters, such as 'U R AN IDIOT & I HATE U!'. Many flaming texts are vicious, horrible, and cruel and are without fact or reason.

vi. Cyberstalking is the utilization of social networks to stalk, hassle, or harass any individual, group, or organization [15]. It might contain false incriminations, accusations, criticism, defamation, maligning, slander, and libel. Cases of cyberstalking can often begin as seemingly harmless interactions. Sometimes, particularly at the beginning, a few strange or maybe distasteful messages may even amuse. Nevertheless, if they turn out to be systematic, it becomes irritating, annoying, and even frightening.

vii. Trolling, also called baiting, attempts to provoke a fight by intentionally writing comments that disagree with other posts in the topic or thread [16]. The poster plans to excite emotions and rouse an argument, while the comments themselves inevitably turn personal, vulgar, enthusiastic, or emotional.

viii. Outing is similar to denigration but requires the bully and the victim to have a close personal relationship, either on social networks or in-person. It includes writing and sharing personal, private, embarrassing, or humiliating information publicly [17]. This information can incorporate stories heard or received from the victim or any personal information such as personal numbers, passwords, or addresses.

ix. Harassment using social networks is equivalent to harassment utilizing more conventional and traditional means [18]. Harassment refers to threatening actions dependent on an individual's age, gender, race, sexual orientation, and so forth.

x. Cyber threats include sending short messages that involve threats of harm, are scary, intimidating, are very aggressive, or incorporate extortion [19]. The dividing line where harassment becomes cyberstalking is obscured; however, one indicator may be when the victim starts to fear for his or her well-being or safety, then the act has to be considered cyberstalking.

According to a more detailed survey from 2021 by PEW Research Center (https://www.pewresearch.org/) (access on 16 February 2023), over 40% of Americans under the age of 30 have experienced online bullying [20]. The most common types of cyberbullying are, as represented in Figure 4, denigration and harassment.

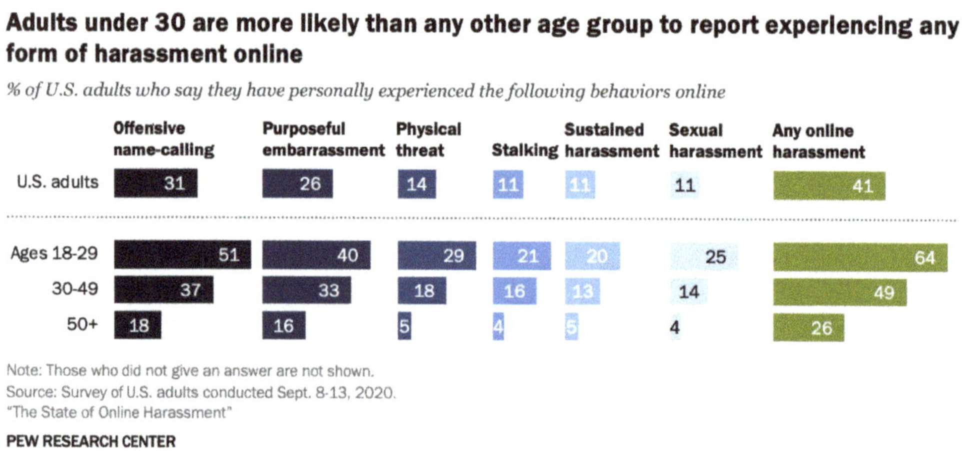


**Figure 4.** Experiences with certain types of online abuse vary by age, gender, race, or ethnicity in the U.S. in 2020.

## 3. Cyber-Hate Speech Datasets

This section presents a compilation of the datasets generated over the last five years for cyber-hate speech detection using different characteristics such as the number of classes, language, size, and availability of the datasets, as shown in Table 1. It covers various types of social network textual content such as Formspring (which contains a teen-oriented Q&A forum), Twitter, Instagram, Facebook (which are considered large microblogging platforms), WhatsApp (which is an application for instant messaging that can run on multiple platform devices) and Wikipedia talk pages (that could be described as a collaborative knowledge repository). Each dataset states a different topic of cyber-hate speech. Twitter datasets comprise examples of offensive, racist, and sexist tweets. Facebook datasets also contain racism and sexism statuses. Instagram and YouTube datasets include examples of personal attacks. However, Formspring datasets are not explicitly about a single topic.

Mangaonkar et al. [21] proposed a binary dataset that contains two subset datasets. The first subset dataset was a balanced dataset with 170 bullying tweets and 170 non-bullying tweets. The second sample was unbalanced, with 177 bullying tweets and 1163 non-bullying tweets. The purpose of developing a balanced and imbalanced dataset is to test the performance of the ML algorithms on various dataset types. These tweets were then manually categorized as "bullying" or "nonbullying" for validation purposes.

Van Hee et al. [22] collected binary cyberbullying data from the social networking site Ask.fm (https://ask.fm/, access on 16 February 2023). They created and implemented a novel method for cyberbullying annotation that describes the existence and intensity of cyberbullying and the role of the author of the post, for example, a harasser, victim, a bystander or not.

Waseem and Hovy [23] gathered a dataset of tweets over a period of two months. They downloaded 136,052 tweets and annotated 16,914 of them, 3383 of which were sexist content sent by 613 users, 1972 of which were racist content sent by 9 users, and 11,559 of which were neither sexist nor racist and were sent by 614 users. Because hate speech is a genuine but restricted phenomenon, they did not balance the data in order to present the most realistic dataset feasible.

Zhao et al. [24] proposed a Twitter dataset that is made up of tweets retrieved from the public Twitter API stream. At least one of the following keywords appears in each tweet: bully, bullied, bullying. Retweets are eliminated by filtering tweets that contain the acronym "RT". Finally, 1762 tweets are randomly selected and manually tagged from the entire twitter collection. It is important to note that labeling is based on bullying traces. Bullying traces are defined as a reaction to a bullying encounter, which includes but vastly outnumbers instances of cyberbullying.

Singh et al. [25] used the Twitter corpus from the Content Analysis for the WEB 2.0 (CAW 2.0) dataset [26]. This corpus comprises around 900,000 postings from 27,135 users (one XML file for each user) from December 2008 to January 2009. They picked this corpus not just because it has been widely used in prior literature but also because it contains information for both textual content and social networks. They chose 800 files at random and kept the comments written with @, which represents direct paths between two people. This yielded a data set of roughly 13,000 messages. Then, they asked three students to categories each message as cyberbullying twice. They designated each post as 'yes' or 'no' based on whether it was believed to entail cyberbullying. This resulted in a data collection with 2150 user pairings and 4865 messages between them.

Al-garadi et al. [27] collected their data via Twitter during January and February of 2015. They have 2.5 million geo-tagged tweets in their data set. To avoid any privacy violations, they extract only publicly available content via the Twitter API and in accordance with Twitter's privacy policy in their study. Their dataset had an uneven class distribution, with just 599 tweets labeled as cyberbullying and 10,007 tweets classified as non-cyberbullying. Such an uneven class distribution can make it difficult for the model to appropriately categories the instances. Learning algorithms that lack class imbalance are prone to be overwhelmed by the major class while ignoring the minor. In real-world applications like fraud detection, instruction detection, and medical diagnosis, data sets frequently contain imbalanced data in which the normal class is the majority, and the abnormal class is the minority. Several solutions to these issues have been offered, including a combination of oversampling the minority (abnormal) class and under-sampling the majority (normal) class.

Hosseinmardi et al. [28] gathered data by using the Instagram API and a snowball sampling technique. They found 41 K Instagram user ids starting from a random seed node. Of these Instagram IDs, 25 K (61%) belonged to users who had public profiles, while the remaining users had private ones.

Zhang et al. [29] gathered data from the social networking site Formspring.me. Almost 3000 messages were collected and labeled by Amazon Mechanical Turk, a web service in which three workers each voted on whether or not a document contained bullying content. As a result, each message receives an equal number of votes from the workers. At least two workers labeled approximately 6.6% of the messages as bullying posts. The authors parsed the original dataset's messages into sentences and relabeled the messages that contained at least one vote. This resulted in 23,243 sentences, with 1623 (or roughly 7%) labeled as bullying messages.

Wulczyn et al. [30] used English Wikipedia to generate a corpus of over 100 k high-quality human-labeled comments. The collected data of debate comments from English Wikipedia discussion pages are generated by computing differences throughout the whole revision history and extracting the new content for each revision. About 10 annotators using Crowdflower (https://www.crowdflower.com/, access on 16 February 2023) annotated the collected corpus into two classes, either attacking or not attacking.

Batoul et al. [31] gathered a massive amount of data. As a result, the decision was made to scrape data from both Facebook and Twitter. This decision was influenced by the fact that those two social media platforms are the most popular among Arabs, particularly Arab youth. After removing all duplicates, this dataset contains 35,273 unique Arabic tweets that were manually labeled as bullying or not bullying.

Davidson et al. [32] gathered tweets containing hate speech keywords using a crowd-sourced hate speech lexicon. They used crowdsourcing to categorize a sample of these tweets into three groups: those containing hate speech, those containing only offensive language, and those containing neither. This dataset resulted in a total of 33,458 tweets.

Sprugnoli et al. [33] presented and distributed an Italian WhatsApp dataset developed through role-playing by three classes of children aged 12–13. The publicly available data has been labeled based on user role and insult type. The Whatsapp chat corpus consists of 14,600 tokens separated into 10 chats. Two annotators used the Celct Annotation Tool (CAT) web-based application [34] to annotate all of the chats.

Founta et al. [35] have published a large-scale crowdsourced abusive tweet dataset of 60 K tweets. An enhanced strategy is applied to effectively annotate the tweets via crowdsourcing. Through such systematic methods, the authors determined that the most appropriate label set in identifying abusive behaviors on Twitter is None, Spam, Abusive, and Hateful, resulting in 11% as 'Abusive,' 7.5% as 'Hateful,' 22.5% as 'Spam,' and 59% as 'None.' They prepare this dataset for a binary classification task by concatenating 'None'/'Spam' and 'Abusive'/'Hateful'.

De Gibert et al. [36] demonstrated the first dataset of textual hate speech annotated at the sentence level. Sentence-level annotation enables dealing with the smallest unit containing hate speech while decreasing noise generated by other clean sentences. About 10,568 sentences were collected from Storm-front and categorized as hate speech or not, as well as two other auxiliary types.

Nurrahmi and Nurjanah [37] gathered information from Twitter. Because the data was unlabeled, the authors created a web-based labeling tool to categorize tweets as cyberbullying or non-cyberbullying. The tool used provided them with 301 cyberbullying tweets and 399 non-cyberbullying tweets.

Albadi et al. [38] investigated the issue of religious hate speech in the Arabic Twittersphere and developed classifiers to detect it automatically. They gathered 6000 Arabic tweets referring to various religious groups and labeled them using crowdsourced workers. They provided a detailed analysis of the labeled dataset, identifying the primary targets of religious hatred on Arabic Twitter. Following the preprocessing of the dataset, they used various feature selection methods to create different lexicons comprised of terms found in tweets discussing religion, as well as scores reflecting their strength in distinguishing sentiment polarity (hate or not hate).

Bosco et al. [39] released a Twitter dataset for the HaSpeeDe (Hate Speech Detection) shared task at Evalita 2018, the Italian evaluation campaign for NLP and speech processing tools. This dataset contains a total of 4000 tweets, with each tweet having an annotation that falls into one of two categories: "hateful post" or "not".

Corazza et al. [40] provided a set of 5009 German tweets manually annotated at the message level with the labels "offense" (abusive language, insults, and profane statements) and "other". More specifically, 1688 messages are tagged as "offense", while 3321 messages are as "other".

Mulki et al. [41] presented the first publicly available Levantine Hate Speech and Abusive (L-HSAB) Twitter dataset, intending to serve as a reference dataset for the automatic identification of online Levantine toxic content. The L-HSAB is a political dataset because the majority of tweets were gathered from the timelines of politicians, social/political activists, and TV anchors. The dataset included 5846 tweets divided into three categories: normal, abusive, and hateful.

Ptaszynski et al. [42] provided the first dataset for the Polish language that included annotations of potentially dangerous and toxic words. The dataset was designed to investigate negative Internet phenomena such as cyberbullying and hate speech, which have recently grown in popularity on the Polish Internet as well as globally. The dataset was obtained automatically from Polish Twitter accounts and annotated by lay volunteers under the supervision of a cyberbullying and hate-speech expert with a total number of 11,041 tweets.

Ibrohim and Budi [43] offered an Indonesian multi-label hate speech and abuse dataset with over 11,292 tweets based on a diverse collection of 126 keywords. These tweets include 6187 non-hate speech tweets and 5105 hate speech tweets and are annotated by 3 annotators.

Basile et al. [44] investigated the detection of hate speech from a multilingual perspective on Twitter. They focused on two specific targets, immigrants and women, in Spanish and English. They made a dataset containing English (13,000) and Spanish (6600) tweets tagged concerning the prevalence of hostile content and its target.

Banerjee et al. [45] investigated the identification of cyberbullying on Twitter in the English language. The dataset used on Twitter consists of 69,874 tweets. A group of human annotators manually labeled the selected tweets as either "0" non-cyberbullying or "1" cyberbullying.

Lu et al. [46] presented a new Chinese Weibo (https://us.weibo.com/index, access on 16 February 2023) comment dataset designed exclusively for cyberbullying detection. They collected a dataset of 17 K comments from more than 20 celebrities with bad reputations or who have been involved in violent incidents. Three members who are familiar with Weibo and have a good understanding of the bloggers manually annotated all data.

Moon et al. [47] offered 9.4 K manually labeled entertainment news comments that were collected from a popular Korean online news portal for recognizing toxic speech. About 32 annotators labeled the comments manually.

Romim et al. [48] created a large dataset of 30,000 comments, 10,000 of which are hate speech, in the Bengali Language. All user comments on YouTube and Facebook were annotated three times by 50 annotators, with the majority vote serving as the final annotation.

Karim et al. [49] offered an 8 K dataset of hateful posts gathered from various sources such as Facebook, news articles, blogs, and so on in the Bengali language. A total of 8087 posts were annotated by three annotators (a linguist, a native Bengali speaker, and an NLP researcher) into political, personal, geopolitical, and religious.

Luu et al. [50] presented the ViHSD, a human-annotated dataset for automatically detecting hate speech on social networks. This dataset contains over 30,000 comments, each of which has one of three labels: CLEAN, OFFENSIVE, or HATE. The ViHSD contains 33,400 comments.

Sadiq et al. [51] presented Data Turks' Cyber-Trolls dataset for text classification purposes. To assist or prevent trolls, this dataset is used to classify tweets. There are two categories: cyber-aggressive (CA) and non-aggressive (NCA). The dataset contains 20,001 items, of which 7822 are cyber-aggressive, and 12,179 are not.

Beyhan et al. [52] compiled a hate speech dataset extracted from tweets in Turkish. The Istanbul Convention dataset is made up of tweets sent out following Turkey's departure from the Istanbul Convention. The Refugees dataset was produced by collecting tweets regarding immigrants and filtering them based on regularly used immigration keywords.

Ollagnier et al. [53] presented the CyberAgressionAdo-V1 dataset, which contains aggressive multiparty discussions in French obtained through a high-school role-playing game with 1210 messages. This dataset is based on scenarios that mimic cyber aggression situations that may occur among teenagers, such as ethnic origin, religion, or skin color. The collected conversations have been annotated in several layers, including participant roles, the presence of hate speech, the type of verbal abuse in the message, and whether utterances use different humor figurative devices such as sarcasm or irony.

ALBayari and Abdallah [54] introduced the first Instagram Arabic corpus (multi-class sub-categorization) concentrating on cyberbullying. The dataset is primarily intended for detecting offensive language in the text. They ended up with 200,000 comments, with three human annotators annotating 46,898 of them manually. They used SPSS (Kapa statistics) to evaluate the labeling agreements between the three annotators in order to use the dataset as a benchmark. The final score was 0.869, with a $p$-value of 103, indicating a near-perfect agreement among the annotators.

Patil et al. [55] investigated hate speech detection in Marathi, an Indian regional language. They presented the L3Cube-MahaHate Corpus, the largest publicly available Marathi hate speech dataset. The dataset was gathered from Twitter and labeled with four fine-grained labels: Hate, Offensive, Profane, and None. The dataset contains over 25,000 samples that have been manually labeled with the classes.

Kumar and Sachdeva [56] developed two datasets FormSpring.me and MySpace. The Formspring.me dataset is an XML file containing 13,158 messages published by 50 different users on the Formspring.me website. The dataset is divided into two categories: "Cyberbullying Positive" and "Cyberbullying Negative". While negative messages represent messages that do not include cyberbullying, positive messages include cyberbullying. There are 892 messages in the Cyberbullying Positive class and 12,266 messages in the Cyberbullying Negative class. The Myspace dataset is made up of messages gathered from Myspace group chats. The dataset's group chats are labeled and organized into ten message groups. If a group conversation contains 100 messages, the first group contains 1–10 messages, the second group contains 2–11 messages, and the final message group contains 91–100 messages. Labeling is done once for each group of ten messages, and it is labeled whether or not those ten messages contain bullying. This dataset contains 1753 message groups divided into 10 groups, each with 357 positive (Bullying) and 1396 negative (Non-Bullying) labels.

Atoum [57] collected two datasets (Dataset-1 and Dataset-2) from Twitter (one month apart). Twitter dataset 1 consists of 6463 tweets, with 2521 cyberbullying tweets and 3942 non-cyberbullying tweets. Twitter dataset 2 consists of 3721 with 1374 cyberbullying tweets and 2347 non-cyberbullying tweets.

Nabiilah et al. [58] proposed a dataset of toxic comments that were manually collected, processed, and labeled. Data is gathered from Indonesian user comments on social media platforms such as Instagram, Twitter, and Kaskus (https://www.kaskus.co.id/, access on 16 February 2023), which have multi-label characteristics and allow for the classification of more than one class. Pornography, Hate Speech, Radicalism, and Defamation are among the 7773 records in the dataset.

Below Table 1 is a brief description of each of the datasets.

**Table 1.** Cyber-hate Speech Datasets.

| Dataset | Category | Number of Classes | Classes | Social Network Platform | Language | Size | Availability | Year |
|---|---|---|---|---|---|---|---|---|
| Mangaonkar et al. [21] | Trolling and Harassment | 2 | Bullying<br>Non-Bullying | Twitter | English | 1340 | N/A | 2015 |
| Van Hee et al. [22] | Cyber Threats and Harassment | 2 | Bullying<br>Non-Bullying | Ask.fm | Dutch | 85,485 | N/A | 2015 |
| Waseem and Hovy [23] | Cyber Threats and Harassment | 3 | Racism<br>Sexism<br>None | Twitter | English | 16 K | [59] | 2016 |
| Zhao et al. [24] | Trolling and Harassment | 2 | Bullying<br>Non-Bullying | Twitter | English | 1762 | N/A | 2016 |
| Singh et al. [25] | Trolling and Harassment | 2 | Bullying<br>Non-Bullying | Twitter | English | 4865 | N/A | 2016 |
| Al-garadi et al. [27] | Trolling and Harassment | 2 | Bullying<br>Non-Bullying | Twitter | English | 10,007 | N/A | 2016 |
| Hosseinmardi et al. [28] | Flaming and Stalking and Harassment | 2 | Bullying<br>Non-Bullying | Instagram | English | 1954 | N/A | 2016 |
| Zhang et al. [29] | Trolling and Harassment | 2 | Bullying<br>Non-Bullying | Formspring | English | 13 K | N/A | 2016 |
| Wulczyn et al. [30] | Denigration and Masquerade and Harassment | 2 | Attacking<br>Non-Attacking | Wikipedia | English | 100 K | [60] | 2017 |
| Batoul et al. [31] | Trolling and Harassment | 2 | Bullying<br>Non-Bullying | Twitter | Arabic | 35,273 | N/A | 2017 |
| Davidson et al. | Trolling and Harassment | 3 | Bullying<br>Non-Bullying<br>Neither | Twitter | English | 33,458 | [61] | 2017 |
| Sprugnoli et al. [33] | Flaming and Stalking and Harassment and Trolling | 10 | Defense<br>General Insult<br>Curse or Exclusion<br>Threat or Blackmail<br>Encouragement to the Harassment<br>Body Shame<br>Discrimination-Sexism<br>Attacking relatives<br>Other<br>Defamation | WhatsApp | Italian | 14,600 | [62] | 2018 |

**Table 1.** *Cont.*

| Dataset | Category | Number of Classes | Classes | Social Network Platform | Language | Size | Availability | Year |
|---|---|---|---|---|---|---|---|---|
| Founta et al. [35] | Cyber Threats and Harassment | 7 | Offensive<br>Abusive<br>Hateful<br>Aggressive<br>Cyberbullying<br>Spam<br>Normal | Twitter | English | 100 K | [63] | 2018 |
| De Gibert et al. [36] | Trolling and Harassment | 2 | Hateful<br>Non-Hateful | Stormfront | English | 10,568 | [64] | 2018 |
| Nurrahmi and Nurjanah [37] | Trolling and Harassment | 2 | Bullying<br>Non-Bullying | Twitter | Indonesian | 700 | N/A | 2018 |
| Albadi et al. [38] | Trolling and Harassment | 2 | Hateful<br>Non-Hateful | Twitter | Arabic | 6 K | [65] | 2018 |
| Bosco et al. [39] | Trolling and Harassment | 2 | Bullying<br>Non-Bullying | Twitter | Italian | 4 K | | 2018 |
| Corazza et al. [40] | Trolling and Harassment | 2 | Bullying<br>Non-Bullying | Twitter | German | 5009 | | 2018 |
| Mulki et al. [41] | Trolling and Harassment | 3 | Normal<br>Abusive<br>Hate | Twitter | Arabic | 6 K | [66] | 2019 |
| Ptaszynski et al. [42] | Trolling and Harassment | 3 | Non-harmful<br>Cyberbullying<br>Hate-speech and other harmful contents | Twitter | Polish | 11,041 | [67] | 2019 |
| Ibrohim and Budi [43] | Trolling and Harassment | 2 | Hateful<br>Non-Hateful | Twitter | Indonesian | 11,292 | [68] | 2019 |
| Basile et al. [44] | Trolling and Harassment | 2 | Hateful<br>Non-Hateful | Twitter | English<br>Spanish | 13,000<br>6600 | [69] | 2019 |
| Banerjee et al. [45] | Trolling and Harassment | 2 | Bullying<br>Non-Bullying | Twitter | English | 69,874 | N/A | 2019 |
| Lu et al. [46] | Cyber Threats and Harassment | 3 | Sexism<br>Racism<br>Neither | Sina Weibo | Chinese | 16,914 | [70] | 2020 |
| Moon et al. [47] | Trolling and Harassment | 3 | Hateful<br>Offensive<br>None | Online News Platform | Korean | 9.4 K | [71] | 2020 |
| Romim et al. [48] | Trolling and Harassment | 2 | Hateful<br>Non-Hateful | Facebook and YouTube | Bengali | 30 K | [72] | 2021 |
| Karim et al. [49] | Trolling and Harassment | 2 | Hateful<br>Non-Hateful | Facebook, YouTube comments, and newspapers | Bengali | 8087 | [73] | 2021 |
| Luu et al. [50] | Trolling and Harassment | 3 | Hate<br>Offensive<br>Clean | Facebook and YouTube | Vietnamese | 33,400 | [74] | 2021 |

**Table 1.** *Cont.*

| Dataset | Category | Number of Classes | Classes | Social Network Platform | Language | Size | Availability | Year |
|---|---|---|---|---|---|---|---|---|
| Sadiq et al. [51] | Trolling and Harassment | 2 | Bullying | Twitter | English | 20,001 | [75] | 2021 |
| | | | Non-Bullying | | | | | |
| Beyhan et al. [52] | Trolling and Harassment | 2 | Hateful | Twitter | Turkish | 2311 | [76] | 2022 |
| | | | Non-Hateful | | | | | |
| Ollagnier et al. [53] | Flaming, Stalking, Harassment and Trolling | 2 | Hateful | WhatsApp | French | 1210 | [77] | 2022 |
| | | | Non-Hateful | | | | | |
| ALBayari and Abdallah [54] | Flaming, Stalking, Harassment and Trolling | 2 | Bullying | Instagram | Arabic | 46,898 | [78] | 2022 |
| | | | Non-Bullying | | | | | |
| Patil et al. [55] | Trolling and Harassment | 4 | Hate | Twitter | Marathi | 25 K | [79] | 2022 |
| | | | Offensive | | | | | |
| | | | Profane | | | | | |
| | | | None | | | | | |
| Kumar and Sachdeva [56] | Trolling and Harassment | 2 | Bullying | Formspring | English | 13,158 | N/A | 2022 |
| | | | Non-Bullying | MySpace | | 1753 | | |
| Atoum [57] | Trolling and Harassment | 2 | Bullying | Twitter Dataset 1 | English | 6463 | N/A | 2023 |
| | | | Non-Bullying | Twitter Dataset 2 | | 3721 | | |
| Nabiilaha et al. [58] | Trolling, Harassment and Flaming | 2 | Bullying | Instagram, Twitter and Kaskus | Indonesian | 7773 | N/A | 2023 |
| | | | Non-Bullying | | | | | |

As demonstrated in Table 1, the majority of the studies and experiments were implemented on Twitter datasets. This is due to the effortless accessibility and availability of tweets that can be crawled utilizing the Twitter API. Out of all, most of the research focuses on the identification of hate speech and differentiating them from non-hate (or offensive) texts. Most of the research into cyber-hate was applied in the English language, while related work in other languages is scarce due to the lack of available datasets or the difficulty of their morphology as in the Arabic language.

## 4. Cyber-Hate Speech Detection Approaches

In recent years, some sentiment-based methods have been published to detect and identify abusive language [80]. These approaches are the machine-learning approach, the lexicon-based approach, and the hybrid approach, as shown in Figure 5.

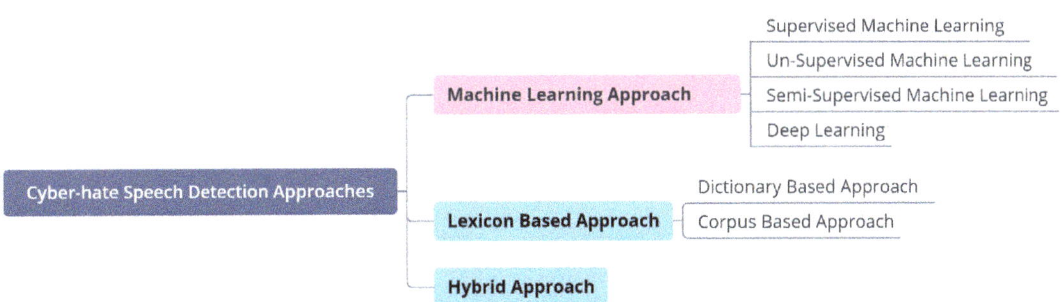


**Figure 5.** Cyber-hate Speech Detection Approaches.

On the one hand, the Machine Learning Approach (MLA) comprises the following methods: supervised machine learning, un-supervised machine learning, semi-supervised machine learning and deep learning. In a supervised machine learning approach, the classifier is built to learn the properties of categories or classes automatically from a set of pre-annotated training textual content. When utilizing the supervised machine learning approach, some main issues and challenges have to be considered, such as the categories to be used to classify the instances, the labeled training data, the extracted and selected features to be used to represent each unknown textual content, and the selected algorithm to be used for categorization [81]. In unsupervised machine learning, the machine attempts to find and understand the hidden structure within unlabeled data [82]. Semi-supervised learning is concerned with how the combination of labeled and unlabeled data will change the behavior of learning and designing algorithms that benefit from such a combination [83]. The Deep Learning approach, inspired by artificial neural networks, is an evolving branch of machine learning [84]. With the aid of the hierarchy of layers, it provides ways of learning data representations in a supervised and unsupervised manner, allowing multiple processing [85].

On the other hand, the Lexicon-Based Approach (LBA) comprises making a list of words that is called the dictionary, which is searched and counted in the textual content. These calculated frequencies can be utilized explicitly as features or to calculate scores for classifying textual content. A potential limitation of this approach regarding its classification efficacy is its dependency on domain-specific words presented in a dictionary; also, it needs an automatic methodology for the classification and scoring of words to reduce the amount of manpower required for the manual scoring of domain-specific words [86]. A corpus-based approach utilizes a collection of sentiment words with pre-defined polarity to recognize new sentiment words and their polarity in a large corpus [87]. A corpus-based approach provides a data-driven approach where one has access not only to sentiment labels but also to a context that one can use. A dictionary-based approach exploits the lexicographical tools such as Artha (https://sourceforge.net/projects/artha/, access on 16 February 2023), Tematres (https://www.vocabularyserver.com/, access on 16 February 2023), Wordhoard (https://wordhoard.northwestern.edu/, access on 16 February 2023), or WordNet (https://wordnet.princeton.edu/, access on 16 February 2023). In these, the key strategy methods are gathering an initial collection of sentiment words and manually orienting them; then, looking in a dictionary to enlarge this collection by finding their synonyms and antonyms [88].

Finally, the Hybrid Approach (HA) is the amalgamation of both machine learning and lexicon-based methods.

## 5. Cyber-Hate Detecting Techniques

Textual data mining and analysis have become an active and attractive research field. The global availability of such data makes text analytics acquire a major consideration. Hate speech detection tasks can be performed as binary or multi-class classifications based on the number of classes in these datasets.

### *5.1. Binary Cyber-Hate Classification*

Cyber-hate detection has been approached as a binary classification task (cyberbullying -vs.- non-cyber bullying) or (Hate–Non-Hate). This section presents a summary of studies in cyber-hate binary classification techniques.

Mangaonkar et al. [21] applied different algorithms to classify tweets, then performed AND and OR parallelism. They combined the output of multiple classifiers to enhance the performance. They classified tweets using a four-node detection system and experimented with homogeneous (all computing nodes use the same classification algorithm), heterogeneous (each node uses a different algorithm), and selective (the best-performing node is chosen as the expert, and all other nodes defer to it) collaborations. Each tweet is processed by all nodes and classified as cyberbullying if more than half of the nodes in the AND configuration flag it as bullying or if any node flags it as bullying in the OR configuration. They discovered that OR parallelism produces the highest recall values at 60%, while AND parallelism produces the highest accuracy at 70%.

Van Hee et al. [22] presented a proposed system in the Dutch language for the automatic detection of cyberbullying. The dataset enclosing cyberbullying posts was gathered from the social networking site Ask.fm. The experimental results showed that Support Vector Machines (SVM) achieved an F1-score of 55.39%.

Nandhini and Sheeba [89] proposed a system for detecting the existence of cyberbullying activity on social networks in the English language in order to help the government take action before more people become cyberbullying victims. The used dataset has a record of almost 4 K, which is gathered from social networks (Formspring.me, Myspace.com) [90]. For this purpose, they used the Naïve Bayes (NB) classifier achieving 92% on the Formspring dataset and 91% on MySpace.me dataset.

Zhao et al. [24] have presented Embedding-enhanced Bag-of-Words (EBoW), a unique representation learning method for cyberbullying detection. EBoW combines bag of words features, latent semantic characteristics, and bullying features. Bullying characteristics are generated from word embeddings, which can capture the semantic information behind words. When the final representation is learned, a linear SVM is used to detect bullying messages with a recall of 79.4%.

Singh et al. [25] employed probabilistic fusion approaches to mix social and text information as the classifier's input. The proposed methodology has been applied to the English Twitter dataset. The accuracy of the obtained results was 89%.

Al-garadi et al. [27] utilized supervised machine learning algorithms such as NB, SVM, Random Forest (RF), and K Nearest Neighbor (KNN) to detect cyberbullying on Twitter in the English language. Based on an evaluation, their model accuracy is 70.4% by NB, 50% by SVM, 62.9% by Random Forest (RF), and 56.8% by KNN.

Hosseinmardi et al. [28] investigated the problem of predicting cyberbullying in the Instagram media-based social network. They demonstrated that non-text features such as image and user metadata were important in predicting cyberbullying, with a Logistic Regression (LR) classifier achieving 72% recall and 78% precision.

Zhang et al. [29] proposed a novel Pronunciation-based Convolutional Neural Network (PCNN) to detect cyberbullying. They assessed the performance of their model using a cyberbullying dataset in English from Formspring.me. Their experiment revealed that PCNN can achieve an accuracy of 88.1%.

Wulczyn et al. [30] demonstrated a methodology in cyberbullying detection by applying LR and Multi-Layer Perceptron (MLP) to Wikipedia, resulting in an open dataset of over 100 k high-quality human-labeled comments. They evaluated their models using Area Under the Receiver Operating Characteristic (AUROC) and achieved 96.18% using LR and 96.59% using MLP.

Batoul et al. [31] proposed a system for detecting Arabic cyberbullying. They worked on an Arabic Twitter dataset that contains 35,273 unique tweets after removing all duplicates. NB and SVM obtained F-measure with 90.5% and 92.7%.

De Gibert et al. [36] conducted a thorough qualitative and quantitative analysis of their dataset, as well as several baseline experiments with various classification models, which are SVM, Convolution Neural Networks (CNN), and Long-Short Term Memory (LSTM). The experiments employ a well-balanced subset of labeled sentences. All of the HATE sentences were collected, and an equal number of NOHATE sentences were randomly sampled, a total of 2 k labeled sentences. Eighty percent of this total has been used for training, with the remaining 20% for testing. The evaluated algorithms, SVM, CNN, and LSTM, achieved 71%, 66%, and 73% accuracy, respectively.

Nurrahmi and Nurjanah [37] studied cyberbullying detection for Indonesian tweets to recognize cyberbullying text and actors on Twitter. The study of cyberbullying has successfully identified tweets that contain cyberbullying with an F1-score of 67% utilizing the SVM algorithm.

Albadi et al. [38] are the first to address the issue of identifying and recognizing speech promoting religious hate on Arabic Twitter. They implemented different classification models utilizing lexicon-based, n-gram-based, and deep-learning-based approaches. They concluded that a straightforward Recurrent Neural Networks (RNN) architecture with Gated Recurrent Units (GRU) and pre-trained word embeddings could sufficiently detect religious hate speech since it gives an AUROC of 84%.

Basile et al. [39] evaluated the SVM on a dataset of 13,000 tweets in English and 6600 tweets in Spanish [45], 60% of which were labeled as hate speech. In terms of performance, the system had an F1-score of 65% for the English tweets and 73% for the Spanish tweets.

Ibrohim and Budi [43] conducted a combination of feature, classifier, and data transformation methods between word unigram, Random Forest Decision Tree (RFDT), and Label Power-set (LP) to identify abusive language and hate speech. Their system achieved an accuracy of 77.36% for the classification of hate speech without identifying the target, categories, and level of hate speech. Moreover, their system identifies abusive language and hate speech, including identifying the target, categories, and level of hate speech with an accuracy of 66.1%.

Banerjee et al. [45] represented an approach for the detection of cyberbullying in the English language. They applied CNN to a Twitter dataset of 69,874 tweets. Their proposed approach achieved an accuracy of 93.97%.

Corazza et al. [4] proposed a neural architecture for identifying the forms of abusive language, which shows satisfactory performance in several languages, namely English [23], Italian [37], and German [40]. Different components were employed in the system, which are Long Short-Term Memory (LSTM), GRU, and Bidirectional Long Short-Term Memory (BiLSTM). For the feature selection, they used n-gram, word embedding, social network-specific features, emotion lexica, and emoji. The results show that LSTM outperforms other used algorithms in multilingual classification with an F1-score of 78.5%, 71.8%, and 80.1% in English, German, and Italian, respectively.

Romim et al. [48] ran a baseline model (SVM) and several deep learning models, as well as extensive pre-trained Bengali word embedding such as Word2Vec, FastTest, and BengFastText, on their collected dataset (Facebook and YouTube comments). The experiment demonstrated that, while all of the deep learning models performed well, SVM achieved the best result with an 87.5% accuracy.

Karim et al. [49] proposed DeepHateExplainer, an explainable approach for detecting hate speech in the under-resourced Bengali language. Bengali texts are thoroughly preprocessed before being classified into political, personal, geopolitical, and religious hatreds using a neural ensemble method of transformer-based neural architectures (i.e., monolingual Bangla BERT-cased/uncased, and XLM-RoBERTa). Before providing human-interpretable explanations for hate speech detection, important (most and least) terms are identified using a sensitivity analysis and layer-wise relevance propagation (LRP). Evaluations against machine learning (linear and tree-based models) and neural networks (i.e., CNN, Bi-LSTM, and Conv-LSTM with word embeddings) baselines produce F1-scores of 78%, 91%, 89%, and 84%, respectively, outperforming both ML and DNN baselines.

Sadiq et al. [51] proposed a system using a combination of CNN with LSTM and CNN with BiLSTM for cyberbullying detection on English tweets of the cyber-troll dataset. Statistical results proved that their proposed model detects aggressive behavior with 92% accuracy.

Beyhan et al. [52] created a hate speech detection system (BERTurk) based on the transformer architecture to serve as a baseline for the collected dataset. The system is evaluated using 5-fold cross-validation on the Istanbul Convention dataset; the classification accuracy is 77%.

ALBayari and Abdallah [54] used the most basic classifiers (LR, SVM, RFC, and Multinomial Naive Bayes (MNB)) for cyberbullying detection using their dataset. As a result, the SVM classifier has a significantly higher F1-score value of 69% than the other classifiers, making it a preferable solution.

Kumar and Sachdeva [56] proposed a hybrid model, Bi-GRU Attention-CapsNet (Bi-GAC), that benefits from learning sequential semantic representations and spatial location information using a Bi-GRU with self-attention followed by CapsNet for cyberbullying detection in social media textual content. The proposed Bi-GAC model is evaluated for performance using the F1-score and the ROC-AUC curve as metrics. On the benchmark Formspring.me and MySpace datasets, the results outperform existing techniques. In comparison to conventional models, the F1-score for MySpace and Formspring.me datasets achieved by nearly 94% and 93%, respectively.

Atoum [57] developed and refined an efficient method for detecting cyberbullying in tweets that uses sentiment analysis and language models. Various machine learning algorithms are examined and compared across two tweet datasets. CNN classifiers with higher n-gram language models outperformed other ML classifiers such as DT, RF, NB, and SVM. The average accuracy of CNN classifiers was 93.62% and 91.03%.

Nabilah et al. [58] used a Pre-Trained Model trained for Indonesian to detect comments containing toxic sentences on social media in Indonesia. The Multilingual BERT (MBERT), IndoBERT, and Indo Roberta Small models were used in this study to perform a multi-label classification and evaluate the classification results. The BERT model with an F1 Score of 88.97% yielded the best results in this study.

### 5.2. Multi-Class Cyber-Hate Classification

Several studies have been conducted for multi-class cyber-hate classification. This section summarizes the studies of multi-class cyber-hate classification techniques in different languages.

Waseem and Hovy [23] investigated the impact of various features in the classification of cyberbullying. They used an LR classifier and 10-fold cross-validation to test and quantify the impact of various features on prediction performance with an F1-score of 73%.

Badjatiya et al. [91] investigate the use of deep neural network architectures for hate speech detection in the English language [23]. They proposed a combination of deep neural network model embeddings and gradient-boosted decision trees, leading to better accuracy values. Embeddings gained from deep neural network models, when joined with gradient-boosted decision trees, prompted the best accuracy values with an F1-score of 93%.

Park and Fung [92] proposed a two-step approach to abusive language classification for detecting and identifying sexist and racist languages. They first classify the language as abusive or not and then classify it into explicit types in a second step. With a public English Twitter corpus [23] that contains 20 thousand tweets of a sexist and racist nature, their approach shows a promising performance of 82.7% F1-score using Hybrid-CNN in the first step and 82.4% F1-score using LR in the second step.

Watanabe et al. [93] presented a methodology to detect hate speech on Twitter in the English language [23]. The proposed approach consequently detects hate speech signs and patterns using unigrams as feature extraction along with sentimental and semantic features to classify tweets into hateful, offensive, and clean. The proposed approach achieves an accuracy of 78.4% for the classification of tweets.

Mulki et al. [41] presented the first publicly available Levantine Hate Speech and Abusive Behavior (L-HSAB) Twitter dataset intending to serve as a benchmark dataset. NB and SVM classifiers were used in machine learning-based classification experiments on L-HSAB. The results showed that NB outperformed SVM in terms of accuracy, with 88.4% and 78.6%, respectively.

Lu et al. [46] proposed an automatic method for determining whether the text in social media contains cyberbullying. It learns char-level features to overcome spelling mistakes and intentional data obfuscation. On the Weibo dataset, the CNN model achieved Precision, F1-score, and Recall values of 79.0%, 71.6%, and 69.8%, respectively.

Moon et al. [47] presented 9.4 K manually labeled entertainment news comments collected from a popular Korean online news platform for identifying Korean toxic speech.

They used three baseline classifiers: a character-level convolutional neural network (Char-CNN), a bidirectional long short-term memory (BiLSTM), and a bidirectional encoder representation from a Transformer (BERT) model. BERT has the best performance, with an F1-score of 63.3%.

Luu et al. [44] created the ViHSD dataset, a large-scale dataset for detecting hate speech in Vietnamese social media texts. The dataset contains 33,400 human-annotated comments and achieves an F1-score of 62.69% using the BERT model.

Patil et al. [55] presented L3CubeMahaHate, a hate speech dataset with 25,000 distinct samples evenly distributed across four classes. They ran experiments on various deep learning models such as CNN, LSTM, BiLSTM, and transformer-based BERT. The BERT model outperformed other models with an accuracy of 80.3%.

Wang et al. [94] proposed a framework for Metamorphic Testing for Textual Content Moderation (MTTM) software. They conducted a pilot study on 2000 text messages from real users and summarized eleven metamorphic relations at three perturbation levels: character, word, and sentence. MTTM uses these metamorphic relations on the toxic textual content to generate test cases that are still toxic but are unlikely to be moderated. When the MTTM is tested, the results show that the MTTM achieves up to 83.9% error-finding rates.

### 5.3. Analysis of the Literature Review

Table 2 presents, for each of the previously described works on binary and multiclass classification, a summary of the dataset used in the experimentation and its number of classes, the language under study, the algorithms tested, and the results obtained.

This comparative study identified that binary classification is the most common task carried out in cyber-hate detection, as shown in Table 2.

Moreover, most of the research on cyber-hate speech detection focuses on English textual content, so most of the resources, assets, libraries, and tools have been implemented for English language use only.

**Table 2.** Cartography of Existing Research in Hate Speech Detection.

| Author | Classes | Dataset | Language | Approach | Algorithm | Evaluation Metric | |
|---|---|---|---|---|---|---|---|
| Mangaonkar et al. 2015 [21] | 2 Classes (Cyberbullying, Non-Cyberbullying) | Twitter | English | MLA | LR (OR parallelism) | Recall | 60% |
| | | | | | LR (AND parallelism) | Accuracy | 70% |
| Van Hee et al. 2015 [22] | 2 Classes (Cyberbullying, Non-Cyberbullying) | Ask.fm | Dutch | MLA | SVM | F1-score | 55.39% |
| | | | | | | Recall | 51.46% |
| | | | | | | Precision | 59.96% |
| Nandhini and Sheeba, 2015 [89] | 2 Classes (Cyberbullying–Non-Cyberbullying) | Formspring | English | MLA | NB | Accuracy | 92% |
| | | MySpace.com | | | | | 91% |
| Waseem and Hovy 2016 [23] | 3 Classes (Sexism, Racism, Neither) | Twitter | English | MLA | LR | F1-score | 73% |
| Zhao et al. 2016 [24] | 2 Classes (Cyberbullying, Non-Cyberbullying) | Twitter | English | MLA | SVM | F1-score | 79.4% |
| Singh et al. 2016 [25] | 2 Classes (Cyberbullying, Non-Cyberbullying) | Twitter | English | LBA | Probabilistic Fusion approach | Accuracy | 89% |
| Al-garadi et al. 2016 [27] | 2 Classes (Cyberbullying, Non-Cyberbullying) | Twitter | English | MLA | NB | Accuracy | 70.4% |
| | | | | | SVM | | 50% |
| | | | | | RF | | 62.9% |
| | | | | | KNN | | 56.8% |
| Hosseinmardi et al. 2016 [28] | 2 Classes (Cyberbullying, Non-Cyberbullying) | Instagram | English | MLA | LR | Recall | 72% |
| | | | | | | Precision | 78% |

**Table 2.** *Cont.*

| Author | Classes | Dataset | Language | Approach | Algorithm | Evaluation Metric | |
|---|---|---|---|---|---|---|---|
| Zhang et al. 2016 [29] | 2 Classes (Cyberbullying, Non-Cyberbullying) | Formspring | English | MLA | PCCN | Accuracy | 88.1%. |
| Wulczyn et al. 2017 [30] | 2 Classes (Attacking, Non-Attacking) | Wikipedia | English | MLA | LR | AUROC | 96.18% |
| | | | | | MLP | | 96.59% |
| Batoul et al. 2017 [31] | 2 Classes (Cyberbullying–Non-Cyberbullying) | Twitter | Arabic | MLA | NB | Precision | 90.1% |
| | | | | | | Recall | 90.9% |
| | | | | | | F1-score | 90.5% |
| | | | | | SVM | Precision | 93.4% |
| | | | | | | Recall | 94.1% |
| | | | | | | F1-score | 92.7% |
| Badjatiya, Pinkesh et al. 2017 [91] | 3 classes (Sexism, Racism, Neither) | Twitter | English | MLA | LSTM | F1-score | 93% |
| Park, Ji Ho et al. 2017 [92] | 3 classes (Sexism, Racism, Neither) | Twitter | English | MLA | CNN | F1-score | 82.7% |
| | | | | | LR | | 82.4% |
| De Gibert et al. 2018 [36] | 2 Classes (Hate, Non-Hate) | Stormfront | English | MLA | SVM | Accuracy | 71% |
| | | | | | CNN | | 66% |
| | | | | | LSTM | | 73% |
| Nurrahmi and Nurjanah, 2018 [37] | 2 Classes (Cyberbullying–Non-Cyberbullying) | Twitter | Indonesian | MLA | SVM | F1-score | 67% |
| N. Albadi et al. 2018 [38] | 2 Classes (Hate, Non-Hate) | Twitter | Arabic | MLA | GRU-based RNN | Precision | 76% |
| | | | | | | Recall | 78% |
| | | | | | | F1-score | 77% |
| | | | | | | AUROC | 84% |
| Watanabe et al. 2018 [93] | 3 Classes (Hateful, Offensive and Clean) | Twitter | English | MLA | J48graft | Precision | 88% |
| | | | | | | Recall | 87.4% |
| | | | | | | F1-score | 87.5% |
| Mulki et al. 2019 [41] | 3 Classes (Normal, Abusive, Hate) | Twitter | English | | NB | Accuracy | 88.4% |
| | | | | | SVM | | 78.6% |
| Ibrohim and Budi, 2019 [43] | 2 Classes (Hateful, Non-Hateful) | Twitter | Indonesian | | RFDT | Accuracy | 77.36% |
| | | | | | LP | | 66.1% |
| Basile et al. 2019 [39] | 2 Classes (Hate, Non-Hate) | Twitter | English | | SVM | F1-score | 65% |
| | | | Spanish | | | | 73% |
| Banerjee et al. 2019 [45] | 2 Classes (Cyberbullying–Non-Cyberbullying) | Twitter | English | MLA | CNN | Accuracy | 93.97% |
| Corazza, Michele et al. 2020 [4] | 2 Classes (Hateful, Non-Hateful) | Twitter | English | MLA | LSTM | F1-score | 78.5% |
| | | | German | | | | 71.8% |
| | | | Italian | | | | 80.1% |
| Lu et al. 2020 [46] | 3 Classes (Sexism, Racism, and Neither) | Sina Weibo | Chinese | MLA | CNN | Precision | 79% |
| | | | | | | F1-score | 71.6% |
| | | | | | | Recall | 69.7% |
| Moon et al. 2020 [47] | 3 Classes (Hate, Offensive, None) | Korean Online News Platform | Korean | MLA | CharCNN | F1-score | 53.5% |
| | | | | | BiLSTM | | 29.1% |
| | | | | | BERT | | 63.3% |

**Table 2.** *Cont.*

| Author | Classes | Dataset | Language | Approach | Algorithm | Evaluation Metric | |
|---|---|---|---|---|---|---|---|
| Romim et al. 2021 [48] | 2 Classes (Hateful, Non-Hateful) | Facebook and YouTube | Bengali | MLA | SVM | Accuracy | 87.5% |
| | | | | | Word2Vec + LSTM | | 83.85% |
| | | | | | Word2Vec + Bi-LSTM | | 81.52% |
| | | | | | FastText + LSTM | | 84.3% |
| | | | | | FastText + Bi-LSTM | | 86.55% |
| | | | | | BengFastText + LSTM | | 81% |
| | | | | | BengFastText + Bi-LSTM | | 80.44% |
| Karim et al. 2021 [49] | 2 Classes (Hateful, Non-Hateful) | Facebook, YouTube comments, and newspapers | Bengali | MLA | LR | F1-score | 67% |
| | | | | | NB | | 64% |
| | | | | | SVM | | 66% |
| | | | | | KNN | | 66% |
| | | | | | RF | | 68% |
| | | | | | GBT | | 68% |
| | | | | | CNN | | 73% |
| | | | | | Bi-LSTM | | 75% |
| | | | | | Conv-LSTM | | 78% |
| | | | | | Bangla BERT | | 86% |
| | | | | | mBERT-cased | | 85% |
| | | | | | XML-RoBERTA | | 87% |
| | | | | | mBERT-uncased | | 86% |
| | | | | | Ensemble * | | 88% |
| Luu et al. 2021 [44] | 3 Classes (Offensive, Hate, None) | Facebook and YouTube | Vietnamese | MLA | BERT | F1-score | 62.69% |
| Sadiq et al. 2021 [51] | 2 Classes (Cyber-aggressive, Non-Cyber-aggressive) | Twitter | English | MLA | CNN + LSTM + Bi-LSTM | Accuracy | 92% |
| Beyhan et al. 2022 [52] | 2 Classes (Hateful, Non-Hateful) | Twitter | Turkish | MLA | BERTurk | Accuracy | 77% |
| ALBayari and Abdallah 2022 [54] | 2 Classes (Cyberbullying–Non-Cyberbullying) | Instagram | Arabic | MLA | MNB | F1-score | 66% |
| | | | | | RF | | 65% |
| | | | | | SVM | | 69% |
| | | | | | LR | | 66% |
| Patil et al. 2022 [55] | 4 Classes (Hate, Offensive, Profane, None) | Twitter | Marathi | MLA | CNN | Accuracy | 75.1% |
| | | | | | LSTM | | 75.1% |
| | | | | | BiLSTM | | 76.1% |
| | | | | | BERT | | 80.3% |
| Kumar and Sachdeva 2022 [56] | 2 Classes (Cyberbullying–Non-Cyberbullying) | Formspring | English | HA | Bi-GAC | F1-score | 94.03% |
| | | MySpace | | | | | 93.89% |
| Wang et al. 2023 [94] | 3 Classes (Cyberbullying–Non-Cyberbullying, Neither) | Twitter | English | HA | MTTM | Error Finding Rates | 83.9% |

**Table 2.** *Cont.*

| Author | Classes | Dataset | Language | Approach | Algorithm | Evaluation Metric | |
|---|---|---|---|---|---|---|---|
| Atoum, 2023 [57] | 2 Classes (Cyberbullying–Non-Cyberbullying) | Twitter Dataset 1 | English | MLA | CNN | Accuracy | 93.62% |
| | | Twitter Dataset 2 | | | | | 91.03% |
| Nabilah et al. 2023 [58] | 2 Classes (Cyberbullying–Non-Cyberbullying) | Instagram and Twitter and Kaskus | Indonesian | MLA | BERT | F1-score | 88.97% |

As previously mentioned, the most common task carried out in cyber-hate detection is binary classification rather than multi-class classification. Cyber-hate texts are known as representatives of a "bullying" class, and all other documents belong to "non-bullying". Twitter is the most commonly studied data source compared to other social media platforms. Most researchers applied and compared many supervised machine learning algorithms in order to determine the ideal ones for cyber-hate detection problems. As for the traditional machine learning algorithms, SVM has been used to build prediction models for cyberbullying and has been found to be accurate and efficient. On the other hand, CNN was the most common deep learning algorithm used in cyber-hate classification for binary or multiple-class classification. Researchers measure the effectiveness of their proposed model to determine how successfully the model can distinguish cyberbullying texts from non-cyber bullying texts by using various evaluation measures such as F1-score, accuracy, recall, and Precision [95,96].

Subsequently, the algorithms used for binary and multi-class classification for the English and Arabic languages are analyzed according to the results obtained with them.

Figure 6 shows the accuracy of binary cyber-hate classification on different English datasets. As illustrated in it, CNN has better accuracy than SVM, NB, and CNN + LSTM + Bi-LSTM. In addition, NB gives an acceptable accuracy on different datasets between 91% and 92%.

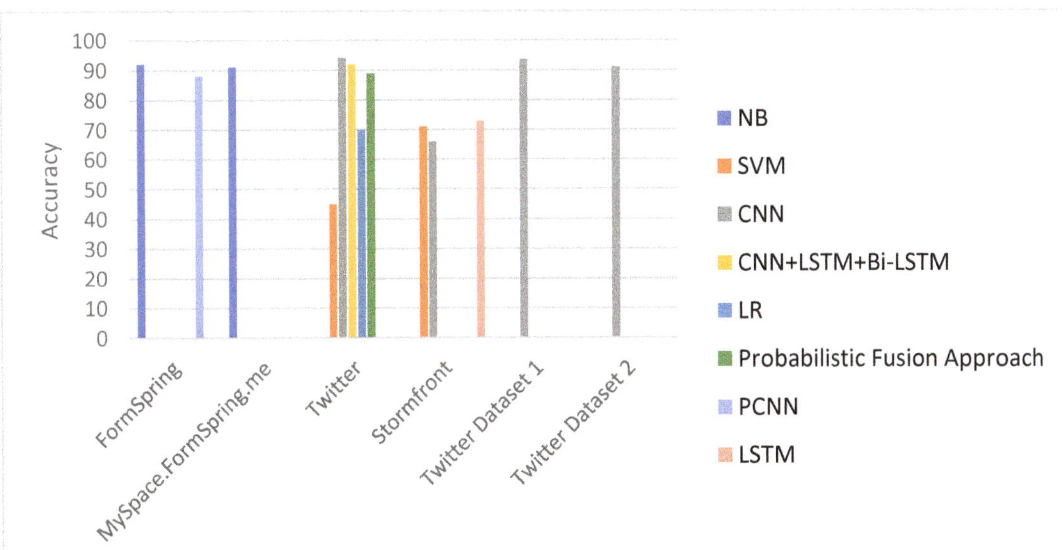


**Figure 6.** Accuracy of Binary Cyber-hate Classification in English.

Figure 7 shows the F1-Score of binary cyber-hate classification in Arabic on different platforms, which are Twitter and Instagram. In this language, the algorithm providing the best results is SVM on different platforms, with more than 92% in terms of F1 score. Otherwise, the combination CNN+LSTM achieves the lowest value of F1 score, which is 73% on Twitter, and RF with 65% on the Instagram platform.

Finally, Figure 8 shows the F1-Score of multiple class cyber-hate classifications on Twitter in English. It illustrates that the best performance result from different machine learning algorithms applied was LSTM, with an F1-Score of 93%.

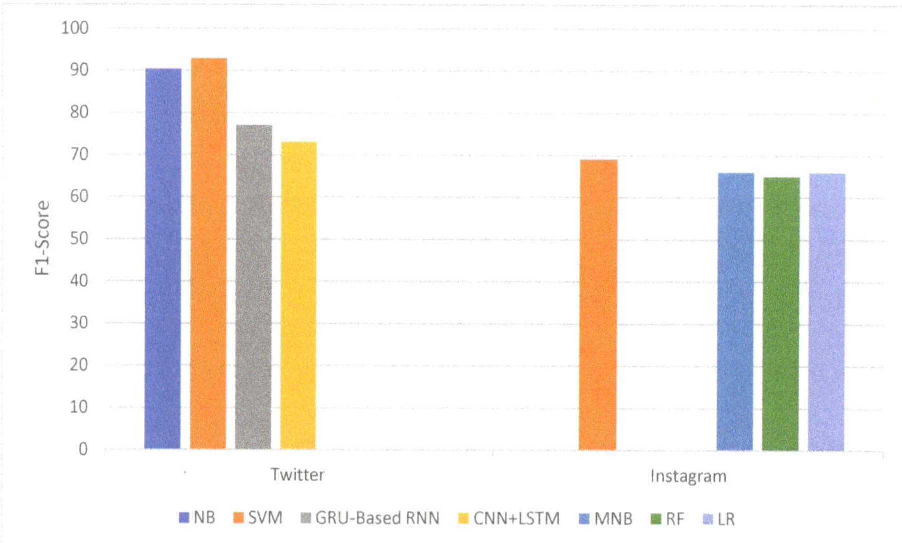


**Figure 7.** F1-Score of Binary Cyber-hate Classification in Arabic.

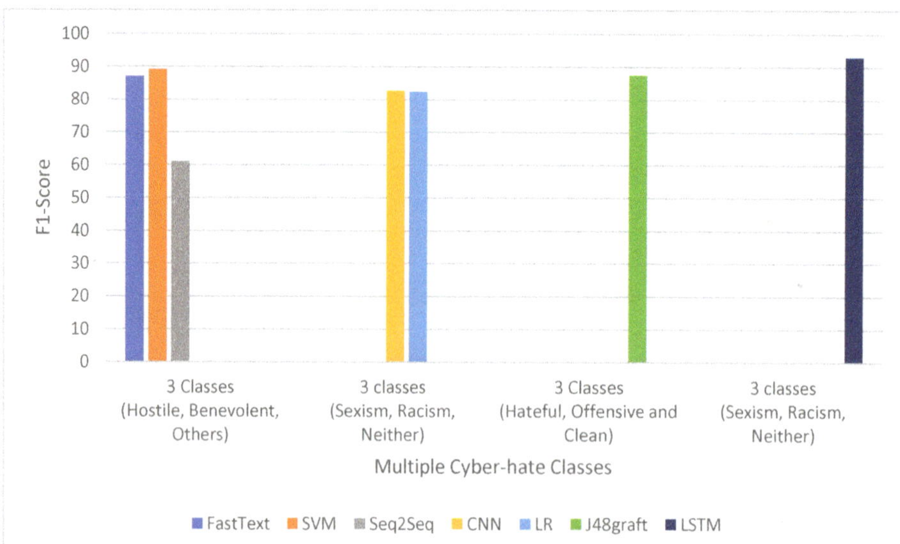


**Figure 8.** F1-Score of Multiple Class Cyber-Hate Classification in English.

## 6. Cyber-Hate Challenges

In this work, several issues were identified that affect the mainstream of the current research on cyber-hate speech detection:

- Data scarcity.
- The ambiguity of the context.
- The complexity of the Arabic language.
- Availability and accessibility to data on social networks.
- Manual Data Labelling.
- The degree of cyberbullying severity.

The field suffers from data scarcity in different languages, such as Arabic, due to the difficulty of collecting accurate cyber-hate speech data in the wild. In addition, discovering the context of a conversation is considered a challenge. The context is significant because numerous words are, in essence, ambiguous. The complexity of the Arabic language poses syntactic, semantic, and figurative ambiguity in terms of its pronunciation, vocabulary, phonetics, and morphology. This challenge could be solved by constructing an Arabic lexicon; the lexicon of offensive words may be useful in other languages to create a benchmark for the Arabic cyber-hate dataset. Current models of cyber-hate detection depend on the accessibility to accurate, relevant information from social media accounts and the experiences of potential victims. However, in actual cases, the availability of this data is influenced by consumer privacy habits and restrictions imposed by social networks. Privacy preservation is considered a challenging point. A proper solution entails the individuals' understanding of privacy preferences. Data labeling is a labor-intensive and time-consuming task, as it is necessary to select appropriate meanings of key terms that would be used during the labeling of ground truth before the process starts. The degree of cyberbullying severity is considered a challenge to be determined. Predicting various degrees of cyber-hate severity involves not only machine learning understanding but also a detailed analysis to identify and categorize the degree of cyber-hate severity from social and psychological experiences.

## 7. Conclusions and Future Work

In this manuscript, we have briefly reviewed the existing research on detecting cyber-hate behavior on different social media websites using various machine-learning approaches. Existing datasets of cyber-hate in different languages have been reviewed. In addition, a comparative study including binary and multiple class cyber-hate classification has been introduced, summarizing the most recent work that has been done during the last five years in different languages. Finally, the main challenges and open research issues were described in detail. Even though this research field in Arabic language is still in its early stages, existing studies confirm the importance of tackling Arabic cyberbullying detection. For future work, we aim to start with the construction of an annotated Arabic Cyber-hate dataset. Then, we will explore and apply different machine and deep learning algorithms for cyber-hate speech detection in Arabic. Future work will also include optimized real-time detection of Arabic cyberbullying.

**Author Contributions:** Conceptualization, D.G.; Validation, D.G., M.A. (Marco Alfonse) and S.M.J.-Z.; Formal analysis, D.G., M.A. (Marco Alfonse) and S.M.J.-Z.; Investigation, D.G.; Resources, D.G.; Writing—original draft, D.G.; Writing—review & editing, D.G., M.A. (Marco Alfonse) and S.M.J.-Z.; Visualization, D.G., M.A. (Marco Alfonse) and S.M.J.-Z.; Supervision, M.A. (Marco Alfonse), S.M.J.-Z. and M.A. (Mostafa Aref); Project administration, M.A. (Mostafa Aref). All authors have read and agreed to the published version of the manuscript.

**Funding:** This research received no external funding.

**Data Availability Statement:** Data sharing not applicable.

**Acknowledgments:** This work has been partially supported by Project CONSENSO (PID2021-122263OB-C21), Project MODERATES (TED2021-130145B-I00) and Project SocialTox (PDC2022-133146-C21) funded by MCIN/AEI/10.13039/501100011033 and by the European Union NextGenerationEU/PRTR, and Big Hug project (P20_00956, PAIDI 2020) and WeLee project (1380939, FEDER Andalucía 2014-2020) funded by the Andalusian Regional Government. Salud María Jiménez-Zafra has been partially supported by a grant from Fondo Social Europeo and Administración de la Junta de Andalucía (DOC_01073).

**Conflicts of Interest:** The authors declare no conflict of interest.

## References

1. Alsayat, A.; Elmitwally, N. A Comprehensive Study for Arabic Sentiment Analysis (Challenges and Applications). *Egypt. Inform. J.* **2020**, *21*, 7–12. [CrossRef]
2. Available online: https://www.Statista.Com/Statistics/278414/Number-of-Worldwide-Social-Network-Users/ (accessed on 3 July 2020).
3. Available online: https://www.Oberlo.Com/Statistics/How-Many-People-Use-Social-Media (accessed on 3 July 2020).
4. Corazza, M.; Menini, S.; Cabrio, E.; Tonelli, S.; Villata, S. A Multilingual Evaluation for Online Hate Speech Detection. *ACM Trans. Internet Technol.* **2020**, *20*, 1–22. [CrossRef]
5. StopBullying.Gov. Available online: https://www.stopbullying.gov (accessed on 3 July 2020).
6. Bisht, A.; Singh, A.; Bhadauria, H.S.; Virmani, J. Detection of Hate Speech and Offensive Language in Twitter Data Using LSTM Model. *Recent Trends Image Signal Process. Comput. Vis.* **2020**, *1124*, 243–264.
7. Miro-Llinares, F.; Rodriguez-Sala, J.J. Cyber Hate Speech on Twitter: Analyzing Disruptive Events from Social Media to Build a Violent Communication and Hate Speech Taxonomy. *Des. Nat. Ecodynamics* **2016**, *11*, 406–415. [CrossRef]
8. Blaya, C. Cyberhate: A Review and Content Analysis of Intervention Strategies. *Aggress. Violent Behav.* **2018**, *45*, 163–172. [CrossRef]
9. Namdeo, P.; Pateriya, R.K.; Shrivastava, S. A Review of Cyber Bullying Detection in Social Networking. In Proceedings of the Inventive Communication and Computational Technologies, Coimbatore, India, 10–11 March 2017; pp. 162–170.
10. Hang, O.C.; Dahlan, H.M. Cyberbullying Lexicon for Social Media. In Proceedings of the Research and Innovation in Information Systems (ICRIIS), Johor Bahru, Malaysia, 2–3 December 2019; IEEE: Piscataway, NJ, USA, 2019; pp. 1–6.
11. Sangwan, S.R.; Bhatia, M.P.S. Denigration Bullying Resolution Using Wolf Search Optimized Online Reputation Rumour Detection. *Procedia Comput. Sci.* **2020**, *173*, 305–314. [CrossRef]
12. Colton, D.; Hofmann, M. Sampling Techniques to Overcome Class Imbalance in a Cyberbullying Context. *Comput. Linguist. Res.* **2019**, *3*, 21–40. [CrossRef]
13. Qodir, A.; Diponegoro, A.M.; Safaria, T. Cyberbullying, Happiness, and Style of Humor among Perpetrators: Is There a Relationship? *Humanit. Soc. Sci. Rev.* **2019**, *7*, 200–206. [CrossRef]
14. Peled, Y. Cyberbullying and Its Influence on Academic, Social, and Emotional Development of Undergraduate Students. *Heliyon* **2019**, *5*, e01393. [CrossRef]
15. Dhillon, G.; Smith, K.J. Defining Objectives for Preventing Cyberstalking. *Bus. Ethics* **2019**, *157*, 137–158. [CrossRef]
16. la Vega, D.; Mojica, L.G.; Ng, V. Modeling Trolling in Social Media Conversations. In Proceedings of the Eleventh International Conference on Language Resources and Evaluation (LREC), Miyazaki, Japan, 7–12 May 2018; pp. 3701–3706.
17. Hassan, S.; Yacob, M.I.; Nguyen, T.; Zambri, S. Social Media Influencer and Cyberbullying: A Lesson Learned from Preliminary Findings. In Proceedings of the 9th Knowledge Management International Conference (KMICe), Miri, Sarawak, Malaysia, 25–27 July 2018; pp. 200–205.
18. Raisi, E.; Huang, B. Weakly Supervised Cyberbullying Detection Using Co-Trained Ensembles of Embedding Models. In Proceedings of the Advances in Social Networks Analysis and Mining (ASONAM), Barcelona, Spain, 28–31 August 2018; pp. 479–486.
19. Willard, N.E. *Cyberbullying and Cyberthreats: Responding to the Challenge of Online Social Aggression, Threats, and Distress*; Research Press: Champaign, IL, USA, 2007.
20. Available online: https://www.Pewresearch.Org/Internet/2021/01/13/Personal-Experiences-with-Online-Harassment/ (accessed on 3 July 2020).
21. Mangaonkar, A.; Hayrapetian, A.; Raje, R. Collaborative Detection of Cyberbullying Behavior in Twitter Data. In Proceedings of the Electro/Information technology (EIT), Dekalb, IL, USA, 21–23 May 2015; IEEE: Piscataway, NJ, USA, 2015; pp. 611–616.
22. Van Hee, C.; Lefever, E.; Verhoeven, B.; Mennes, J.; Desmet, B.; De Pauw, G.; Daelemans, W.; Hoste, V. Detection and Fine-Grained Classification of Cyberbullying Events. In Proceedings of the Recent Advances in Natural Language Processing (RANLP), Hissar, Bulgaria, 1–8 September 2015; pp. 672–680.
23. Waseem, Z.; Hovy, D. Hateful Symbols or Hateful People? Predictive Features for Hate Speech Detection on Twitter. In Proceedings of the NAACL Student Research Workshop, San Diego, CA, USA, 1 June 2016; pp. 88–93.
24. Zhao, R.; Zhou, A.; Mao, K. Automatic Detection of Cyberbullying on Social Networks Based on Bullying Features. In Proceedings of the 17th International Conference on Distributed Computing and Networking, New York, NY, USA, 4 January 2016; pp. 1–6.
25. Singh, V.K.; Huang, Q.; Atrey, P.K. Cyberbullying Detection Using Probabilistic Socio-Textual Information Fusion. In Proceedings of the Advances in Social Networks Analysis and Mining (ASONAM), San Francisco, CA, USA, 18–21 August 2016; IEEE: Piscataway, NJ, USA, 2016; pp. 884–887.

26. Available online: https://www.Ra.Ethz.Ch/Cdstore/Www2009/Caw2.Barcelonamedia.Org/Index.Html (accessed on 3 July 2020).
27. Al-garadi, M.A.; Varathan, K.D.; Ravana, S.D. Cybercrime Detection in Online Communications: The Experimental Case of Cyberbullying Detection in the Twitter Network. *Comput. Hum. Behav.* **2016**, *63*, 433–443. [CrossRef]
28. Hosseinmardi, H.; Rafiq, R.I.; Han, R.; Lv, Q.; Mishra, S. Prediction of Cyberbullying Incidents in a Media-Based Social Network. In Proceedings of the Advances in Social Networks Analysis and Mining (ASONAM), San Francisco, CA, USA, 18–21 August 2016; IEEE: Piscataway, NJ, USA, 2016; pp. 186–192.
29. Zhang, X.; Tong, J.; Vishwamitra, N.; Whittaker, E.; Mazer, J.P.; Kowalski, R.; Hu, H.; Luo, F.; Macbeth, J.; Dillon, E. Cyberbullying Detection with a Pronunciation Based Convolutional Neural Network. In Proceedings of the 15th IEEE International Conference on Machine Learning and Applications (ICMLA), Anaheim, CA, USA, 18–20 December 2016; IEEE: Piscataway, NJ, USA, 2016; pp. 740–745.
30. Wulczyn, E.; Thain, N.; Dixon, L. Ex Machina: Personal Attacks Seen at Scale. In Proceedings of the 26th International Conference on World Wide Web, Perth, Australia, 3 April 2017; pp. 1391–1399.
31. Haidar, B.; Chamoun, M.; Serhrouchni, A. Multilingual Cyberbullying Detection System: Detecting Cyberbullying in Arabic Content. In Proceedings of the 1st Cyber Security in Networking Conference (CSNet), Rio de Janeiro, Brazil, 18–20 October 2017; pp. 1–8.
32. Davidson, T.; Warmsley, D.; Macy, M.; Weber, I. Automated Hate Speech Detection and the Problem of Offensive Language. *Int. AAAI Conf. Web Soc. Media* **2017**, *11*, 512–515. [CrossRef]
33. Sprugnoli, R.; Menini, S.; Tonelli, S.; Oncini, F.; Piras, E. Creating a Whatsapp Dataset to Study Pre-Teen Cyberbullying. In Proceedings of the 2nd Workshop on Abusive Language Online (ALW2), Brussels, Belgium, 31 October 2018; pp. 51–59.
34. Bartalesi Lenzi, V.; Moretti, G.; Sprugnoli, R. Cat: The Celct Annotation Tool. In Proceedings of the Eight International Conference on Language Resources and Evaluation (LREC'12), Istanbul, Turkey, 21–27 May 2012; pp. 333–338.
35. Founta, A.-M.; Djouvas, C.; Chatzakou, D.; Leontiadis, I.; Blackburn, J.; Stringhini, G.; Vakali, A.; Sirivianos, M.; Kourtellis, N. Large Scale Crowdsourcing and Characterization of Twitter Abusive Behavior. In Proceedings of the Weblogs and Social Media (ICWSM), Palo Alto, CA, USA, 25–28 June 2018; pp. 491–500.
36. de Gibert, O.; Perez, N.; García-Pablos, A.; Cuadros, M. Hate Speech Dataset from a White Supremacy Forum. In Proceedings of the 2nd Workshop on Abusive Language Online (ALW2), Brussels, Belgium, 31 October 2018; pp. 11–20.
37. Nurrahmi, H.; Nurjanah, D. Indonesian Twitter Cyberbullying Detection Using Text Classification and User Credibility. In Proceedings of the Information and Communications Technology (ICOIACT), Yogyakarta, Indonesia, 6–7 March 2018; pp. 543–548.
38. Albadi, N.; Kurdi, M.; Mishra, S. Are They Our Brothers? Analysis and Detection of Religious Hate Speech in the Arabic Twittersphere. In Proceedings of the IEEE/ACM International Conference on Advances in Social Networks Analysis and Mining (ASONAM), Barcelona, Spain, 28–31 August 2018; IEEE: Piscataway, NJ, USA, 2018; pp. 69–76.
39. Bosco, C.; Felice, D.; Poletto, F.; Sanguinetti, M.; Maurizio, T. Overview of the Evalita 2018 Hate Speech Detection Task. *Ceur Workshop Proc.* **2018**, *2263*, 1–9.
40. Michele, C.; Menini, S.; Pinar, A.; Sprugnoli, R.; Elena, C.; Tonelli, S.; Serena, V. Inriafbk at Germeval 2018: Identifying Offensive Tweets Using Recurrent Neural Networks. In Proceedings of the Germ Eval Workshop, Vienna, Austria, 21 September 2018; pp. 80–84.
41. Mulki, H.; Haddad, H.; Ali, C.B.; Alshabani, H. L-HSAB: A Levantine Twitter Dataset for Hate Speech and Abusive Language. In Proceedings of the Third Workshop on Abusive Language Online, Florence, Italy, 1–2August 2019; pp. 111–118.
42. Ptaszynski, M.; Pieciukiewicz, A.; Dybała, P. Results of the PolEval 2019 Shared Task 6: First Dataset and Open Shared Task for Automatic Cyberbullying Detection in Polish Twitter. In Proceedings of the Pol Eval 2019 Workshop, Warsaw, Poland, 31 May 2019; pp. 89–110.
43. Ibrohim, M.O.; Budi, I. Multi-Label Hate Speech and Abusive Language Detection in Indonesian Twitter. In Proceedings of the Third Workshop on Abusive Language Online, Florence, Italy, 1–2 August 2019; pp. 46–57.
44. Basile, V.; Bosco, C.; Fersini, E.; Nozza, D.; Patti, V.; Pardo, F.M.R.; Rosso, P.; Sanguinetti, M. Semeval-2019 Task 5: Multilingual Detection of Hate Speech against Immigrants and Women in Twitter. In Proceedings of the 13th International Workshop on Semantic Evaluation, Minneapolis, MN, USA, 6–7 June 2019; pp. 54–63.
45. Banerjee, V.; Telavane, J.; Gaikwad, P.; Vartak, P. Detection of Cyberbullying Using Deep Neural Network. In Proceedings of the 5th International Conference on Advanced Computing & Communication Systems (ICACCS), Piscataway, NJ, USA, 15 March 2019; pp. 604–607.
46. Lu, N.; Wu, G.; Zhang, Z.; Zheng, Y.; Ren, Y.; Choo, K.R. Cyberbullying Detection in Social Media Text Based on Character-level Convolutional Neural Network with Shortcuts. *Concurr. Comput. Pract. Exp.* **2020**, *32*, 1–11. [CrossRef]
47. Moon, J.; Cho, W.I.; Lee, J. BEEP! Korean Corpus of Online News Comments for Toxic Speech Detection. In Proceedings of the Eighth International Workshop on Natural Language Processing for Social Media, Online, 10 July 2020; pp. 25–31.
48. Romim, N.; Ahmed, M.; Talukder, H.; Islam, S. Hate Speech Detection in the Bengali Language: A Dataset and Its Baseline Evaluation. In Proceedings of the International Joint Conference on Advances in Computational Intelligence, Singapore, 23–24 October 2021; pp. 457–468.

49. Karim, M.R.; Dey, S.K.; Islam, T.; Sarker, S.; Menon, M.H.; Hossain, K.; Hossain, M.A.; Decker, S. Deephateexplainer: Explainable Hate Speech Detection in under-Resourced Bengali Language. In Proceedings of the 2021 IEEE 8th International Conference on Data Science and Advanced Analytics (DSAA), Porto, Portugal, 6–9 October 2021; pp. 1–10.
50. Luu, S.T.; Van Nguyen, K.; Nguyen, N.L.-T. A Large-Scale Dataset for Hate Speech Detection on Vietnamese Social Media Texts. In Proceedings of the International Conference on Industrial, Engineering and Other Applications of Applied Intelligent Systems, Kitakyushu, Japan, 19–22 July 2021; pp. 415–426.
51. Sadiq, S.; Mehmood, A.; Ullah, S.; Ahmad, M.; Choi, G.S.; On, B.-W. Aggression Detection through Deep Neural Model on Twitter. *Futur. Gener. Comput. Syst.* **2021**, *114*, 120–129. [CrossRef]
52. Beyhan, F.; Çarık, B.; İnanç, A.; Terzioğlu, A.; Yanikoglu, B.; Yeniterzi, R. A Turkish Hate Speech Dataset and Detection System. In Proceedings of the Thirteenth Language Resources and Evaluation Conference, Marseille, France, 20–25 June 2022; pp. 4177–4185.
53. Ollagnier, A.; Cabrio, E.; Villata, S.; Blaya, C. CyberAgressionAdo-v1: A Dataset of Annotated Online Aggressions in French Collected through a Role-Playing Game. In Proceedings of the Thirteenth Language Resources and Evaluation Conference, Marseille, France, 20–25 June 2022; pp. 867–875.
54. ALBayari, R.; Abdallah, S. Instagram-Based Benchmark Dataset for Cyberbullying Detection in Arabic Text. *Data* **2022**, *7*, 83. [CrossRef]
55. Patil, H.; Velankar, A.; Joshi, R. L3cube-Mahahate: A Tweet-Based Marathi Hate Speech Detection Dataset and Bert Models. In Proceedings of the Third Workshop on Threat, Aggression and Cyberbullying (TRAC 2022), Gyeongju, Republic of Korea, 17 October 2022; pp. 1–9.
56. Kumar, A.; Sachdeva, N. A Bi-GRU with Attention and CapsNet Hybrid Model for Cyberbullying Detection on Social Media. *World Wide Web* **2022**, *25*, 1537–1550. [CrossRef]
57. Atoum, J.O. Detecting Cyberbullying from Tweets Through Machine Learning Techniques with Sentiment Analysis. In *Advances in Information and Communication*; Arai, K., Ed.; Springer Nature: Cham, Switzerland, 2023; pp. 25–38.
58. Nabiilah, G.Z.; Prasetyo, S.Y.; Izdihar, Z.N.; Girsang, A.S. BERT Base Model for Toxic Comment Analysis on Indonesian Social Media. *Procedia Comput. Sci.* **2023**, *216*, 714–721. [CrossRef]
59. Hate Speech Twitter Annotations. Available online: https://github.com/ZeerakW/hatespeech (accessed on 9 August 2020).
60. Wikipedia Detox. Available online: https://github.com/ewulczyn/wiki-detox (accessed on 20 August 2020).
61. Used a Crowd-Sourced Hate Speech Lexicon to Collect Tweets Containing Hate Speech Keywords. We Use Crowd-Sourcing to Label a Sample of These Tweets into Three Categories: Those Containing Hate Speech, Only Offensive Language, and Those with Neither. Available online: https://arxiv.org/abs/1703.04009 (accessed on 16 February 2023).
62. WhatsApp-Dataset. Available online: https://github.com/dhfbk/WhatsApp-Dataset (accessed on 18 August 2020).
63. Hate and Abusive Speech on Twitter. Available online: https://github.com/ENCASEH2020/hatespeech-twitter (accessed on 22 August 2020).
64. Hate Speech Dataset from a White Supremacist Forum. Available online: https://github.com/Vicomtech/hate-speech-dataset (accessed on 18 November 2022).
65. Available online: https://Github.Com/Nuhaalbadi/Arabic_hatespeech (accessed on 18 November 2022).
66. L-HSAB Dataset: Context and Topics. Available online: https://github.com/Hala-Mulki/L-HSAB-First-Arabic-Levantine-HateSpeech-Dataset (accessed on 18 November 2022).
67. Dataset for Automatic Cyberbullying Detection in Polish Laguage. Available online: https://github.com/ptaszynski/cyberbullying-Polish (accessed on 15 August 2020).
68. Multi-Label Hate Speech and Abusive Language Detection in the Indonesian Twitter. Available online: https://github.com/okkyibrohim/id-multi-label-hate-speech-and-abusive-language-detection (accessed on 6 February 2022).
69. HatEval. Available online: http://hatespeech.di.unito.it/hateval.html (accessed on 18 November 2022).
70. BullyDataset. Available online: https://github.com/NijiaLu/BullyDataset (accessed on 6 January 2020).
71. Korean HateSpeech Dataset. Available online: https://github.com/kocohub/korean-hate-speech (accessed on 10 February 2022).
72. Available online: https://www.Kaggle.Com/Datasets/Naurosromim/Bengali-Hate-Speech-Dataset (accessed on 10 February 2022).
73. Available online: https://Github.Com/Rezacsedu/DeepHateExplainer (accessed on 10 February 2022).
74. Available online: https://Github.Com/Sonlam1102/Vihsd (accessed on 10 February 2022).
75. Available online: https://www.Kaggle.Com/Datasets/Dataturks/Dataset-for-Detection-of-Cybertrolls (accessed on 10 February 2022).
76. Available online: https://Github.Com/Verimsu/Turkish-HS-Dataset (accessed on 10 February 2022).
77. CyberAgressionAdo-v1. Available online: https://Github.Com/Aollagnier/CyberAgressionAdo-V1 (accessed on 10 February 2022).
78. Available online: https://Bit.Ly/3Md8mj3 (accessed on 10 February 2022).
79. Available online: https://Huggingface.Co/L3cube-Pune/Mahahate-Bert (accessed on 10 February 2022).
80. Lingiardi, V.; Carone, N.; Semeraro, G.; Musto, C.; D'amico, M.; Brena, S. Mapping Twitter Hate Speech towards Social and Sexual Minorities: A Lexicon-Based Approach to Semantic Content Analysis. *Behav. Inf. Technol.* **2019**, *39*, 711–721. [CrossRef]
81. Alloghani, M.; Al-Jumeily, D.; Mustafina, J.; Hussain, A.; Aljaaf, A.J. *A Systematic Review on Supervised and Unsupervised Machine Learning Algorithms for Data Science*; Springer: Cham, Switzerland, 2020.
82. Alsharif, M.H.; Kelechi, A.H.; Yahya, K.; Chaudhry, S.A. Machine Learning Algorithms for Smart Data Analysis in Internet of Things Environment: Taxonomies and Research Trends. *Symmetry* **2020**, *12*, 88. [CrossRef]

83. Rout, J.K.; Dalmia, A.; Choo, K.-K.R.; Bakshi, S.; Jena, S.K. Revisiting Semi-Supervised Learning for Online Deceptive Review Detection. *IEEE Access* **2017**, *5*, 1319–1327. [CrossRef]
84. Li, Z.; Fan, Y.; Jiang, B.; Lei, T.; Liu, W. A Survey on Sentiment Analysis and Opinion Mining for Social Multimedia. *Multimed. Tools Appl.* **2019**, *78*, 6939–6967. [CrossRef]
85. Ay Karakuş, B.; Talo, M.; Hallaç, İ.R.; Aydin, G. Evaluating Deep Learning Models for Sentiment Classification. *Concurr. Comput. Pract. Exp.* **2018**, *30*, 1–14. [CrossRef]
86. Asghar, M.Z.; Khan, A.; Ahmad, S.; Qasim, M.; Khan, I.A. Lexicon-Enhanced Sentiment Analysis Framework Using Rule-Based Classification Scheme. *Peer-Rev. Open Access Sci. J. (PLoS ONE)* **2017**, *12*, e0171649. [CrossRef] [PubMed]
87. Khan, F.H.; Qamar, U.; Bashir, S. Lexicon Based Semantic Detection of Sentiments Using Expected Likelihood Estimate Smoothed Odds Ratio. *Artif. Intell. Rev.* **2017**, *48*, 113–138. [CrossRef]
88. Ahmed, M.; Chen, Q.; Li, Z. Constructing Domain-Dependent Sentiment Dictionary for Sentiment Analysis. *Neural Comput. Appl.* **2020**, *32*, 14719–14732. [CrossRef]
89. Nandhini, B.S.; Sheeba, J.I. Cyberbullying Detection and Classification Using Information Retrieval Algorithm. In Proceedings of the International Conference on Advanced Research in Computer Science Engineering & Technology (ICARCSET), Tamilnadu, India, 15–16 March 2015; pp. 1–5.
90. Reynolds, K.; Kontostathis, A.; Edwards, L. Using Machine Learning to Detect Cyberbullying. In Proceedings of the 2011 10th International Conference on Machine Learning and Applications and Workshops, NW Washington, DC, USA, 18–21 December 2011; Volume 2, pp. 241–244.
91. Badjatiya, P.; Gupta, S.; Gupta, M.; Varma, V. Deep Learning for Hate Speech Detection in Tweets. In Proceedings of the 26th International Conference on World Wide Web Companion, Republic and Canton of Geneva, Switzerland, 3–7 April 2017; pp. 759–760.
92. Park, J.H.; Fung, P. One-Step and Two-Step Classification for Abusive Language Detection on Twitter. In Proceedings of the First Workshop on Abusive Language Online, Vancouver, BC, Canada, 4 August 2017; pp. 41–45.
93. Watanabe, H.; Bouazizi, M.; Ohtsuki, T. Hate Speech on Twitter: A Pragmatic Approach to Collect Hateful and Offensive Expressions and Perform Hate Speech Detection. *IEEE Access* **2018**, *6*, 13825–13835. [CrossRef]
94. Wang, W.; Huang, J.-t.; Wu, W.; Zhang, J.; Huang, Y.; Li, S.; He, P.; Lyu, M. MTTM: Metamorphic Testing for Textual Content Moderation Software. In Proceedings of the International Conference on Software Engineering (ICSE), Lisbon, Portugal, 14–20 May 2023; pp. 1–13.
95. Roy, P.K.; Tripathy, A.K.; Das, T.K.; Gao, X.-Z. A Framework for Hate Speech Detection Using Deep Convolutional Neural Network. *IEEE Access* **2020**, *8*, 204951–204962. [CrossRef]
96. Yadav, A.; Vishwakarma, D.K. Sentiment Analysis Using Deep Learning Architectures: A Review. *Artif. Intell. Rev.* **2020**, *53*, 4335–4385. [CrossRef]

big data and
cognitive computing

MDPI

*Article*

# The Value of Web Data Scraping: An Application to TripAdvisor

**Gianluca Barbera [1], Luiz Araujo [2] and Silvia Fernandes [2,3,*]**

[1] School of Political Sciences "Cesare Alfieri", University of Florence, 50127 Florence, Italy; gianluca.barbera@stud.unifi.it

[2] Faculty of Economics, University of Algarve, 8005-139 Faro, Portugal; a78318@ualg.pt

[3] CinTurs—Research Centre for Tourism, Sustainability and Well-being, University of Algarve, 8005-139 Faro, Portugal

* Correspondence: sfernan@ualg.pt

**Abstract:** Social Media Analytics (SMA) is more and more relevant in today's market dynamics. However, it is necessary to use it wisely, either in promoting any kind of product/brand, or interacting with customers. This requires its effective understanding and monitoring. One way is through web data scraping (WDS) tools that allow to select sites and platforms to compare them in their performances. They can optimize extraction of big data published on social media. Due to current challenges, a sector that can particularly take advantage of this source is tourism (and its related sectors). This year has the hope of tourism's revival after a pandemic whose impacts are still affecting several activities. Many traders and entrepreneurs have already used these versatile tools. However, do they really know their potential? The present study highlights the use of WDS to collect data from TripAdvisor's social pages. Besides comparing competitors' performance, companies also gain new knowledge of unnoticed preferences/habits. This contributes to more interesting innovations and results for them and for their customers. The approach used here is based on a project for smart tourism consultancy, from the identification of a gap in our region, to aid tourism organizations to enhance their digital presence and business model. Many things can be detected in this big source of unstructured data very quickly and easily without programming. Moreover, exploring code, either to refine the web scraper or connect it with other platforms/apps, can be an object of future research to leverage consumer behavior prediction for more advanced interactions.

**Keywords:** social media; data scraping; tourism; smart consultancy; cognitive system

## 1. Introduction

Social networks/channels are the internet tools most increasingly used. Their influence is seen in many sectors, but it is necessary to use them wisely. They are necessary to promote products, brands, or ideas; to consolidate one's own company or launch a new one; or to build loyalty with customers and seek new ones [1].

This year seems to be prepared for a social and economic revival after a 2022 that saw the world emerge from a heavy situation due to the pandemic. Many companies have been able to proceed through their own boldness or conscious use of online platforms, with emphasis on social media. These have many times been revealed as the best quality/price promotion tools. The results obtained also stem from a conscious investment in them. Those who invested wisely now see a return and will certainly acknowledge that one can no longer do without using social media and related apps.

One of the most affected sectors, given the movements dictated by government decrees to limit human mobility and circulation, has been tourism. This is a key sector for many regions and countries that claim to depend heavily on it [2]. Tourism is a sector that can also affect several other sectors, such as hospitality and traveling. Besides these, it can influence handicraft, environmental, gastronomic, and artistic–cultural sectors. It is therefore decisive for the economic health of many people.

Hence, this present study is based on a project for smart tourism consultancy. It aims at analyzing, on one hand, the potential of web data scraping (WDS), and on the other hand the application of a web scraper to TripAdvisor's social pages. The period chosen is the whole year of 2022 due to its post-pandemic conditions with signs of revival. Moreover, specific related objectives deal with discerning unanswered trends and data-scraping levels. The web scraper used is Fanpage Karma (FpK), which will be compared with other similar tools to point out key features and choice impact. This approach elected TripAdvisor (TA) since it is an important travel site that provides information and reviews of tourism-related user-generated content. Finally, the results obtained are discussed in order to discern lessons for tourism companies and users.

This paper is thus organized as follows: in Section 2, the potential of WDS and initial view of the web scraper to discern relevant metrics and KPI; in Section 3, the application to TA's main social pages to discern main trends, issues, and lessons for better tourism management and forecasting; in Section 4, a comparison between WDS tools to discuss the importance of choosing the right solution by looking at the goals of three organizations in Algarve (interested in the original project), along with a holistic view toward advanced cognitive tools; and in Section 5 the conclusions, other trends, implications, and considerations for future research are provided.

## 2. Materials and Methods

The referred goal of a project for smart tourism consultancy was very welcomed by important organizations in Algarve (a very touristic region in Portugal) such as Dengun (one of the most successful digital marketing companies); CRIA (the innovation accelerator of the University of Algarve); and Tilia (an innovative hostel located in the capital's center). Their managers/owners are even interested in investing in the idea as there is nothing similar in the region.

### *2.1. WDS: A Big Data Source*

As mentioned, TA is one of the largest websites for hotel, bed-and-breakfast, and restaurant reviews, accommodation bookings, and other travel-related content. It also includes interactive travel forums. This site was an early adopter of user-generated content and is supported by an advertising business model. As tourism is a key sector for many economies, it is important that such a platform be monitored in order to better predict and decide upon disruption events. Comparing their content quickly and easily, the level of engagement and other metrics can guide strategies and feedbacks on a timely, regular basis. For example, current concerns in the area are related to safety, sustainability, and well-being.

No company or entrepreneur can think of not being active on social networks today, since this medium is one of the largest data sources most widely used in the world [1]. Consequently, businesses and brands have been progressively investing in them as relevant communication channels [3]. Therefore, a practice in which many are investing to enhance performance through the analysis of desired markets/competition is WDS. Also known as web extraction, this technique extracts data from the internet and saves them to a file system or a database for analysis [4]. It is also widely acknowledged as a powerful technique for collecting big data [5,6]. Thus, web scraping is a practice that captures a large amount of data on the internet [7] to obtain knowledge about competing firms or websites.

Besides web scrapers, one of the most widely used resources for doing WDS is Python language. It is particularly used for the analysis of Twitter pages, through a dedicated library—Tweepy [8,9]. Another way of applying Python is through the Text Blob Tool which analyzes the sentiments of tweets [10]. Other tools in Python enable web-scraping activities such as Scrapy and Pandas [11], SpaCy [12], and Python's Natural Language Toolkit [13]. However, to extract information from a website by only having knowledge of

Python is not sufficient. It is necessary to know HTML because web scraping is a task that needs to be divided into subtasks. Once this marking language is known, which describes the elements of a page, Python libraries can be used more easily (such as Beautiful Soup, Scrapy, or Selenium) [14,15]. However, a software tool such as FpK is much more intuitive and easier to use by the end-user as it does not require knowing the programming code.

Nevertheless, programming language is crucial to create functions that SMA (Social Media Analytics) tools may lack, such as personalizing the scraper or connecting it with other tools or platforms for more advanced goals. An example of this kind of connection is sentiment analysis [16]. Figure 1 illustrates an example of text networks where, from the most prominent terms and their relations, firms might discern unanswered trends and prepare more assertive and timely strategies to cope with them.

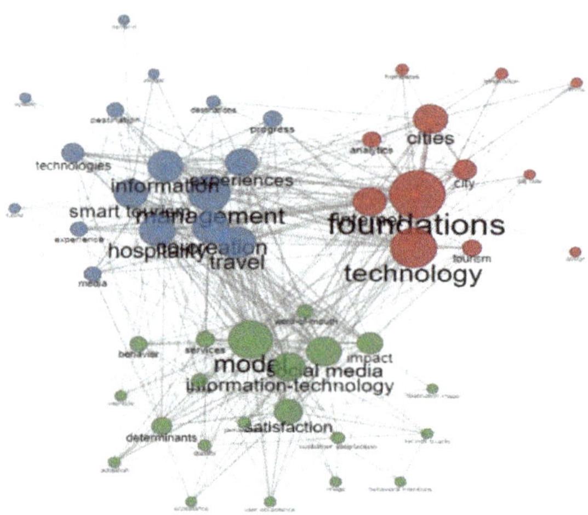


**Figure 1.** Example of text networks that can be used for sentiment analysis (own elaboration).

Other possibilities can be envisioned by integrating WDS with other tools or systems, which can be the object of future research on predicting consumer's behaviors. These discoveries can revolutionize tourism patterns and services [17]. Other aspects related to time and space can be captured for more data granularity (e.g., discover what happens in a certain specific place, etc.). This is relevant in today's context of these critical dimensions. These experiments can contribute to developing AI (artificial intelligence) or cognitive-driven decision-making to help firms be better equipped to fight disasters.

### *2.2. FpK: An Ease of Use Tool*

FpK was launched in 2012 by two online marketing experts from Berlin. Their main motivation was to interpret and analyze Facebook pages, or other social media, in order to optimize them. It helps to manage social media presence and performance. It enables organizations to collect web data in order to compare their own social pages with the competitors' pages. Data about prices practiced, follower growth, and best time to post can be easily obtained. Key features are its ease-of-use and variety of functions related with web data extraction to produce insights. Besides calculating metrics, it has other actions such as benchmarking, tagging, history, live (see data in real-time), etc. (Figure 2). For example, it can check hotel room prices and other information, collect tweets, monitor reputation and SEO (Search Engine Optimization), and extract emails and addresses from map and business websites.

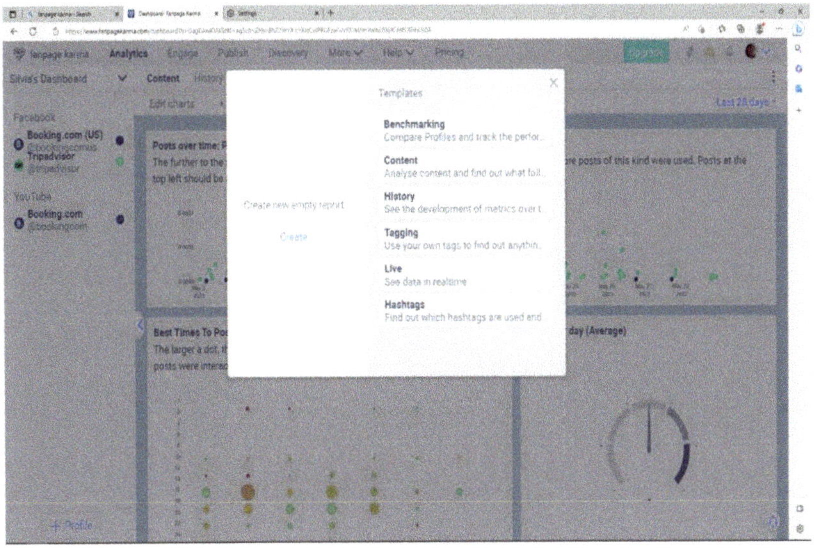


**Figure 2.** Creating a report of the analyses made (own elaboration).

This tool also allows visualizing several metrics such as number of followers in a competitor's page, its growth rate, engagement, response time, influencers, etc. Since it returns the results in a friendly design (including charts), their comparison is much easier. This is very interesting because the information about the competition's strategic options can be more easily discerned and understood. Moreover, this big source of unstructured data can then be accessed and used for deeper analyses of behavioral patterns and trends that went unnoticed before [18]. Currently in FpK there are more than 350 different metrics to choose and arrange in a dashboard, including the websites that can be added to it (Figure 3).

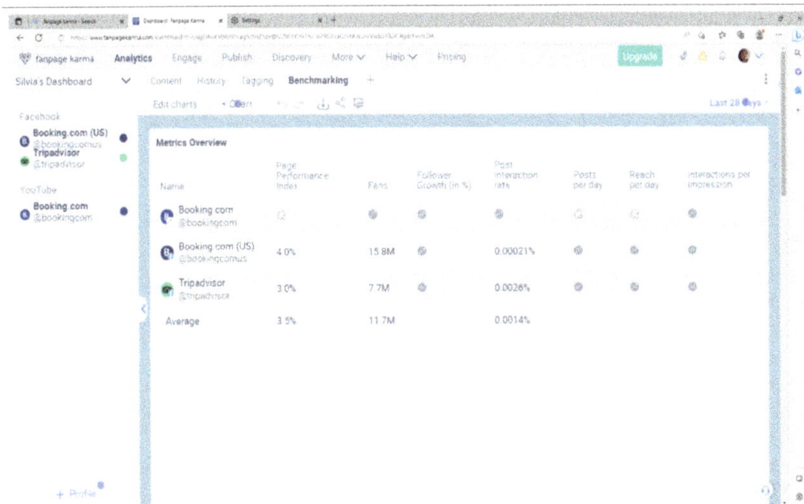


**Figure 3.** Function Analytics –> Benchmarking (own elaboration).

The dashboard is like a matrix of results that are the "raw data" for a graphical customized report. This report can help think of other metrics and combine them to obtain key performance indicators (KPI) that best fit the firm's goals. In addition, the function Analytics can create the dashboard for the last month (or year), or for comparing several months (or entire years) within the period needed.

For instance, two main KPIs of SMA are "follower growth" and "fan interactions" (engagement). The resulting chart (Figure 4) shows which site is the best. In this case, it is Booking.com (upper right corner) compared with TA (lower left corner). This is important to know in order to identify the factors behind it and better plan the structure and strategy of social media portfolios. However, these two social channels have different goals: TripAdvisor has more information because, besides hosting data, it shows reviews and recommendations for various places and activities, such as sights and restaurants.

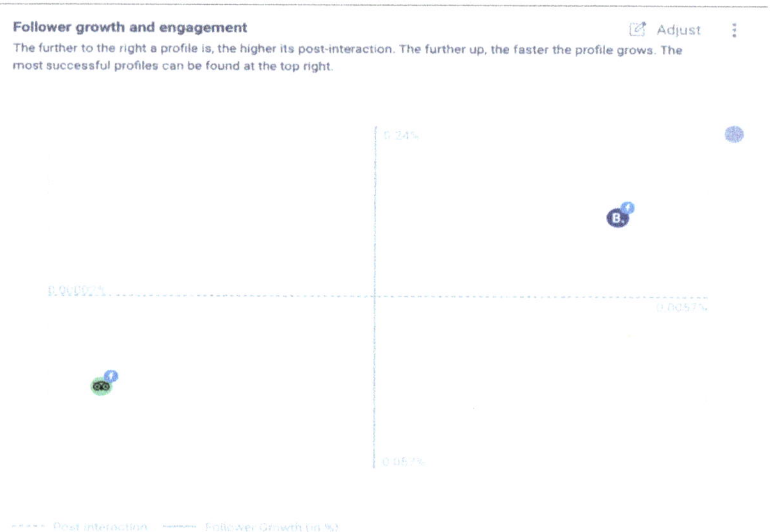


**Figure 4.** Main social performance metrics (own elaboration).

Effectively the web scraper FpK has been applied in several studies to investigate a wide variety of topics, which include political communication [19–21]; public communication [22]; healthcare [23]; customer behavior [24–26]; and tourism [27–30].

## 3. Results

Considering the main aim of this work—analyzing the potential of web data scraping (WDS) applied to TA's social pages—some trends of post-pandemic tourism can be discerned. As a 'sample', we have searched them among the main social channels of TA. This famous platform allows tourists to describe their experiences with hotels, restaurants, and services, among other aspects. Regarding the impact of the pandemic on people's travel habits, it is important to understand how this source of user-generated content has dealt with the pandemic and discern lessons for future tourism management.

Thus, this study has an exploratory nature and follows a comparative approach. It can then contribute to the tourism knowledge base to endure as one of the most important sectors worldwide. Through FpK, we collected data from TA's most active social networks in 2022 (Facebook, Instagram, Twitter, Pinterest, and TikTok). Then, we analyzed the resulting tables and charts. Several metrics were considered, such as page performance index, fans, follower growth, post interaction, and posts per day (Table 1).

**Table 1.** Performance of Tripadvisor's social pages (own elaboration). Period: 1 January to 31 December 2022.

| TA Profiles | Page Performance Index | Fans | Follower Growth (%) | Post Interaction | Posts Per Day |
|---|---|---|---|---|---|
| TikTok | 10.0% | 1127 | 1843.1% | 1.31% | 0.052 |
| Pinterest | 10.0% | 228,048 | 8.6% | 0.0006% | 1.038 |
| Twitter | 8.0% | 3,424,208 | 0.97% | 0.0005% | 6.917 |
| Instagram | 14.0% | 2,769,720 | 12.38% | 0.06% | 1.057 |
| Facebook | 9.0% | 7,626,233 | 0.94% | 0.0008% | 4.608 |
| Average | 10.2% | 2,809,867.2 | 373.2% | 0.27% | 2.734 |

Looking at the page performance index, which combines fans' engagement with page growth, we can see that Instagram had the highest value (14%). This is a platform that had around 1.28 billion users in 2022 (Statista, 2022). However, the figure that stands out the most is the % of follower growth on TikTok. However, TA only has had its TikTok page since the beginning of 2022 (having started posting only in October). It is followed by Instagram with 12.38% follower growth. Regarding post interaction, TikTok was also the best, but this is a biased metric for the same reason mentioned above. In conclusion, Instagram is confirmed as the social network where followers most interacted with the posts published. Finally, the social network that posts the most was Twitter, but it still has the least interaction level.

Therefore, this first scraping query shows at least two things: that more metrics are needed to understand these issues and that tourists/visitors engage more with visual social media (rich in pictures, short videos, and storytelling). This trend is augmenting what requires more experience to be prepared for future VR (virtual reality) or AI-based tourism/hospitality projects. Then, we pursued the analysis to discover which specific content got the most engagement from users. This information is important in order to change or improve campaigns and product/service innovation. Figure 5 shows which post features were published the most (left side of chart) and which ones had the most interactions (right side of chart) according to the following categories: links, pictures, status, carousels, reels, and videos.

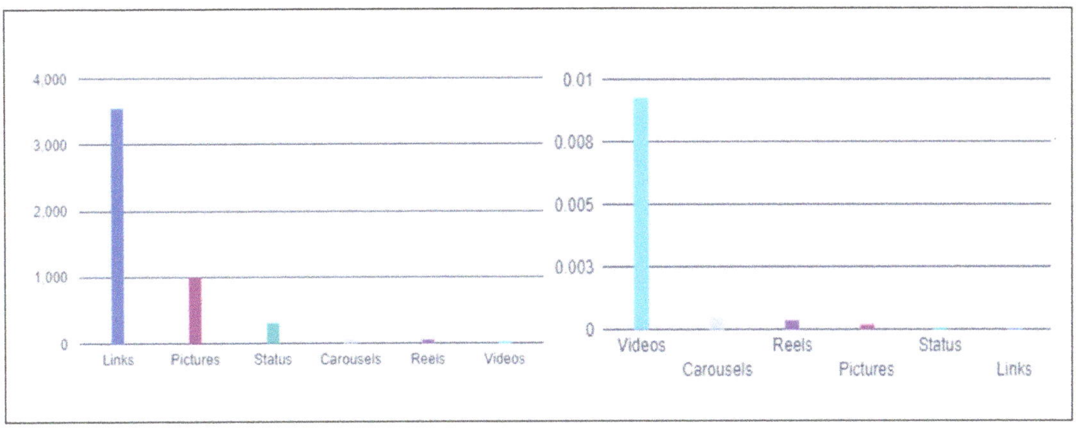


**Figure 5.** Number of posts and post-interaction in TA's social media (own elaboration).

Effectively, video posts had the highest level of interaction against pictures, which had the lowest level. The low results in the other categories (reels: full-screen vertical videos; carousels: ads that combine videos and images; status; and links) suggest (re) thinking the contents and the modern ways of presenting them. For instance, a study on these issues [31] revealed an increasing relevance of interacting with influencers and of co-creating (i.e., including users' ideas in product innovation). Travel, hospitality, and tourism organizations should consider data from WDS tools along with case studies in this area to manage their social media more smartly. Their digital marketing plans must break into these trends to keep up with, or anticipate, the digital transformation.

## 4. Discussion

Regarding the results obtained about the potential of WDS for smart tourism consultancy, through an application to the TA case, we can acknowledge that more metrics of SMA are necessary. To release managers (and other users in the company) from the need to program them, a way of exploring either more metrics or more functions is by comparing web scrapers such as the nine best ones referred in the work of [32] (Table 2).

**Table 2.** The 9 best WDS tools (own elaboration, based on [32]).

| Web Scraper | Key Features |
|---|---|
| Octoparse | Anonymous web data scraping behind login forms |
| ParseHub | Efficiency of data extraction from complex web pages |
| Mozenda | Job sequencers to collect web data in real-time; highly scalable |
| Webhose.io | Fast content indexing; get machine-readable data sets |
| Content Grabber | Allows to build web apps and offers a wide variety of formats |
| Common Crawl | Support for non-code usage; has resources to teach data analysis |
| Scrapy | Open-source tool and easily extensible. Middleware modules available for integrating useful tools |
| Scraperapi | Easy to integrate; allows price scraping, search engine scraping |
| Scrape-it.Cloud | Easy integration in other systems; scrapers for popular sites |

In fact, this study uses FpK as an example of this type of specialized software tool. These are non-code web scrapers that enable users to extract data without the knowledge of HTML structures and elements [32]. However, each one has limitations and choosing the right one is a crucial task. For example, the previously mentioned organizations interested in this project—Dengun, CRIA, and Tilia Hostel—have different goals in mind. Dengun aims to use WDS insights to accelerate potential business in the phases of customer validation and proof of concept; CRIA has both academic and commercial interests: academic in terms of being able to absorb this type of information and transfer it to other sectors, and commercial as an enterprise incubator; and Tilia's owner envisions anticipating the demands of potential customers, maintaining the maximum capacity of its rooms, understanding new ways to approach customers at the right time, and realizing new marketing actions to interact with its publics. The other goals these three interveners mentioned are more assertiveness of investment in new businesses; hiring staff in advance of the high season; forecasting tourists with potential consumption in the region; forecasting segments to be explored for meeting new demands; and perceiving more assertively the desires of consumers.

### 4.1. Towards Cognitive Scraping

This work has shown that many things can be discovered through this big data source. Nevertheless, deeper aspects can be detected, especially by programming, in order to develop connections either within WDS tools or between them and other platforms. This issue raises the discussion around discerning unanswered trends and data-scraping levels.

Currently, 80% of available data are unstructured [33] such as those on the web. This wealth of data contains valuable information about customer habits, preferences, dislikes, intentions, and much more [34]. Well-extracted and analyzed, they may help identify innovation opportunities. Using the right tool(s), firms can start capturing value from this big data source [35]. The knowledge absorbed can give them a competitive advantage.

Considerable amounts of unstructured data are generated every minute [36]. Applying advanced cognitive computing to them will create more complete models of customers' behavior with the relationships and environments in which they act (Figure 6). In turn, the massive volume of these actions will enable better prediction of consumer behaviors [35,37].

For instance, a company can provide branded content to a customer on a social network platform, then track the person/group with whom he shares that content and how they respond. This can help to understand how consumers feel about its brands to find gaps in the marketplace and develop brand positions.

Analyzing data can be a time-consuming practice, but good cognitive analyses can leverage this big amount of data [38]. As this problem can become unsustainable to process, there is a need for better computer tools and developing languages [39]. The developments involved in this holistic view (Figure 6) should be further explored in tourism and hospitality since they have been focused on areas such as education, healthcare, commerce, and human augmentation [40].

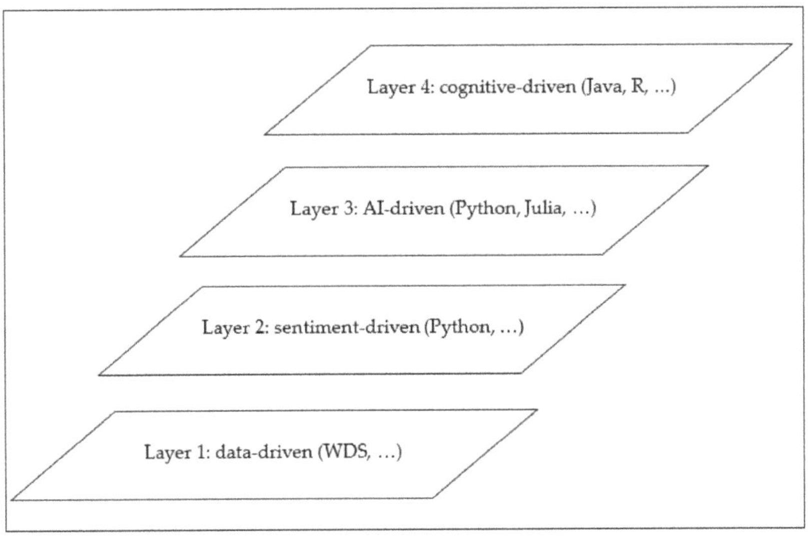


**Figure 6.** Holistic view toward advanced cognitive tools (own elaboration, based on [40,41]).

## 5. Conclusions

This study can contribute, through its innovative approach, to the discussion of web scraping's potential for scalable big data cognitive analytics. WDS tools can search for web data in a very fast and intuitive way, without running into legal issues, such as lack of consent to access private information, as well as wasting software/hardware resources. These tools can aid consultancy in tourism, among other segments, as they can optimize the extraction of competitor's data and other relevant information. This knowledge and its real-time monitoring can give companies a competitive advantage. They provide data quickly and clearly on the performance of any web page. As a case, we decided to explore which of TA's social pages performed best in 2022, the most recent post-pandemic year for tourism as a key sector for our region/country.

We saw that Instagram is the best performing social network, even though it does not have the highest number of followers. Its modern structure and interactive content certainly facilitate user engagement. Moreover, it is precisely through the right interactions (shares, comments, sales, etc.) that the firms obtain value and returns. Another interesting social service is TikTok, whose adoption is augmenting especially among young people. TA has noticed this trend and launched its own page in 2022. Regarding Twitter, the social network that posted the most, it still has the least interaction level. It is mainly used to answer individual questions from users, and often reposts what has been published on other social media.

Given the pace of current challenges, web scraping can be a differential and timesaving resource. WDS tools can easily capture and compare this big data source on a regular basis. There are different ways of exploring their potential, some more dependent on programming and others more independent and intuitive. Several studies made use of both ways to extract and compare data from web and/or social media pages. However, we emphasize that web scrapers usually do not require knowledge of programming languages, which makes them easier to use/explore by managers and marketers. Many metrics can then be obtained and combined in versatile KPI with minimal effort.

### 5.1. Other Trends

A groundbreaking technology that has already started to influence the internet is augmented reality (AR). It may have a strong impact on social media apps in the coming years. It lies between the virtual and real world and is gradually gaining popularity as the virtual elements brought into the real world provide an immersive experience for the user. With the help of AR, 3D content will replace 2D content on social networks. A strong usage of interactive content will reshape social media as 3D content has the power to take the users to an entirely new and unimaginable world. Several social media players are quickly joining AR and VR such as Snapchat, Instagram, and Facebook [42].

Besides AR possibilities, other trends involve audio features. For example, HearMeOut is a voice-based social network that allows users to record and share audio posts. Moreover, the use of hands- and eyes-free social media allows users to do both multitasking and experiment greater 'reality' [43,44]. The social media landscape tends to witness significant trends and changes. With the continued technological innovation and changing user demands, platforms will focus more on personalized experiences, integration of AR and VR, and the enrichment of video content [45]. Entrepreneurs and marketers should stay up to date with these trends to be ahead of the curve and effectively engage their target audiences [46].

According to recent articles and reports [47,48] other trends to be considered are related with the increasing role of influencers; the need for more dynamic and interactive content; conditions for developing social media commerce/shopping as these platforms have wider target audiences; and usage of social media for mobile purposes and hybrid events among others. Companies that offer richer experiences to their customers will be much better equipped to face major challenges and stand up alongside competition.

### 5.2. Implications and Future Work

WDS can be used to gain insights toward key performance indicators. For instance, post interaction can be a good variable for analyzing which content, times, and places are more attractive and then rethink destination marketing strategies. Moreover, with these data, other more advanced analyses can be developed especially through programming connections either between WDS tools or between them and other platforms. Text networks are an example to build sentimental analyses and predict customer behavior more assertively.

Today, tourism organizations can choose from a rich set of marketing instruments. Marketers can now deal with paid (content they pay such as an ad or sponsorship), owned (content they create and control, like own Facebook page or website), and earned (content that others create like reviews or mentions) media [49]. However, these instruments can be better gathered and managed through the dashboard of the right WDS tool.

Currently any digital business model should approach SMA needs [50]. The focus of this study is relevant for advanced consultancy in tourism as it highlights web data extraction optimization to enhance the knowledge base for tourism innovation and revival. Given the importance of social media throughout the entire customer journey, tourism managers need to know and understand the repercussions of their efforts. In fact, measuring social media performance has been considered as a major challenge of modern management [18,50]. In contrast to traditional media, social media resemble 'living' organisms [51] as these platforms involve content, motives, structure, roles and interactions. Therefore, more metrics should be developed and tested to understand and monitor their dynamics. Tourism associations should invest in hiring qualified staff, both on tourism dynamics and social media behavior. Consequently, invest in tools designed to provide a workspace and think-tank where unstructured data can be quick and easily analyzed. Moreover, consider and explore the developments toward advanced cognitive tools. The way people acquire and broaden their knowledge is changing toward cognitive acceleration. Cognitive computing models and systems can decipher unstructured data and draw valuable conclusions. These systems are capable of reasoning, decision-making, and experience-based learning [41]. They can communicate with people in natural language, comprehend real situations, and give personalized answers.

**Author Contributions:** G.B.—Conceptualization, Sections 1 and 3, investigation; L.A.—Section 2.1, Section 2.2 and charts; S.F.—Sections 4.1 and 5, writing, funding acquisition. All authors have read and agreed to the published version of the manuscript.

**Funding:** This research was funded by National Funds provided by the FCT—Foundation for Science and Technology under the project UIDB/04020/2020.

**Institutional Review Board Statement:** Not applicable.

**Informed Consent Statement:** Not applicable.

**Data Availability Statement:** All data supporting results can be found in the references reported.

**Conflicts of Interest:** The authors declare no conflict of interest.

## References

1. González-Padilla, D.; Tortolero-Blanco, L. Social media influence in the COVID-19 Pandemic. *Int. Braz. J. Urol.* **2020**, *46*, 120–124. [CrossRef] [PubMed]
2. Duro, J.; Perez-Laborda, A.; Turrion-Prats, J.; Fernández-Fernández, M. COVID-19 and tourism vulnerability. *Tour. Manag. Perspect.* **2021**, *38*, 100819. [CrossRef]
3. Ashley, C.; Tuten, T. Creative Strategies in Social Media Marketing: An Exploratory Study of Branded Social Content and Consumer Engagement. *Psychol. Mark.* **2015**, *32*, 15–27. [CrossRef]
4. Zhao, B. *Web Scraping. Encyclopedia of Big Data*; Springer: Berlin/Heidelberg, Germany, 2017; pp. 1–3. [CrossRef]
5. Kaisler, S.; Armour, F.; Espinosa, J.; Money, W. Big Data: Issues and Challenges Moving Forward. In Proceedings of the 46th Hawaii International Conference on System Sciences 2013, Wailea, HI, USA, 7–10 January 2013; pp. 995–1004. [CrossRef]
6. Bar-Ilan, J. Data collection methods on the Web for infometric purposes-A review and analysis. *Scientometrics* **2001**, *50*, 7–32. [CrossRef]
7. Mitchell, R. *Web Scraping with Python: Collecting More Data from the Modern Web*, 2nd ed.; O'Reilly: Sebastopol, CA, USA, 2018.
8. Kusumasari, B.; Prabowo, N. Scraping social media data for disaster communication: How the pattern of Twitter users affects disasters in Asia and the Pacific. *Nat. Hazards* **2020**, *103*, 3415–3435. [CrossRef]
9. Kaburuan, E.; Lindawati, A.; Surjandy; Sinswantini; Putra, M.; Utama, D. A Model Configuration of Social Media Text Mining for Projecting the Online-Commerce Transaction (Case: Twitter Tweets Scraping). In Proceedings of the 7th International Conference on Cyber and IT Service Management (CITSM) 2019, Jakarta, Indonesia, 6–8 November 2019; pp. 1–4. [CrossRef]
10. Kaur, C.; Sharma, A. Social Issues Sentiment Analysis using Python. In Proceedings of the 5th International Conference on Computing, Communication and Security (ICCCS) 2020, Patna, India, 14–16 October 2020; pp. 1–6. [CrossRef]
11. Raman, D.; Jayalakshmi, S.; Arumugam, K.; Raj, A.; Balaji, D.; Brightsingh, R. Implementation of Data Analysis and Document Summarization in Social Media Data Using R and Python. In Proceedings of the 4th International Conference on Inventive Research in Computing Applications (ICIRCA) 2022, Coimbatore, India, 21–23 September 2022; pp. 1457–1464. [CrossRef]
12. Bhardwaj, B.; Ahmed, S.; Jaiharie, J.; Dadhich, R.; Ganesan, M. Web Scraping Using Summarization and Named Entity Recognition (NER). In Proceedings of the 7th International Conference on Advanced Computing and Communication Systems (ICACCS) 2021, Coimbatore, India, 19–20 March 2021; pp. 261–265. [CrossRef]
13. Dansana, D.; Adhikari, J.; Mohapatra, M.; Sahoo, S. An Approach to Analyse and Forecast Social media Data using Machine Learning and Data Analysis. In Proceedings of the International Conference on Computer Science, Engineering and Applications (ICCSEA) 2020, Gunupur, India, 13–14 March 2020; pp. 1–5. [CrossRef]

14. Camargo-Henríquez, I.; Núñez-Bernal, Y. A Web Scraping based approach for data research through social media: An Instagram case. In Proceedings of the V Congreso Internacional en Inteligencia Ambiental, Ingeniería de Software y Salud Electrónica y Móvil (AmITIC) 2022, San Jose, Costa Rica, 14–16 September 2022; pp. 1–4. [CrossRef]
15. Zou, J.; Le, D.; Thoma, G. Locating and parsing bibliographic references in HTML medical articles. *Int. J. Doc. Anal. Recognit.* **2010**, *13*, 107–119. [CrossRef]
16. Korab, P. Text Network Analysis: Generate Beautiful Network Visualisations. 2022. Available online: https://towardsdatascience.com/text-network-analysis-generate-beautiful-network-visualisations-a373dbe183ca (accessed on 21 May 2023).
17. Alaei, A.R.; Becken, S.; Stantic, B. Sentiment Analysis in Tourism: Capitalizing on Big Data. *J. Travel Res.* **2019**, *58*, 175–191. [CrossRef]
18. Boegershausen, J.; Datta, H.; Borah, A.; Stephen, A.T. Fields of Gold: Scraping Web Data for Marketing Insights. *J. Mark.* **2022**, *86*, 1–20. [CrossRef]
19. Màrquez-Domínguez, C.; López López, P.; Arias, T. Social networking and political agenda: Donald Trump's Twitter accounts. In Proceedings of the 12th Iberian Conference on Information Systems and Technologies (CISTI) 2017, Lisbon, Portugal, 21–24 June 2017; pp. 1–6. [CrossRef]
20. Tarai, J.; Kant, R.; Finau, G.; Titifanue, J. Political Social Media Campaigning in Fiji's 2014 Elections. *J. Pac. Stud.* **2015**, *35*, 89–114.
21. Rullo, L.; Nunziata, F. "Sometimes the Crisis Makes the Leader?" A Comparison of Giuseppe Conte Digital Communication before and during the COVID-19 Pandemic. *Comun. Politica* **2021**, *3*, 309–332. [CrossRef]
22. Mabillard, V.; Zumofen, R.; Pasquier, M. Local governments' communication on social media platforms: Refining and assessing patterns of adoption in Belgium. *Int. Rev. Adm. Sci.* **2022**, 1–17. [CrossRef]
23. Martínez, T. Comunicación y diabetes, un camino para la reflexión. *RedMarka-Rev. De Mark. Apl.* **2022**, *26*, 96–113. [CrossRef]
24. Jayasingh, S.; Venkatesh, R. Customer Engagement Factors in Facebook Brand Pages. *Asian Soc. Sci.* **2015**, *11*, 19. [CrossRef]
25. Huertas, A.; Marine-Roig, E. User reactions to destination brand contents in social media. *Inf. Technol. Tour.* **2016**, *15*, 291–315. [CrossRef]
26. Caldevilla-Domínguez, D.; Barrientos-Báez, A.; Padilla-Castillo, G. Dilemmas Between Freedom of Speech and Hate Speech: Russophobia on Facebook and Instagram in the Spanish Media. *Politics Gov.* **2022**, *11*, 1–13. [CrossRef]
27. Martínez-Fernández, V.; Amboage, E.; Burneo, M.; Benitez, V. La gestión de los medios sociales en la dinamización de destinos turísticos termales: Análisis crosscultural de modelos aplicados en España, Portugal y Ecuador. *Hologramática* **2015**, *2*, 47–60.
28. Sánchez-Jiménez, M.; Matos, N.; Correia, M. Evolution of the presence and engagement of official social networks in promoting tourism in Spain. *J. Spat. Organ. Dyn.* **2019**, *7*, 210–225.
29. Sánchez-Jiménez, M. Análisis de la comunicación digital oficial en la promoción turística de Brasil. *3c TIC-Cuad. De Desarro. Apl. A Las TIC* **2020**, *9*, 17–39. [CrossRef]
30. Lee, M. Evolution of hospitality and tourism technology research from Journal of Hospitality and Tourism Technology: A computer-assisted qualitative data analysis. *J. Hosp. Tour. Technol.* **2021**, *13*, 62–84. [CrossRef]
31. Pereira, P. Social Media Influencers in Travel and Tourism. Master's Thesis, Master Course in Information Management. Nova Information Management School, Lisbon, Portugal, 2023.
32. Phaujdar, A. 9 Best Web Scraping Tools. 2021. Available online: https://hevodata.com/learn/web-scraping-tools/ (accessed on 22 May 2023).
33. Rizkallah, J. The Big (Unstructured) Data Problem. 2017. Available online: https://www.forbes.com/sites/forbestechcouncil/2017/06/05/the-big-unstructured-data-problem/?sh=cd00fa3493a3 (accessed on 23 March 2023).
34. Selz, D. Unstructured Data Is Key to True Customer Insight. 2017. Available online: https://www.linkedin.com/pulse/unstructured-data-key-true-customer-insight-dorian-selz (accessed on 23 March 2023).
35. Chen, S.; Kang, J.; Liu, S.; Sun, Y. Cognitive computing on unstructured data for customer co-innovation. *Eur. J. Mark.* **2020**, *54*, 570–593. [CrossRef]
36. Marr, B. How Much Data Do We Create Every Day? *The Mind-Blowing Stats Everyone Should Read.* 2018. Available online: https://www.forbes.com/sites/bernardmarr/2018/05/21/how-much-data-do-we-create-every-day-the-mind-blowing-stats-everyone-should-read/?sh=4de1a9aa60ba (accessed on 12 March 2023).
37. Ruan, Z.; Siau, K. Digital Marketing in the Artificial Intelligence and Machine Learning Age. Americas Conference on Information Systems. 2019. Available online: https://www.semanticscholar.org/paper/Digital-Marketing-in-the-Artificial-Intelligence-Ruan-Siau/5d0764dbe4cb3beb6c194b49a4eae1a991a72cd8 (accessed on 13 March 2023).
38. Kim, H.; Chan, H.; Gupta, S. Examining information systems infusion from a user commitment perspective. *Inf. Technol. People* **2016**, *29*, 173–199. [CrossRef]
39. Changchit, C.; Chuchuen, C. Cloud computing: An examination of factors impacting users' adoption. *J. Comput. Inf. Syst.* **2018**, *58*, 1–9. [CrossRef]
40. Biedrzycki, N. Cognitive Computing. What Can It Be Used for? 2020. Available online: https://towardsdatascience.com/cognitive-computing-what-can-it-be-used-for-8af4721928f5 (accessed on 26 May 2023).
41. Frackiewicz, M. The Role of NLP in Cognitive Computing. 2023. Available online: https://ts2.space/en/the-role-of-nlp-in-cognitive-computing/ (accessed on 26 May 2023).
42. Rao, L. Instagram Copies Snapchat Once again with Face Filters. 2017. Available online: https://tinyurl.com/ybcuxxdv (accessed on 5 June 2023).

43. Perry, E. Meet HearMeOut: The Social Media Platform Looking to Bring Audio Back into the Mainstream. 2018. Available online: https://tinyurl.com/y8yxbzah (accessed on 5 June 2023).
44. Katai, L. 3 Reasons Why Audio Will Conquer All Social Media. 2018. Available online: https://www.adweek.com/performance-marketing/3-reasons-why-audio-will-conquer-social-media/ (accessed on 5 June 2023).
45. Shahid, M.Z.; Li, G. Impact of Artificial Intelligence in Marketing: A Perspective of Marketing Professionals of Pakistan. *Glob. J. Manag. Bus. Res.* **2019**, *19*, 27–33.
46. Dwivedi, Y.; Ismagilova, E.; Hughes, D.; Carlson, J.; Filieri, R.; Jacobson, J.; Jain, V.; Karjaluoto, H.; Kefi, H.; Krishen, A.; et al. Setting the future of digital and social media marketing research: Perspectives and research propositions. *Int. J. Inf. Manag.* **2021**, *59*, 102168. [CrossRef]
47. Zoho Social. Social Media Marketing Trends for 2022. 2021. Available online: https://www.zoho.com/social/journal/social-media-marketing-trends-2022.html (accessed on 5 June 2023).
48. NBBJ. Social Media Is Evolving Quickly, and Your Business Needs to Also. 2022. Available online: https://www.northbaybusinessjournal.com/article/industrynews/social-media-is-evolving-quickly-and-your-business-needs-to-also/ (accessed on 5 June 2023).
49. Corcoran, S. Defining Earned, Owned and Paid Media. 2009. Available online: https://www.forrester.com/blogs/09-12-16-defining_earned_owned_and_paid_media/ (accessed on 12 March 2023).
50. Wozniak, T.; Stangl, B.; Schegg, R.; Liebrich, A. Do Social Media Investments Pay Off? Preliminary Evidence from Swiss Destination Marketing Organizations. In Proceedings of the ENTER eTourism Conference 2016, Bilbao, Spain, 2–5 February 2016.
51. Peters, K.; Chen, Y.; Kaplan, A.; Ognibeni, B.; Pauwels, K. Social media metrics-A framework and guidelines for managing social media. *J. Interact. Mark.* **2013**, *27*, 281–298. [CrossRef]

*big data and cognitive computing*

MDPI

*Article*

# Visual Explanations of Differentiable Greedy Model Predictions on the Influence Maximization Problem

**Mario Michelessa [1], Christophe Hurter [2], Brian Y. Lim [1], Jamie Ng Suat Ling [3], Bogdan Cautis [4] and Carol Anne Hargreaves [1,*]**

1 Department of Statistics and Data Science, Faculty of Science, National University of Singapore, Singapore 117546, Singapore; mario.michelessa@u.nus.edu (M.M.); brianlim@comp.nus.edu.sg (B.Y.L.)

2 ENAC, Université de Toulouse, 31400 Toulouse, France; christophe.hurter@enac.fr

3 Institute for Infocomm Research, A*STAR, Singapore 138632, Singapore; jamie@i2r.a-star.edu.sg

4 Department of Computer Science, University of Paris-Sud, 91405 Orsay, France; bogdan.cautis@u-psud.fr

* Correspondence: carol.hargreaves@nus.edu.sg

**Abstract:** Social networks have become important objects of study in recent years. Social media marketing has, for example, greatly benefited from the vast literature developed in the past two decades. The study of social networks has taken advantage of recent advances in machine learning to process these immense amounts of data. Automatic emotional labeling of content on social media has, for example, been made possible by the recent progress in natural language processing. In this work, we are interested in the influence maximization problem, which consists of finding the most influential nodes in the social network. The problem is classically carried out using classical performance metrics such as accuracy or recall, which is not the end goal of the influence maximization problem. Our work presents an end-to-end learning model, SGREEDYNN, for the selection of the most influential nodes in a social network, given a history of information diffusion. In addition, this work proposes data visualization techniques to interpret the augmenting performances of our method compared to classical training. The results of this method are confirmed by visualizing the final influence of the selected nodes on network instances with edge bundling techniques. Edge bundling is a visual aggregation technique that makes patterns emerge. It has been shown to be an interesting asset for decision-making. By using edge bundling, we observe that our method chooses more diverse and high-degree nodes compared to the classical training.

**Keywords:** influence maximization; end-to-end learning; decision-focused learning; graph visualization; edge bundling; differentiable greedy

## 1. Introduction

The rapid growth of social networks in recent years has sparked extensive research on understanding the dynamics of these networks. The diffusion of information within social networks has significant implications in various fields, including marketing [1], politics [2], and surveillance [3]. Social networks have emerged as powerful platforms for mass information diffusion, influencing major events such as elections and social movements such as the Arab Spring. Exploring the mechanisms of information diffusion is crucial for detecting manipulation attempts, mitigating terrorist risks, and optimizing product advertising.

The problem of information diffusion centers around how information spreads and propagates among users. One fundamental problem in this domain is the Influence Maximization problem, which involves identifying a set of users that maximizes the spread of information. However, this problem is known to be NP-hard, posing computational challenges for classical statistical approaches to studying information cascades [4].

Due to the sudden growth of social networks, the quantity of data to process became unmanageable for classical statistical studies of information diffusion instances, called information 'cascades'. However, due to the rapid development of these networks, the quantity of data to process became a challenge to handle. In recent years, machine learning models have shown promise in directly estimating the influence that users exert on each other within social networks. By leveraging these predictions, it becomes possible to identify the most influential users in the network. Traditionally, these two steps of prediction and influence maximization are performed sequentially. However, recent research has suggested the joint execution of these steps for improved influence maximization [5].

In this paper, we propose an end-to-end learning approach using machine learning models to predict diffusion probabilities in social networks. Additionally, we developed a novel visualization method to evaluate the quality of our solution compared to existing methods. We argue that visualizing the graph provides deeper insights into diffusion mechanisms than conventional numerical metrics. Building on the concept highlighted in Anscombe's work [6], where statistics fail to capture the full patterns of data, social networks exhibit a similar phenomenon. Applying edge bundling techniques to visualize dense and cluttered graphs allows us to better understand edge connections.

The contributions of this paper are twofold. Firstly, we introduce a novel graph optimization algorithm for maximizing influence in social networks. Secondly, we employ visualization techniques to validate our model's performance against two baseline models. The visualization model enhances our understanding and evaluation of the proposed solution.

The remainder of this paper is organized as follows. Section 1 provides the background and context, while Section 2 reviews and summarizes related work on the influence maximization problem. In Section 3, we describe the data and methods employed in this study. Section 4 presents the results and findings, and Section 5 concludes the paper by summarizing the key findings and discussing future research opportunities.

## 2. Related Work

In this section, we first review the influence maximization problem and existing end-to-end learning methods. Next, we review existing visual simplification techniques for dense data visualization.

### 2.1. *Influence Maximization*

The study of information diffusion in social networks began in the early 2000s with the seminal work of Kempe et al. [4], in which they propose a greedy framework to find approximations of the optimal subset of nodes maximizing influence with theoretical guarantees. The recent advances in the optimization of submodular functions allowed the improvement of greedy algorithms in a Cost-Effective Lazy Forward algorithm (CELF) [7] and then CELF++ [8], which are much faster.

However, these algorithms require knowing the diffusion probabilities between users, which is problematic on real social networks since this information is not available. To solve this problem, machine learning algorithms have recently been used to learn the influence of content on social networks, forecast the future bursts of popularity of content, or generate new cascades. Decision-trees-based models, support vector machines, and clustering algorithms have been used for a decade to predict the influence of content on social networks. However, since DeepCas [9], they have been progressively replaced by deep learning models.

### 2.2. *End-to-End Learning*

Recently, a new method for training machine learning models arose for solving complex data pipeline problems. End-to-end learning can be used on "prediction-optimization" problems, where the prediction of a model is then used to optimize a certain quantity. The "prediction-optimization" problems are classically solved in two stages. In the first stage, the model is trained to maximize its accuracy. The outputs of the model tend to be close to the historical data. In the second stage, an optimization algorithm is executed on the predicted values of the model. End-to-end learning differs in that the model is not trained to maximize the accuracy but is directly optimized to maximize the final influence of the optimal solution found using the predicted probabilities. This framework has recently been applied to recommendation systems [5] but has never been applied to influence maximization on social networks.

### 2.3. Edge-Bundling

Several initiatives related to the visualization of social networks have been developed in recent years; however, there has been limited focus on influence maximization or information cascades [10]. The large quantity of data involved in information diffusion in social networks makes visualization both important and challenging to develop.

Due to the large number of edges and the high density of edges in the graph, edge-bundling techniques can be used to facilitate the interpretation of results. Edge bundling techniques have in common to cluster close edge paths together, thus increasing the number of white spaces and reducing the clutter in layouts of large graphs. Bundling can be seen as sharpening the edge density in the layout, making areas of high density even denser and areas with a lesser edge density appear sparser or white.

As opposed to graph simplification techniques where edges considered unimportant are simply removed from the layout, no edges are removed during edge bundling, and the overall topology of the graph is conserved.

Edge bundling can be used to identify the links between groups of nodes that would be invisible in a large, dense graph due to the clutter of edges. The identification of clusters is made easier by the white spaces separating the clusters. Edge bundling, however, does not conserve the direction of the edges. In certain cases, the direction of the edges can be important, such as in trajectories or geographical data. The direction of the edge is reduced to a small set of main directions, thus losing the initial directions information.

Recent work has tried to conserve the initial edge directions for automobile traffic and airplane trajectories [11]. In our case, the direction of the edges is not important.

Since 2006, with the seminal work of Gasner and Koren, various edge bundling techniques have been developed. Initially, edge bundles were drawn as straight lines based on spatial proximity [12]. Qu et al. added NURBS splines to replace the original straight-line bundles [13]. Most notably, hierarchical edge bundling was developed by Holten [14] to bundle large graphs of several thousands of nodes easily. After this, variants spurred, adapted to whether the graph is static or dynamic, directed or undirected, 3D or 2D [15].

## 3. Data and Methods

In this section, we briefly present the Weibo dataset used to train the models. Furthermore, we detail the specificity of the end-to-end approach we used to train the SGREEDYNN model.

### 3.1. Dataset

Our approach uses data scraped from real social networks in order to predict the best influencers. The dataset used is scraped from the Chinese micro-blogging social media Weibo [16]. It contains examples of information cascades stored as lists of reposts. Information about user profiles, topic classification of the messages, and the social graph are also available.

Table 1 above provides a summary of our dataset. From this dataset, we extract and use 24 features describing the link $(u, v)$. These features are extracted from the users' profiles, the topology of the social graph, the topic modeling of the posts of the users, and the cascade information. The detailed list of the features is given in Appendix A.

**Table 1.** Dataset summary.

| Dataset | # Cascades | # Users | # Reposts |
|---|---|---|---|
| Weibo | 300 K | 1.7 M | 200 M |

The goal is then to use the end-to-end learning method to train machine learning models to predict the diffusion probabilities between influencers and targets. The ground truth diffusion probabilities used are extracted from the previous examples of information diffusion cascades.

Figure 1 above provides the end-to-end process for the learning framework for maximizing the influence. The social network and its content are preprocessed to create an instance X containing the features vectors $X[u, v]$ of all the pairs of nodes $(u, v)$ and a matrix P containing the ground-truth diffusion probabilities. The instance X is fed to our SGREEDYNN model. This model predicts a diffusion probability matrix, which is then fed to the optimization algorithm. Given these diffusion probabilities, the algorithm chooses the best subset of k influencers among the users in the social network that maximizes the information propagation. The influence function $\sigma$ then returns the real influence of this subset. The weights of the model are then updated to maximize this final influence.

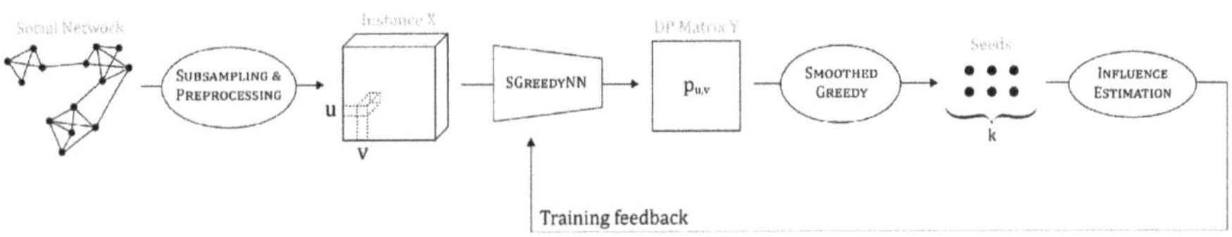


**Figure 1.** General pipeline of the learning framework.

Given a history of actions $\{(u, v)\}$ where $u$ influences $v$, the ground truth diffusion probability $p_{u,v}$ is defined as

$$p_{u,v} = \frac{A_{u,v}}{A_{u,\cdot} + A_{\cdot,v}} \tag{1}$$

where $A_{u,v}$ is the number of times $u$ influenced $v$, $A_{u,\cdot}$ is the number of posts posted by $u$, and $A_{\cdot,v}$ is the total number of reactions of $v$ [17].

The obtained diffusion probabilities are, in reality, very small. The estimated ground truth probabilities are mapped to higher values to facilitate the training. For positive probabilities, the first two deciles are mapped to a probability of 1, the next 3 deciles are mapped to a value of 0.5, the next 3 deciles are mapped to 0.2, and the rest is mapped to 0.1. The probabilities equal to 0 do not change. We chose these values.

To increase the number of positive examples in the diffusion probability matrix during the training, only the targets participating in more than 150 cascades are considered, and only the top 20% of influencers on this induced graph are considered. This gives a subset of influencers $I \subset V$ and $T \subset V$.

Given this sub-sample, $I$, $T$, the training dataset D is created by randomly drawing $n$ influencers in $I$ and $m$ targets in $T$ and creating a matrix X of size $(n, m, 24)$ where $X[u, v, :]$ is the feature vector of size 24 associated with the potential influence link $(u, v)$. At the same time, a matrix $Y$ of size $(n, m)$ is created with $Y[u, v] = p_{u,v}$ where $p_{u,v}$ has been previously defined. An example of a $Y$ matrix is provided in Appendix A.

### *3.2. End to End Learning*

The end-to-end machine learning model is trained on the dataset D = $\{(X, Y)\}$. The model is then optimized to minimize the following function by stochastic gradient descent.

The use of the Smoothed Greedy algorithm (SGREEDY) as an optimization algorithm allows an easy estimation of the objective function's gradient.

$$J(\theta) = -\sum_{(X,Y)\in D} \sigma(\mathrm{SG{\small REEDY}}(m(X;\Theta)), Y) \tag{2}$$

The model $m$ is parameterized by $\Theta$, takes an instance $X$ of size $(n, m, 24)$ as an input, and returns a diffusion probability matrix of size $(n, m)$. The Smoothed Greedy algorithm takes a diffusion probability matrix as an input and returns the index of the best seeds in the network. The influence spread function $\sigma$ takes a seed set and a diffusion probability matrix and returns a positive number. Here,

$$\sigma(S) = \sum_{v=1}^{n} \left(1 - \prod_{u\in S}(1 - p_{u,v})\right) \tag{3}$$

The model has been trained on 50 instances of size 500 $\times$ 500, on 100 epochs, with a batch size of 4, a learning rate $\lambda = 5e^{-4}$. The temperature of the Smoothed Greedy algorithm is $\epsilon = 0.1$, and the sample size of the Smoothed Greedy algorithm is 20. The hyperparameters were chosen to maximize the number of targets influenced in the test dataset using the Bayesian optimization and hyperband method [18].

### 3.3. Visual Aggregation

In addition to the other performance metrics, we explain and compare the results found by the models using edge bundling techniques on the different networks studied. These visualizations give new insights into the behavior of our method and how it surpasses the other methods.

The edge bundling algorithm used is the kernel-based estimation edge bundling algorithm [19]. The first step of the algorithm is to estimate the edge density map using the kernel density estimation. Then, the normalized gradient direction is estimated, and the edges are moved in the gradient direction and smoothed by using Laplacian filtering. These steps are repeated with a decreasing kernel size until the result is convincing. Tuning these parameters typically involves a combination of manual experimentation and automated optimization. The following parameters are chosen such that only a few main edge bundles appear while retaining most of the information of individual edges:

- number $n$ of iterations of the algorithm
- tension $t$ of the edges
- initial bandwidth $bw$ of the kernel
- decay rate $d$ of the kernel's size

We apply this technique to an instance from the training dataset. The instance is a graph containing 500 influencers and 500 targets. The graph contains the edge $(u, v)$ if $v$ is present in a cascade initiated by $u$. We call this graph the cascade graph. This graph is bipartite, and the only possible information diffusion is between an influencer and a target. The dense instances contain tens of thousands of edges; thus, the use of edge bundling on this graph is appropriate.

### 3.4. Choice of Graph Layouts

Edge bundling techniques are very dependent on the layout used. We investigated several layouts, as shown in Figure 2 below. We color the edges according to the position of the origin node to better see the direction of the bundled edges. In circular layouts, the color of the edge depends on the angular coordinate of the origin node. In lateral layouts, the color depends on the y-component of the origin node.

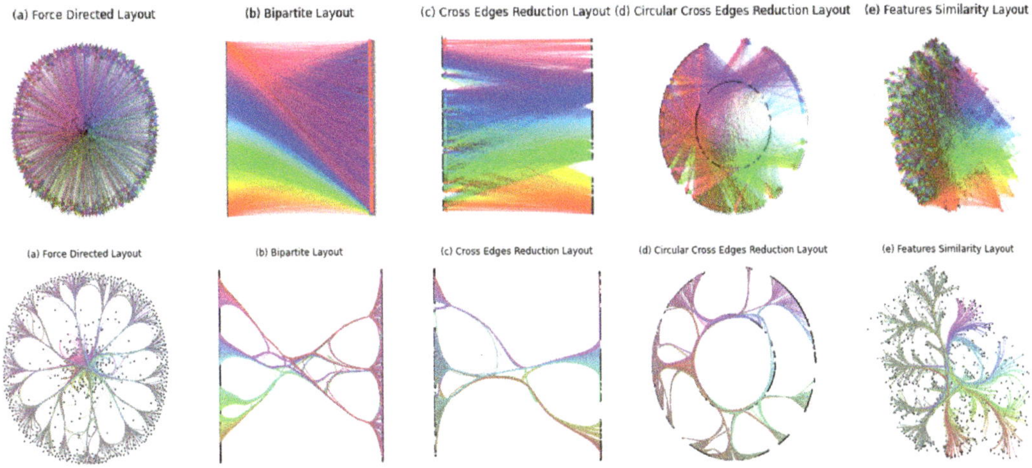


**Figure 2.** Comparison of 5 different layouts in the bundled and unbundled case. Edge bundling is applied on the second row. The five different layouts are, from left to right, (**a**) force-directed, (**b**) bipartite, (**c**) cross-edge reduced bipartite, (**d**) circular cross-edge reduced bipartite, (**e**) similarity. The unbundled graphs are cluttered, which makes any interpretation very difficult. Edge bundling makes the visualization clearer.

Several layouts have been explored. We explain briefly their specificity.

#### 3.4.1. Force-Directed Layout

The principle of the force-directed layout is to consider the edges as strings and to minimize the potential energy of the system [20]. The forces acting on the edges tend to group the nodes in clusters. In our case, the nodes having a higher degree are in the middle; they link the peripheral nodes having a lesser degree. By applying edge bundling on this layout and coloring the edges according to the directions, we can notice that the edges arriving in the outer layer of the figure are colored grey. The outer layer mainly consists of the targets, and the grey color comes from all the colors mixing together. This means that the influencers do not spread their influence in a particular direction, and all targets can receive influence from the targets of any color. This layout may thus not be suited to give significant insights when bundled.

#### 3.4.2. Bipartite Layout

We also take advantage of the fact that the graph is bipartite by considering a bipartite layout. The bipartite layout consists of displaying the two groups of nodes (influencers and targets) on two parallel lines.

The order of the nodes is important, and the initial order does not give significant insights.

#### 3.4.3. Cross Edge Reduced Bipartite Layout

To counter the issue mentioned above, we apply a cross-edge reduction algorithm [21]. As shown in Figure 2c, the edge bundling technique displays bundles of the same color, which is consistent with the behavior of the edge cross-reduction algorithm.

#### 3.4.4. Circular Cross-Edge Reduced Bipartite Layout

To better see the edges between influencers and targets, we spread the targets around the influencers and then organized these two in a circular layout. The influencers are positioned in the inner circle, and the targets are in the outer circle.

However, the placement of the nodes does not depend on the features the model used to predict the diffusion probability. This placement only depends on the original cascade graph topology, which is not accessible to the network.

3.4.5. Similarity Layout

This observation motivates us to consider a layout where two nodes are close if they have similar behavior. The principle is to generate an undirected weighted graph having the same nodes as the social network. The edges' weights are defined as the cosine similarity between the feature vectors of the two nodes. This dense graph is then pruned by removing the edges having a weight less than a certain threshold, and we then consider its force-directed layout.

In these conditions, two nodes having a very similar feature vector will be close in this layout. The bundling on this graph shows that the graph is clustered into two separate groups, i.e., the influencers and the targets, even if this separation is not performed in the features. This first observation shows that the features of the influencers and the targets are significantly different. To visualize the effect of the features on the layout, we display the values of the features according to the position of the nodes in the layout.

The following observations can be made. The difference between influencers and targets in a social network mainly comes from the number of followers, the number of cascades initiated, and the PageRank metric. The number of friends is higher for targets than for influencers. This may be due to the subsampling of the targets $T$ explained in the methodology. The number of likes reaches its highest value in the boundary between the influencers and the targets.

3.4.6. Influence Maximization

The problem of influence maximization consists of finding the subset of nodes maximizing the influence on the social network's population. To do that, we can apply brushing on the selected nodes by the algorithm. If an algorithm returns a seed set $S$, it is possible to see the influence coverage on the edge bundling graph by only plotting the influence edges $(u, v)$ if $u \in S$.

An example of such brushing is shown in Figure 3 below. This figure shows the influence of the number of seeds on the information diffusion. The seeds are selected by the Oracle Greedy algorithm, detailed in the next section. Logically, by adding more seeds to the seed set, the number of nodes reached by the seed set increases. This can be seen by the increasing number of plotted edges in the bundled case.

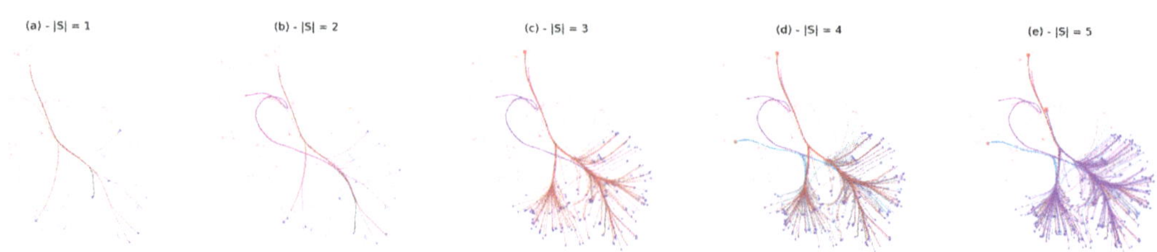


**Figure 3.** Influence of the number of seeds selected on the targets influenced. The red nodes are the influencers, and the blue nodes are the targets. The size of the nodes corresponds to the degree of the nodes in the cascade graph. The color of the edges depends on the position of the origin node, as in Figure 2. The different columns correspond to different numbers of selected seeds $|S|$. The layout is a similarity layout with a weight_threshold of 0.8, an inter-node distance of $k = 0.04$, and 50 iterations. The spread weakly increases when $|S| < 3$ before suddenly increasing with the addition of important nodes to the seed set.

The first observation that can be made on all layouts is the direction of the spread on the first seeds. What can be seen on the edge bundling graph is that the first seeds already influence the target through all the largest bundles of the diagram. The seed set reaches targets in all directions.

The second observation that can be made thanks to the edge bundling visualization is the difference in spread between the added seeds. In Figure 3b, we can see that the added node does not participate much in the increase of the number of influenced nodes.

The following node added in Figure 3c spreads its influence in all branches and adds more nodes.

## 4. Results & Findings

### 4.1. Introduction

Different layouts have been explored. Due to the weaknesses of the first four layouts, we developed a layout based on the similarity between the feature vectors of the nodes. Figure 2 compares the graph visualization with and without edge bundling. Acquiring insights on the graph topology on the visualization without edge bundling is impossible. The edge bundling techniques give some insights into the 3 cases.

The graph layouts in the unbundled cases are cluttered, which makes any interpretation impossible. The edge bundling technique applied to the different layouts gives clearer visualizations of the relations between the nodes.

### 4.2. Comparison between Algorithms

We compare the results of the three following algorithms. Selecting the best-performing nodes can be separated into two different tasks. An estimation task and an optimization task.

SGREEDYNN: This is our proposed method. This model is trained in the manner explained in Figure 1. The model then infers the diffusion probability matrix of the network.

2STAGE: To prove the performances of the end-to-end learning on influence maximization, we train this model in a classical way. This model is trained to minimize the mean square loss between the predicted and the ground-truth diffusion probability matrices.

ORACLEGREEDY: We also evaluate how well the two estimated matrices match up with the actual diffusion probabilities. These actual probabilities are detailed in Equation (1). To compare the performances, we directly execute the greedy algorithm on the ground-truth diffusion probability matrix.

The optimization part of the task is the same in the three methods. The $k$ nodes constituting the seed set $S$ are selected by a greedy algorithm. Thus, the difference in the three methods is only in how to estimate the diffusion probability matrices.

We compare this visualization using edge bundling on the similarity layout in Figure 4 below. The edge bundling helps to determine which nodes the influencers are targeting. The edge bundling shows that the first influencers to be added are targeting in every direction. However, when the number of seeds increases, the behavior of our method varies from the classical greedy algorithm. While the greedy algorithm continues to add nodes influencing in every direction, the additional seeds added by our model seem to focus on areas where the first seeds had low coverage. The difference between our method and the two other models is visible in Figure 5 below.

For $|S| = 3$, our method chooses three seeds having a high degree and far away on the layout from each other (Figure 4b). The distance between the influencers informs us that our model chooses diverse types of influencer profiles. The number of targets influenced by the seed set is thus increased. Indeed, similar influencers may influence the same targets; thus, different influencers have different sets of targets. This observation contrasts with Figure 5a (OracleGreedy), where the selected nodes are small and close to each other, creating an overlap in the influenced nodes.

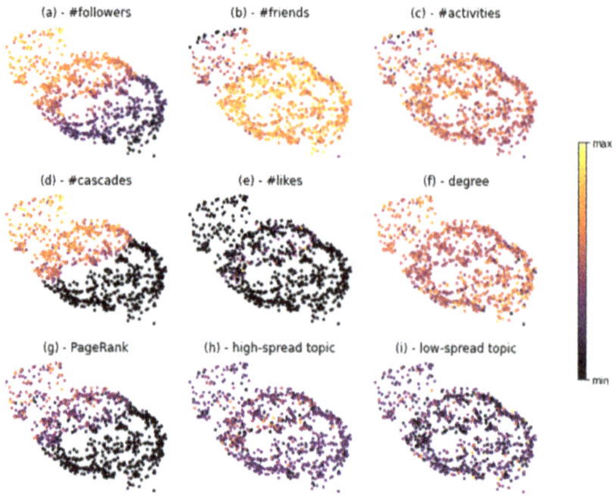


**Figure 4.** Distribution of nine different features on the similarity layout. For each figure, the brightest color corresponds to the highest value, and the darkest color corresponds to the lowest value. The layout used is the same as in Figure 3.

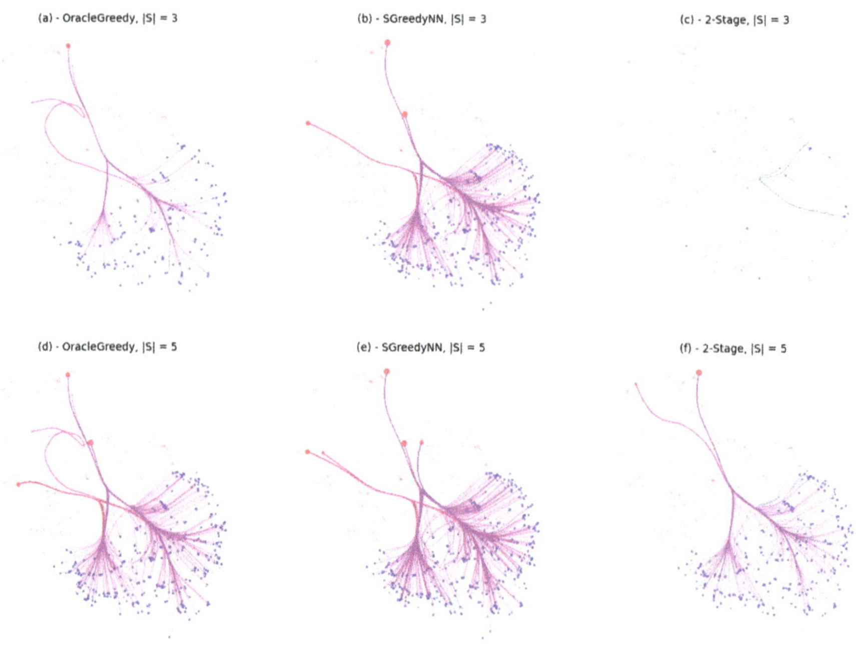


**Figure 5.** Comparison between the three methods. The layout used is the same similarity layout used in Figure 3. The three columns correspond to the three methods compared, and the two rows correspond to two different values of initial number of seeds. They are the Oracle Greedy algorithm in (**a**), our method SGREEDYNN in (**b**), and the classical 2-staged training method in (**c**). The red nodes are the influencers, and the blue nodes are the targets. The size of the nodes correspond to the degree of the nodes in the cascade graph. The color of the edges depends on the position of the origin node. The two rows of the figure correspond to two different numbers of seeds selected. We can see that the performances can greatly vary among the models. With one seed, the decision-focused model already reaches a high number of targets, whereas the 2-Staged and the Oracle greedy algorithms mainly choose small peripheral nodes.

On the next line, corresponding to $|S| = 5$, the differences between the performances of the algorithm are less visible. The gap of influenced targets thus rapidly decreases.

The visualization techniques confirm the superiority of the SGREEDYNN model compared to Oracle Greedy and the 2-staged classical learning. When the number of seeds is very low, the edge bundling helps visualize the direction in which the seeds influence the targets.

## 5. Conclusions & Future Work

In this study, we implemented an end-to-end method to learn the information diffusion probabilities between users in a social network. Our model SGREEDYNN performs better than the classical learning method. In addition, we developed visualization methods to better compare and understand the influence of the social network. The performances of our models have been confirmed by the visualization of the social network using edge bundling.

The method presented here has different advantages compared to the numerical performance metrics normally used in influence maximization. After a high-level analysis of different layouts, we showed that edge bundling techniques applied to the training dataset validated our method compared to the classical training.

Several limitations have been noted and could be investigated in future works. The model can be tested with different types of data. In this work, the preprocessing of the social network's data modified the original topology by subsampling the influencers and targets available. It may be interesting to test the visualization techniques on data subsampled differently. In addition, similar social networks, such as Twitter for example, can also be investigated to verify the results.

Different architectures of the model can be investigated. In this work, we only considered artificial neural networks, but this choice can be extended to different types of machine learning models.

Concerning the visualization of the instances, interactivity could be added to facilitate the brushing and the exploration of the graph.

**Author Contributions:** Conceptualisation: C.H.; Methodology: B.C.; Software: M.M.; Formal Analysis: M.M.; Resources: B.C.; Writing original draft preparation: M.M.; Writing—review and editing: C.A.H., C.H. and B.Y.L.; Visualization: M.M.; Supervision: C.A.H., C.H., B.Y.L. and J.N.S.L.; Project Administration: C.A.H. All authors have read and agreed to the published version of the manuscript.

**Funding:** This research was funded by the National Research Foundation, Prime Minister's Office, Singapore, under its Campus for Research Excellence and Technological Enterprise (CREATE).

**Data Availability Statement:** Data can be provided upon request.

**Acknowledgments:** This research project was conducted during an internship at CNRS@CREATE. I would like to acknowledge their support and provision of necessary resources.

**Conflicts of Interest:** The authors declare no conflict of interest.

## Appendix A. Features Used by the Models

### *Appendix A.1. List of Features*

To estimate the information diffusion probabilities, our SGREEDYNN model uses the following features. The first column contains $u$ if the feature comes from the influencer, $v$ if it comes from the target, and $u, v$ if it depends on both the influencer and the target. "#" represents, "the number of".

### *Appendix A.2. Example of Y*

Here is an example of a $Y$ matrix containing the diffusion probabilities $p_{u,v}$ between influencers $u$ and targets $v$ for a subsampling of 500 influencers and targets.

**Table A1.** Features used for estimating the diffusion probability between two users $(u, v)$.

| User | Feature |
|---|---|
| $u$ | # followers |
| $u$ | # friends |
| $u$ | # activities |
| $u$ | verified |
| $u$ | gender |
| $u$ | # cascades |
| $u$ | # likes |
| $u$ | # reposts |
| $u$ | out-degree |
| $u$ | PageRank |
| $u$ | high-spread topic |
| $u$ | med-spread topic |
| $u$ | low-spread topic |
| $v$ | # followers |
| $v$ | # friends |
| $v$ | # reposts |
| $v$ | verified |
| $v$ | gender |
| $v$ | in-degree |
| $v$ | PageRank |
| $v$ | high-spread topic |
| $v$ | med-spread topic |
| $v$ | low-spread topic |
| $u,v$ | social edge |

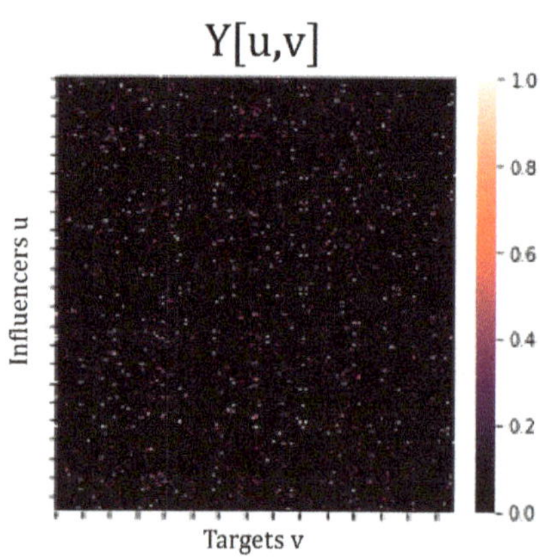


**Figure A1.** Example of a $Y$ matrix used for visualization in this paper.

## References

1. Chen, W.; Wang, C.; Wang, Y. Scalable Influence Maximization for Prevalent Viral Marketing in Large-Scale Social Networks. In Proceedings of the 16th ACM SIGKDD International Conference on Knowledge Discovery and Data Mining, KDD '10, Washington, DC, USA, 25–28 July 2010; pp. 1029–1038. [CrossRef]
2. Bond, R.; Fariss, C.; Jones, J.; Kramer, A.; Marlow, C.; Settle, J.; Fowler, J. A 61-Million-Person Experiment in Social Influence and Political Mobilization. *Nature* **2012**, *489*, 295–298. [CrossRef] [PubMed]
3. Bedi, M. Social networks, government surveillance, and the Fourth Amendment Mosaic Theory. *BUL Rev.* **2014**, *94*, 1809.
4. Kempe, D.; Kleinberg, J.; Tardos, E. Maximizing the Spread of Influence through a Social Network. In Proceedings of the Ninth ACM SIGKDD International Conference on Knowledge Discovery and Data Mining, KDD '03, Washington, DC, USA, 24–27 August 2003; pp. 137–146. [CrossRef]

5. Sakaue, S. Differentiable Greedy Algorithm for Monotone Submodular Maximization: Guarantees, Gradient Estimators, and Applications. In Proceedings of the 24th International Conference on Artificial Intelligence and Statistics, San Diego, CA, USA, 13–15 April 2021; Banerjee, A., Fukumizu, K., Eds.; Proceedings of Machine Learning Research (PMLR). UK, 2021; Volume 130, pp. 28–36.
6. Anscombe, F.J. Graphs in Statistical Analysis. *Am. Stat.* **1973**, *27*, 17–21. [CrossRef]
7. Leskovec, J.; Krause, A.; Guestrin, C.; Faloutsos, C.; VanBriesen, J.; Glance, N. Cost-Effective Outbreak Detection in Networks. In Proceedings of the 13th ACM SIGKDD International Conference on Knowledge Discovery and Data Mining, KDD '07, New York, NY, USA, 12–15 August 2007; pp. 420–429. [CrossRef]
8. Goyal, A.; Lu, W.; Lakshmanan, L.V. CELF++: Optimizing the Greedy Algorithm for Influence Maximization in Social Networks. In Proceedings of the 20th International Conference Companion on World Wide Web, WWW '11, Hyderabad, India, 28 March–1 April 2011; pp. 47–48. [CrossRef]
9. Li, C.; Ma, J.; Guo, X.; Mei, Q. DeepCas: An End-to-End Predictor of Information Cascades, Republic and Canton of Geneva, CHE. In Proceedings of the 26th International Conference on World Wide Web, WWW '17, Perth, Australia, 3–7 April 2017; pp. 577–586. [CrossRef]
10. Wu, Y.; Cao, N.; Gotz, D.; Tan, Y.P.; Keim, D.A. A survey on visual analytics of social media data. *IEEE Trans. Multimed.* **2016**, *18*, 2135–2148. [CrossRef]
11. Lyu, Y.; Liu, X.; Chen, H.; Mangal, A.; Liu, K.; Chen, C.; Lim, B. OD Morphing: Balancing simplicity with faithfulness for OD bundling. *IEEE Trans. Vis. Comput. Graph.* **2019**, *26*, 811–821. [CrossRef] [PubMed]
12. Gansner, E.R.; Hu, Y.; North, S.C.; Scheidegger, C.E. Multilevel agglomerative edge bundling for visualizing large graphs. In Proceedings of the IEEE Pacific Visualization Symposium, Hong Kong, China, 1–4 March 2011; pp. 187–194.
13. Qu, H.; Zhou, H.; Wu, Y. Controllable and progressive edge clustering for large networks. In Proceedings of the International Symposium on Graph Drawing, Karlsruhe, Germany, 18–20 September 2006; Springer: Berlin/Heidelberg, Germany, 2006; pp. 399–404.
14. Holten, D. Hierarchical edge bundles: Visualization of adjacency relations in hierarchical data. *IEEE Trans. Vis. Comput. Graph.* **2006**, *12*, 741–748. [CrossRef] [PubMed]
15. Lhuillier, A.; Hurter, C.; Telea, A. State of the Art in Edge and Trail Bundling Techniques. *Comput. Graph. Forum* **2017**, *36*, 619–645. [CrossRef]
16. Zhang, J.; Liu, B.; Tang, J.; Chen, T.; Li, J. Social Influence Locality for Modeling Retweeting Behaviors. In Proceedings of the Twenty-Third International Joint Conference on Artificial Intelligence, IJCAI '13, Beijing, China, 3–9 August 2013; AAAI Press: Washington, DC, USA, 2013; pp. 2761–2767.
17. Goyal, A.; Bonchi, F.; Lakshmanan, L.V. Learning influence probabilities in social networks. In Proceedings of the Third ACM International Conference on Web Search and Data Mining, New York, NY, USA, 3–6 February 2010; pp. 241–250.
18. Falkner, S.; Klein, A.; Hutter, F. BOHB: Robust and efficient hyperparameter optimization at scale. In Proceedings of the International Conference on Machine Learning, PMLR, Stockholm, Sweden, 10–15 July 2018; pp. 1437–1446.
19. Hurter, C.; Ersoy, O.; Telea, A. Graph Bundling by Kernel Density Estimation. *Comput. Graph. Forum* **2012**, *31*, 865–874. [CrossRef]
20. Fruchterman, T.M.J.; Reingold, E.M. Graph drawing by force-directed placement. *Softw. Pract. Exp.* **1991**, *21*, 1129–1164. [CrossRef]
21. Sugiyama, K.; Tagawa, S.; Toda, M. Methods for visual understanding of hierarchical system structures. *IEEE Trans. Syst. Man Cybern.* **1981**, *11*, 109–125. [CrossRef]

big data and cognitive computing

MDPI

*Article*

# MM-EMOR: Multi-Modal Emotion Recognition of Social Media Using Concatenated Deep Learning Networks

**Omar Adel *, Karma M. Fathalla and Ahmed Abo ElFarag**

Department of Computer Engineering, Faculty of Engineering and Technology, Arab Academy for Science, Technology and Maritime Transport (AAST), Alexandria 1029, Egypt
* Correspondence: omar95adel@aast.edu; Tel.: +20-111-439-4765

**Abstract:** Emotion recognition is crucial in artificial intelligence, particularly in the domain of human–computer interaction. The ability to accurately discern and interpret emotions plays a critical role in helping machines to effectively decipher users' underlying intentions, allowing for a more streamlined interaction process that invariably translates into an elevated user experience. The recent increase in social media usage, as well as the availability of an immense amount of unstructured data, has resulted in a significant demand for the deployment of automated emotion recognition systems. Artificial intelligence (AI) techniques have emerged as a powerful solution to this pressing concern in this context. In particular, the incorporation of multimodal AI-driven approaches for emotion recognition has proven beneficial in capturing the intricate interplay of diverse human expression cues that manifest across multiple modalities. The current study aims to develop an effective multimodal emotion recognition system known as MM-EMOR in order to improve the efficacy of emotion recognition efforts focused on audio and text modalities. The use of Mel spectrogram features, Chromagram features, and the Mobilenet Convolutional Neural Network (CNN) for processing audio data are central to the operation of this system, while an attention-based Roberta model caters to the text modality. The methodology of this study is based on an exhaustive evaluation of this approach across three different datasets. Notably, the empirical findings show that MM-EMOR outperforms competing models across the same datasets. This performance boost is noticeable, with accuracy gains of an impressive 7% on one dataset and a substantial 8% on another. Most significantly, the observed increase in accuracy for the final dataset was an astounding 18%.

**Keywords:** classification; MobileNet; Roberta; multimodal; emotion; recognition; IEMOCAP; MELD; social media

## 1. Introduction

Correctly perceiving different people's emotions is a key factor in proper communication. Emotional understanding makes social networking more natural by eliminating ambiguity and helps interpret the conveyed messages effectively. Due to the importance and complexity of emotions in conversations, emotion recognition has become one of the vital fields of study applying artificial intelligence (AI) techniques. In addition, the ubiquitous use of social networking services created further demand for automated content and emotion analysis using machine learning, as it has become infeasible to analyze them otherwise [1].

Emotion recognition has a wide range of applications in various fields such as robotics, security, healthcare, automated identification, customer support call review, and lie detection. It also plays an essential role in improving human–computer interaction (HCI) [2]. It can help the machines to take user feedback to enhance the user experience. The diversity of the applications and the availability of big data volumes stipulate the significance of developing novel approaches for emotion recognition. Emotional expression can be verbal and non-verbal. The human perception of emotion involves capturing and analyzing facial expressions, speech text, and voice examination. Hence, automated emotion recognition can bridge the gap between humans and machines [3].

Speech recognition has gained a lot of focus over the last decades. It is an important research area for human-to-machine communication. Early methods focused on manual feature extraction and conventional techniques such as Gaussian mixture models (GMM) [4], and Hidden Markov models (HMM) [5]. More recently, neural networks such as recurrent neural networks (RNNs) [6], and convolutional neural networks (CNNs) [7] have been applied to speech recognition and have achieved great performance.

In addition, emotion extraction from text is of huge importance and is an uprising area of research in natural language processing. Recognition of emotions from text has high practical utilities for quality improvement like in the field of HCI. Most works on it have proposed solutions based on neural network techniques like LSTM [8], and CNN [9] with adequate results.

Despite the extensive development of unimodal learning models, they still cannot cover all the aspects of human interpretation of emotion. People present their emotions with various modes of expression. The combination of speech and text analysis carries the potential for improving emotion recognition [10]. Hence, in this study, a multimodal emotion (MM-EMOR) analysis system is proposed to examine audio and text data.

While the MM-EMOR system uses well-established technologies such as MobileNet for image data and ROBERTA for text data, we feel that its originality comes in the pragmatic merging of various modalities. This integration, while seemingly simple, is motivated by the desire for a more thorough knowledge of human emotions. Furthermore, our motivation goes beyond the technological aspects. We envision MM-EMOR as a versatile tool with far-reaching implications in disciplines such as mental health, human–computer interaction, and beyond. The capacity to effectively recognize and interpret emotions can improve our ability to communicate, empathize, and, eventually, improve the human experience in an increasingly digital environment.

The proposed MM-EMOR is a collaborative effort to process and synthesize information from a variety of modalities [11]. MM-EMOR uses a multimodal approach to bridge the gap between these modalities, ushering in a new era of sophisticated emotion recognition. The meticulously designed unimodal learning models are central to the architecture of MM-EMOR, each of which is tailored to perform well within its respective domain while still yielding competitive overall performance. Notably, the audio-based model stands out as a standout component, harnessing the subtle nuances of emotions via the complex interplay of Mel Spectrogram and Chromagram features. These audio representations give each emotion a distinct identity, adding depth to our understanding of emotional states. Furthermore, modalities are integrated using a seemingly simple yet effective concatenation technique that harmoniously blends the insights gained from audio and other modalities to achieve a robust multimodal emotion classification. The proposed MM-EMOR system stands as a testament to its efficacy in the grand tapestry of emotion recognition, as evidenced by its noteworthy performance that outperforms existing benchmarks in the field.

The rest of this paper is organized as follows. In Section 2, we clarify the related work which motivated and inspired the current study. In Section 3, we define the materials and methods used in this study. In Section 4, the results of MM-EMOR and its comparison with the state of the work are presented. Finally, conclusions are given in Section 5.

## 2. Related Works

Due to the importance of emotion recognition, many studies have been directed toward improving the performance of emotion recognition systems. The studies employed various deep learning methods and algorithms to distinguish human emotions [10,12,13]. Recently, Huddar et al. [14] proposed a cascaded approach for merging modalities, where textual features were extracted using CNN, audio features using the openSMILE toolkit [15], and visual features using 3D-CNN. They extracted the unimodal context using biLSTM and merged two modalities at a time to get the bimodal features. Similarly, biLSTM extracts the bimodal context and merges it to get the trimodal features. The adopted approach makes the complexity of the algorithm high. The IEMOCAP dataset [16] was used to train and test their model, with four classes (happy, angry, sadness, and neutral).

Kumar et al. [17] made a combination of three kinds of acoustic features (Mel spectrogram, MFCC, and chroma vectors) and trained it by RNN. For lexical features, they used the Bert model (bidirectional encoder representations from transformers), which is a transformer-based machine-learning technique for natural language processing. After that, they concatenated these features and used drop-out and dense layers to make the prediction. The IEMOCAP dataset was used to verify the performance of their model, with four classes (happy merged with excited, angry, sadness, and neutral).

Guo et al. [10] used BLSTM for text modality and audio modality, where the weights between different modalities needed to be considered to obtain a richer comprehensive emotional representation. As such, they used a weighted fusion layer. The IEMOCAP dataset was used with four classes (happy, angry, sadness, and neutral).

Singh et al. [18] extracted 33 audio features (16 prosody-based, 16 spectral-based, and one voice quality-based feature), in addition to the MFCC feature. For text, they used ELMo, which is a state-of-the-art NLP framework developed by AllenNLP and trained on the one Billion Word Benchmark. After being concatenated, they used the DNN model for prediction. To test their model performance, they used the IEMOCAP dataset with classes (happy merged with excited, angry merged with frustration, sadness, and neutral).

Wang et al. [19] proposed a new method called multimodal transformer augmented fusion (MTAF) for SER. MTAF uses a hybrid fusion strategy that combines feature-level fusion and model-level fusion. A model-fusion module composed of three cross-transformer encoders was proposed to generate a multimodal emotional representation for modal guidance and information fusion. Specifically, the multimodal features obtained by feature-level fusion and text features were used to enhance speech features. Experiments were implemented on the IEMOCAP dataset and the MELD dataset.

Zaidi et al. [20] proposed a multimodal dual attention transformer for cross-language speech emotion recognition. The multimodal dual attention transformer combined two attention mechanisms: one for the acoustic features of the speech signal called Roberta, and one for the textual features called wav2vec. The proposed approach was evaluated on the IEMOCAP dataset.

Canal et al. [21] provided a significant contribution to the field of facial expression analysis. This study investigated novel ways of facial expression identification using state-of-the-art machine learning algorithms and computer vision methodology. The research looked at the development of deep learning models, namely convolutional neural networks (CNNs), for extracting relevant characteristics from facial photos and accurately detecting and classifying a wide range of emotional states. In addition, it covered the use of these techniques in a variety of domains, highlighting their potential impact on fields such as human–computer interaction, affective computing, and emotion-aware systems. The insights and approaches described in this research are a significant resource for researchers and practitioners interested in advancing the subject of facial expression analysis and its practical applications.

The outlined related work shows promising results in emotion recognition, but some limitations exist due to the nature of the used datasets like multimodal integration challenges. Combining information from different modalities (e.g., text, audio, video) into a unified model can be complex and may introduce additional challenges [22]. This limitation needs a powerful model to overcome this. Our approach involves training Roberta, a highly performant natural language processing model, to extract textual features. Simultaneously, for audio data, we applied a feature extraction process to obtain a set of powerful multiple features from the acoustic domain. These features are then merged to capture complex relationships between text and audio data.

Another possible limitation is overfitting of emotion recognition models, particularly deep learning models, which may overfit the training data, leading to high training accuracy but poor generalization to new, unseen data [23]. We tried to overcome this by adding a dropout layer and regularizer to avoid overfitting in our model.

Another characteristic challenge of the problem is continuous emotions where some emotions are continuous and can vary in intensity, making it difficult to assign discrete labels [24]. We used the Roberta model which can overcome this. Therefore, we propose MM-EMOR an effective approach that can overcome these limitations.

## 3. Materials and Methods

MM-EMOR aims to analyze multimodal speech for emotion recognition. It merges textual and audio modalities for this task. The proposed system's architecture is described in Figure 1, with textual- and audio-based emotion recognition modules. The system is explained in detail in the following sections.

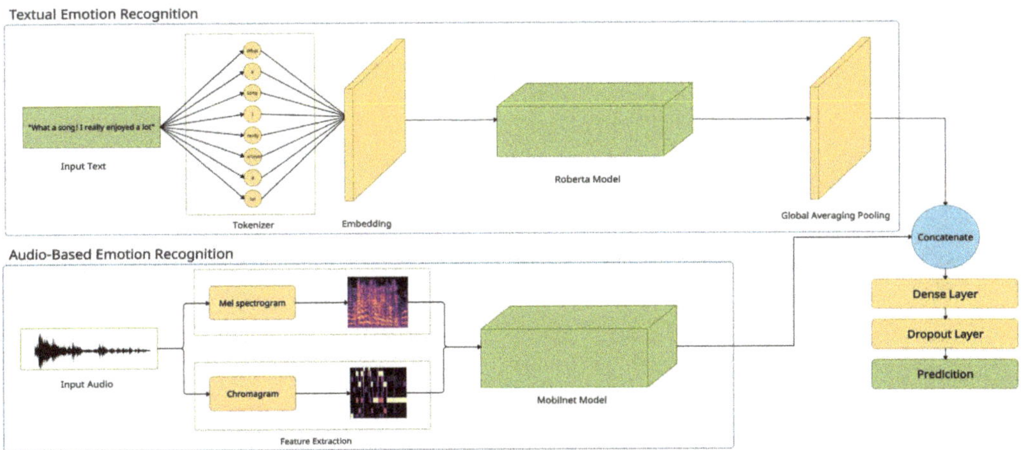


**Figure 1.** The proposed MM-EMOR system.

### 3.1. *Textual Emotion Recognition*

Roberta

In this part, the textual emotion recognition model is described. The core pivotal component of this module is the Roberta model. Facebook AI Research (FAIR) unveiled the robust language model known as Roberta in 2019 [25].

The Bidirectional Encoder Representations from the Transformers (BERT) model [26], another well-known NLP model, is referenced in the model's name since it incorporates many of BERT's advantages. Its bidirectional context allows it to comprehend text completely, and it employs transfer learning to rapidly adapt to varied NLP tasks. BERT's contextual embeddings efficiently capture nuances, and its large-scale architecture allows it to represent long-term dependencies.

To improve its functionality and sturdiness, Roberta is an improved version of the BERT model that has several advantages. It employs a better training methodology, a larger dataset, and data augmentation techniques. Roberta eliminates the next sentence prediction (NSP) assignment from BERT's training procedure and concentrates on successfully predicting masked tokens. Typically, the model is trained on a larger dataset, resulting in stronger contextual embeddings and increased performance on downstream NLP tasks. Roberta requires less fine-tuning and is more generalizable across domains. Its repeatability and consistent outstanding performance has made it a preferred choice for a wide range of NLP applications.

The training strategy is one significant difference between Roberta and BERT. BERT was trained with a masked language modeling (MLM) objective, whereas Roberta uses a version known as "dynamically masked language modeling" (DMLM). Instead of randomly masking words during training, DMLM trains the model on a vast volume of unlabeled text with no masks. This allows for improved generalization by exploiting the entire context of the text.

Roberta also has a larger training corpus than BERT, which includes both in-domain and out-of-domain data. It is trained using large amounts of text data from books, websites, and Wikipedia. Due to the thorough pretraining, the model can learn a wide range of linguistic patterns and semantic correlations. We used the Roberta-Base model which has 12 layers of the transformer model as shown in Figure 2. We selected a token length of 1000 due to the inherent characteristics of our datasets, characterized by the presence of lengthy text sentences. In addition, longer token lengths can lead to longer training times, so there is often a trade-off between training duration and model performance. A token length of 1000 tokens might strike a reasonable balance.

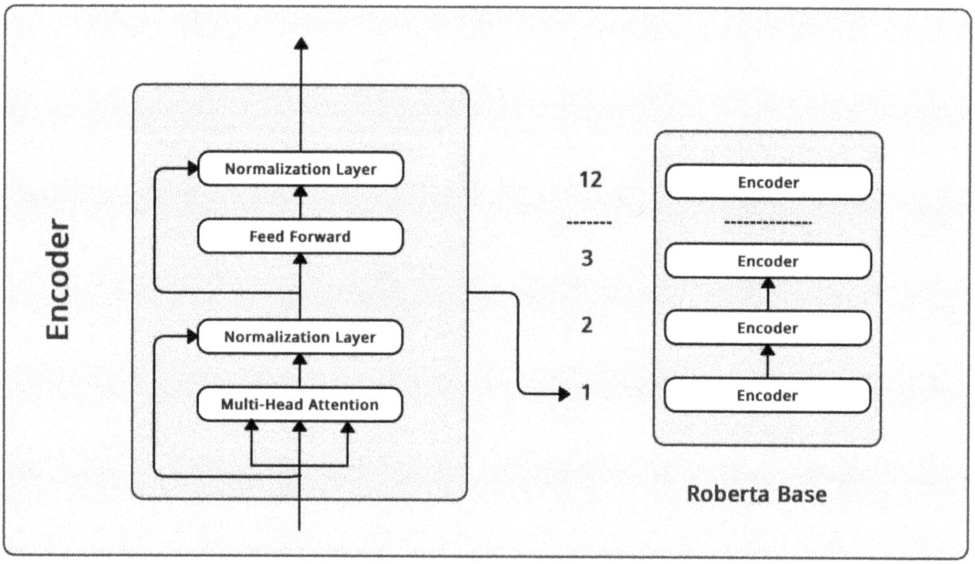


**Figure 2.** Roberta model architecture.

### 3.2. *Audio-Based Emotion Recognition*

#### 3.2.1. Feature Extraction

Audio signals are dynamic and multidimensional; they are difficult to analyze and comprehend. They are comprised of temporal, spectral, and frequently harmonic data. Variations in pitch, timbre, rhythm, and dynamics, as well as the existence of background noise and transitory occurrences, add to the complications. Audio signals can also transmit emotions, intentions, and cultural nuances, adding another degree of complication to their interpretation. Two main feature representations were emphasized to capture different features of these signals: spectrograms and chromagrams.

Mel spectrogram features are extracted from the audio signals. The extracted features are well suited to the task of emotion recognition [27,28]. Each frame of the spectrum in the Mel spectrum contains a short-time Fourier transform (STFT), which maps the signal from a linear frequency scale to a logarithmic Mel scale. After that, it is applied to the filter bank to produce the eigenvector, whose eigenvalues can be roughly described as the distribution of signal energy on the Mel-scale frequency. To extract audio features, we used Librosa [29] to extract the Mel spectrogram features, which were converted into a 2D image.

To train the convolutional neural networks (CNN) for recognition, the generated Mel spectrogram-based 2D images were input to the network. In Figure 3, we show some examples of Mel-spectrogram images of various emotions. As can be seen from the figure below, there are evident differences between various types of emotions. After representing the audio data as an image, MobileNet CNN was used to model the data and perform the emotion classification task.

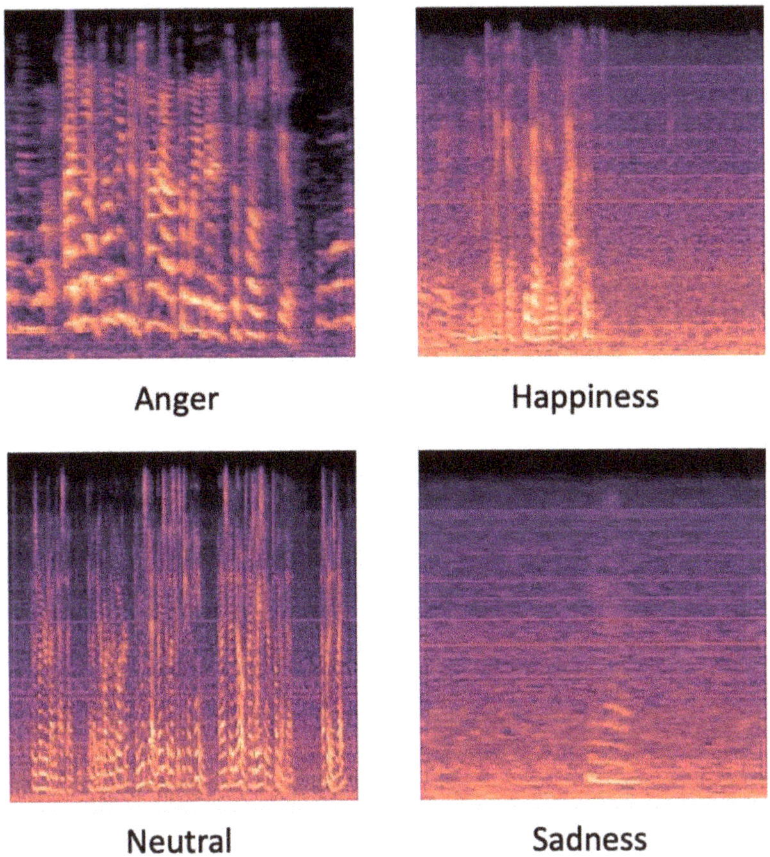


**Figure 3.** Differences between classes Mel spectrogram.

The chromagram is a characteristic that is extensively utilized in audio signal processing, including AI and machine learning applications. It depicts the distribution of energy in different frequency bands or pitches in an audio signal over time.

A chromagram is a two-dimensional representation of an audio signal in which time is represented horizontally and frequency bands are shown vertically. It is frequently estimated using the short-time Fourier transform (STFT) approach, which analyzes an audio signal's spectrum over short overlapping time periods.

The chromagram's vertical axis is often quantized into a given number of bins, each representing a specific pitch or frequency range.

To generate a chromagram, an audio signal is separated into frames or windows, and the frequency content of each frame is analyzed using the STFT. The generated spectrum is then mapped to the correct chroma bins by grouping frequencies in the same pitch.

In AI and machine learning problems such as audio classification, chromagrams are frequently employed as a feature representation. Chromagrams can capture the underlying musical structure, tonality, and chord progressions in an audio signal by describing it in terms of its harmonic content.

Once computed, the chromagram can be utilized as an input feature for machine learning methods such as convolutional neural networks (CNNs) as an image. These models can learn to recognize patterns and extract meaningful information from chromagram representations.

Figure 4 shows the difference between emotions. We used Librosa to extract chromagram features and merged them with the Mel-spectrogram to use it to train the MobileNet CNN model.

**Figure 4.** Differences between classes chromagram.

3.2.2. MobileNet

For the purpose of audio classification, we opted for a 2D image-based approach over 1D audio-based classification due to several compelling reasons. Firstly, 2D image-based classification is useful when working with audio data that can be converted into spectrograms. Spectrograms provide a comprehensive representation of essential temporal and frequency domain information, making them ideal for image-based classification applications. This approach is commonly used in speech recognition and sound analysis applications. Secondly, using 2D image-based techniques has the great advantage of utilizing pre-trained image classification models, particularly convolutional neural networks (CNNs). This not only saves time and computing resources but also takes advantage of the amount of knowledge contained inside well-established image models. It is important to note that Solovyev et al. [30] conducted a comparative study and similarly advocated for the 2D image-based approach, achieving better results in comparison to 1D audio-based methods. We chose to use mobilNet model for audio classification.

MobileNet was first introduced in 2017 by Howard et al. [31], MobileNet is an efficient and portable CNN architecture, which is a sub-category of neural networks and is currently one of the most efficient image classification models. MobileNet is used to build lighter models by using depth-wise separable convolutions in place of the standard convolutions used in earlier architectures. MobileNet introduces two new global hyperparameters (width multiplier and resolution multiplier) that allow model developers to trade off latency or accuracy for speed and small size depending on their requirements.

The MobileNet model is primarily based on depth-wise separable convolutions that are a shape of standard convolution right into a depth-wise convolution, which is a type of convolution where we apply a single convolutional filter for each input channel and a $1 \times 1$ convolution known as a pointwise convolution. The pointwise convolution then applies a $1 \times 1$ convolution to combine the outputs of the depth-wise convolution as shown in Figure 5.

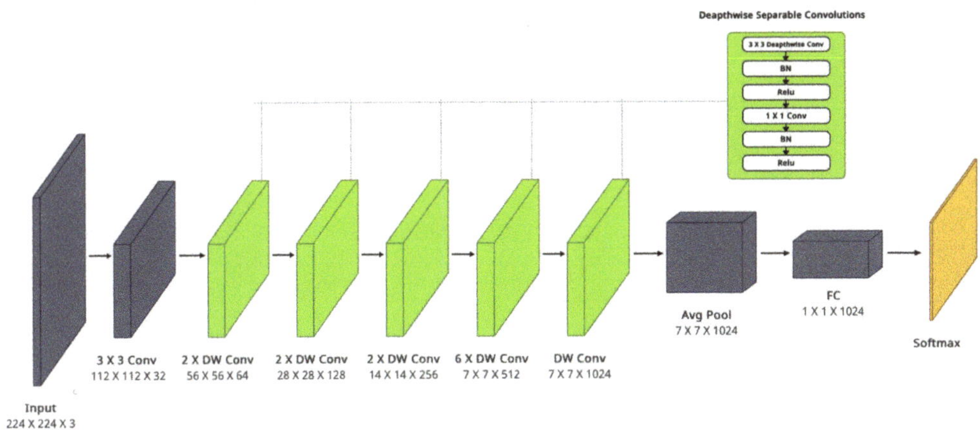


**Figure 5.** MobileNet architecture.

A standard convolutional layer takes as input a $D_F \times D_F \times M$ feature map F and produces a $D_F \times D_F \times N$ feature map $G$ where:

- $D_F$ is the spatial width and height of a square input feature map.
- $M$ is the number of input channels (input depth).
- $D_G$ is the spatial width and height of a square output feature map.
- $N$ is the number of output channels (output depth).
- The standard convolutional layer is parameterized by convolution kernel $K$ of size $D_K \times D_K \times M \times N$ where:
- $D_K$ is the spatial dimension of the kernel assumed to be square.
- $M$ is the number of input channels.
- $N$ is the number of output channels as defined previously.

Standard convolutions have the computational cost of $D_K \cdot D_K \cdot M \cdot N \cdot D_F \cdot D_F$

Depthwise convolution has a computational cost of $D_K \cdot D_K \cdot M \cdot D_F \cdot D_F$

The combination of depthwise convolution and $1 \times 1$ (pointwise) convolution is called depthwise separable convolution, which was originally introduced in [32].

Depthwise separable convolution costs are $D_K \cdot D_K \cdot M \cdot D_F \cdot D_F + M \cdot N \cdot D_F \cdot D_F$ which is the sum of the depthwise and $1 \times 1$ pointwise convolutions.

By expressing convolution as a two-step process of filtering and combining, we get a reduction in the computation of:

$$\frac{D_K \cdot D_K \cdot M \cdot D_F \cdot D_F + M \cdot N \cdot D_F \cdot D_F}{D_K \cdot D_K \cdot M \cdot N \cdot D_F \cdot D_F} = \frac{1}{N} + \frac{1}{D_K^2} \quad (1)$$

MobileNet uses $3 \times 3$ depthwise separable convolutions which use between 8 to 9 times less computation than standard convolutions.

*3.3. Concatenation and Prediction*

After taking the output from the Roberta model, we used a global averaging pooling 1D layer to preprocess the data to concatenate with the MobileNet model. After concatenation, we used a dense layer with a ReLU activation function.

We also added a regularizer to avoid overfitting in our model. regularizers are techniques used to prevent overfitting, which occurs when a model fits the training data too closely and fails to generalize well to new, unseen data. Regularization helps improve a model's ability to generalize by adding a penalty term to the loss function, discouraging overly complex or extreme parameter values. This makes the model more robust and less prone to overfitting.

After this, we added a dropout layer with a drop probability of 0.3 to avoid overfitting. The outputs of the dense and dropout layers were finally used to predict the probabilities of the emotion classes. The proposed model was trained using the Adam optimizer [33] and relies on categorical cross-entropy for the loss metric.

## 4. Results and Discussion

In this section, a detailed evaluation of MM-EMOR performance on different datasets is performed.

*4.1. Dataset*

In this study, publicly available well-studied datasets were chosen to test the performance of the presented multimodal approach, namely Tweeteval [34], Multimodal Emotion Lines Dataset (MELD) [35], and the Interactive Emotional Dyadic Motion Capture (IEMOCAP) [16]. The chosen datasets allow comparison with state-of-the-art performance.

4.1.1. Tweeteval Dataset

Tweeteval [34] is used to analyze people's emotions on social networks and determine the success of the proposed model to recognize emotions from textual data. Tweeteval [34] consists of seven heterogeneous tasks in Twitter, all framed as multi-class tweet classifications. The tasks include irony, hate, offensive, stance, emoji, emotion, and sentiment. All tasks have been unified into the same benchmark, with each dataset presented in the same format and with fixed training, validation, and test split. The number of labels and instances in training, validation, and test sets for each task is shown in Table 1.

**Table 1.** Tweeteval dataset stratification for the used classes.

| Task | Number of Classes | Train | Val | Test |
|---|---|---|---|---|
| Emotion | 4 | 3257 | 374 | 1421 |
| Hate | 2 | 9000 | 1000 | 2970 |
| Irony | 2 | 2862 | 955 | 784 |
| Offensive | 2 | 11,916 | 1324 | 860 |
| Sentiment | 3 | 45,389 | 2000 | 11,906 |
| Stance | 3 | 2620 | 294 | 1249 |
| Emoji | 20 | 45,000 | 5000 | 50,000 |

#### 4.1.2. IEMOCAP Dataset

The Interactive Emotional Dyadic Motion Capture (IEMOCAP) database [16] is an acted multimodal and multi-speaker database, recently collected at the SAIL lab at the University of Sothern California (USC). It contains approximately 12 h of audiovisual data, including video, speech, motion capture of face, and text transcriptions. Our experiments will focus on speech and text transcriptions modalities. The dataset contains 9 classes namely anger, happiness, excitement, sadness, frustration, fear, surprise, other, and neutral.

The most commonly used classes in the literature are anger, happiness, sadness, and neutral. Hence, they will be used in our experiments. In addition, due to the small number of happiness class records, some of the related studies merged happiness with excitement. As such, we will be reporting the merged results. The merging of emotion classes has been done based on Plutchik's wheel of emotions [36]. The number of records in each class is shown in Table 2.

**Table 2.** IEMOCAP dataset stratification for the used classes.

| Classes | Happiness | Anger | Sadness | Neutral |
|---|---|---|---|---|
| Happiness, anger, sadness, and neutral | 595 | 1103 | 1084 | 1708 |
| Happiness + excitement, anger, sadness, and neutral | 1636 | 1103 | 1084 | 1708 |

#### 4.1.3. MELD Dataset

The Multimodal Emotion Lines Dataset (MELD) [35] has more than 1400 dialogues and 13,000 utterances from the Friends TV series. Multiple speakers participated in the dialogues. Each utterance in dialogue has been labeled by any of these seven emotions: anger, disgust, sadness, joy, neutral, surprise, and fear. MELD also has sentiment (positive, negative, and neutral) annotations for each utterance. The number of records in each class is shown in Table 3.

**Table 3.** MELD dataset stratification for the used classes.

| Classes | Surprise | Neutral | Fear | Joy | Sadness | Disgust | Anger |
|---|---|---|---|---|---|---|---|
| Train | 1205 | 4710 | 268 | 1743 | 683 | 271 | 1109 |
| Test | 281 | 1256 | 50 | 402 | 208 | 68 | 345 |
| Validation | 150 | 470 | 40 | 163 | 111 | 22 | 153 |

The proposed model was evaluated on these datasets. For all training and testing purposes, networks from the Keras library for Python were implemented. For IEMOCAP, tests were performed using 5-fold cross-validation, for Tweeteval and MELD, we used test records on the dataset.

### *4.2. Performance Measures*

The performance of MM-EMOR was evaluated using the following evaluation metrics: accuracy, precision, recall, and F1-score. The expression of these metrics is given as follows:

- Unweighted accuracy is just the proportion of correctly predicted observations to all observations. $\frac{TP+TN}{TP+FP+TN+FN}$
- Weighted accuracy takes into account the class-specific accuracy and assigns different weights to each class based on their importance or prevalence in the dataset. $\sum(W_i \times acc_i)/\sum W_i$
- Recall is a metric for how well a model detects true positives. $\frac{TP}{TP+FN}$
- Precision is the ratio of accurately anticipated positive observations to all actual class observations. $\frac{TP}{TP+FP}$
- F1-score is the weighted average of precision and recall, weighted. $2\frac{Precision \times Recall}{Precision+Recall}$

### 4.3. *Performance Evaluation*

The training of the models takes a maximum of 30 epochs. The results are reported as the best mode reached.

#### 4.3.1. IEMOCAP Dataset

Figure 6 and Table 4 show the confusion matrix and the results of (anger, happiness, neutral, and sadness) classes on the IEMOCAP dataset with our performance metrics. Each metric for a given modality generates different values for every emotion. The anger emotion has the best value of precision of 90%. For recall, neutral has the best value of 82%. Otherwise, in the F1-score anger has the best value of 85%.

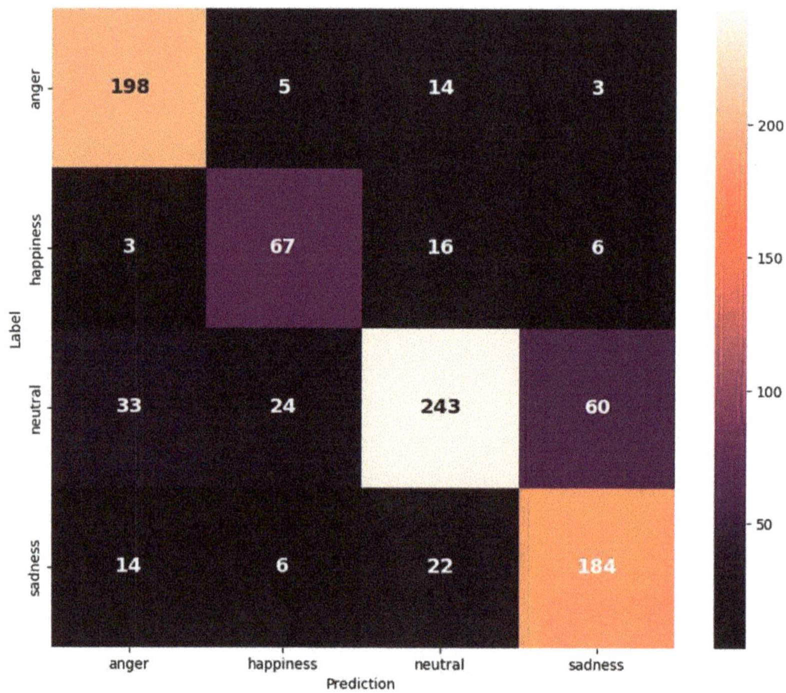


**Figure 6.** Confusion matrix for IEMOCAP (anger, happiness, neutral, and sadness) on the text and audio modalities.

**Table 4.** Classification report results for the IEMOCAP (anger, happiness, neutral, and sadness) on the text and audio modalities in the function of (precision, recall, and F1-score).

| Classes | Precision | Recall | F1-score |
|---|---|---|---|
| Anger | 0.90 | 0.80 | 0.85 |
| Happiness | 0.73 | 0.66 | 0.69 |
| Neutral | 0.68 | 0.82 | 0.74 |
| Sadness | 0.81 | 0.73 | 0.77 |

Figure 7 and Table 5 show the confusion matrix and the results of (anger, happiness merged with excitement, neutral, and sadness) classes on the IEMOCAP dataset. Happiness and sadness emotions have the same value of precision of 84%. Recall anger has the best value of 82%. In the F1-score, anger also has the best value of 82%.

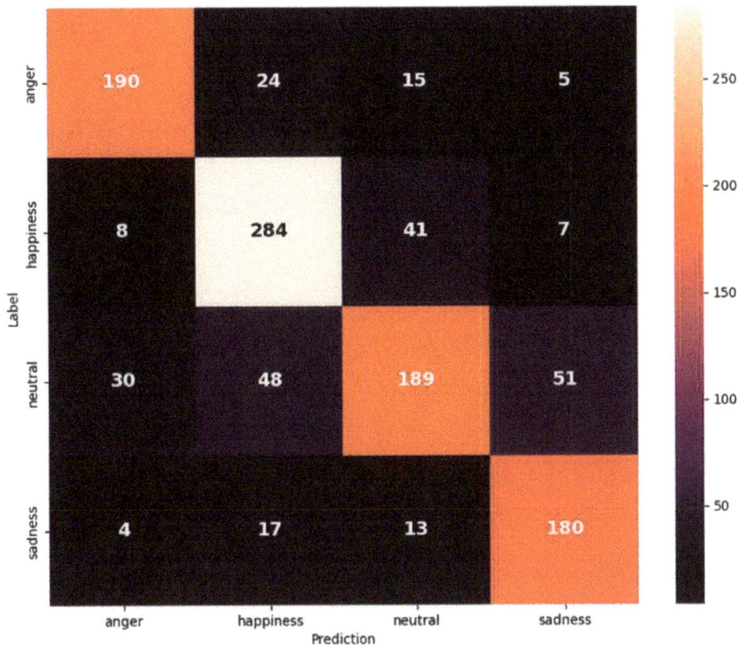


**Figure 7.** Confusion matrix of IEMOCAP (anger, happiness merged with excitement, neutral, and sadness) on the text and audio modalities.

**Table 5.** Classification report results for the IEMOCAP (anger, happiness merged with excitement, neutral, and sadness) on the text and audio modalities in the function of (precision, recall, and F1-score).

| Classes | Precision | Recall | F1-score |
|---|---|---|---|
| Anger | 0.81 | 0.82 | 0.82 |
| Happiness | 0.84 | 0.76 | 0.80 |
| Neutral | 0.59 | 0.73 | 0.66 |
| Sadness | 0.84 | 0.74 | 0.79 |

According to Table 6, the model has an accuracy of 77.06% in (happiness, anger, sadness, and neutral) classes, otherwise it has an accuracy of 76.22% in (happiness merged with excitement, anger, sadness, and neutral) classes.

**Table 6.** Accuracy results of the IEMOCAP dataset.

| Classes | UA | WA |
|---|---|---|
| Happiness, anger, sadness, and neutral | 0.7706 | 0.7792 |
| Happiness + excitement, anger, sadness, and neutral | 0.7622 | 0.7705 |

In Figure 8 the training accuracy versus epochs of the IEMOCAP dataset in using happiness, anger, sadness, and neutral classes is visualized. In Figure 9, the training accuracy versus epochs is illustrated, but for happiness and excitement, anger, sadness, and neutral classes. The figures elucidate improving accuracy with the increasing number of epochs, indicating the potential of the approach to model the data.

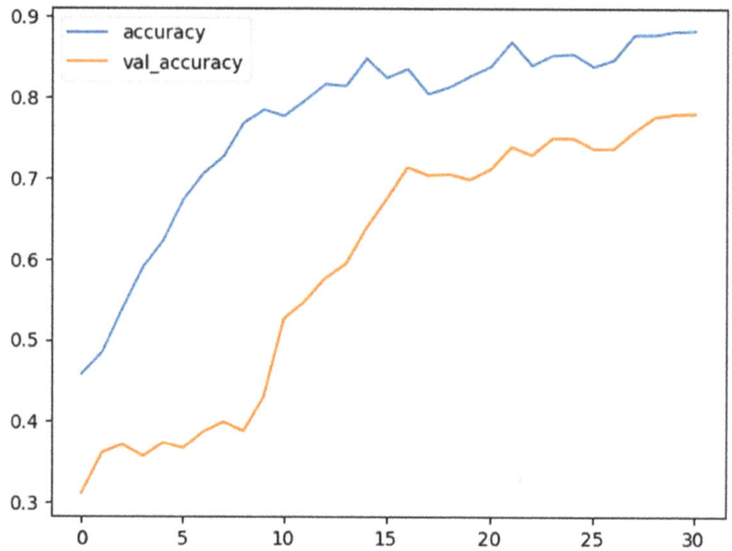


**Figure 8.** Training and validation accuracy versus epochs in happiness, anger, sadness, and neutral classes of IEMOCAP.

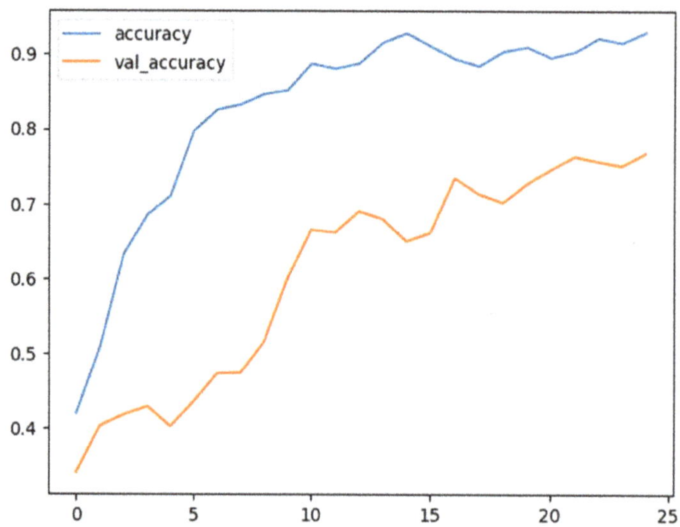


**Figure 9.** Training and validation accuracy versus epochs in happiness merged with excitement, anger, sadness, and neutral classes of IEMOCAP.

#### 4.3.2. MELD Dataset

For the MELD dataset, Figure 10 and Table 7 show its confusion matrix and results. Neutral emotion has the best value of precision of 92%. Recall sadness has the best value of 92%. Otherwise, the F1-score for neutral has the best value of 74%, and the model has an accuracy of 63.33%. Figure 11 illustrates the training accuracy versus epochs on the MELD dataset, where the validation accuracy continues to rise with the training epochs.

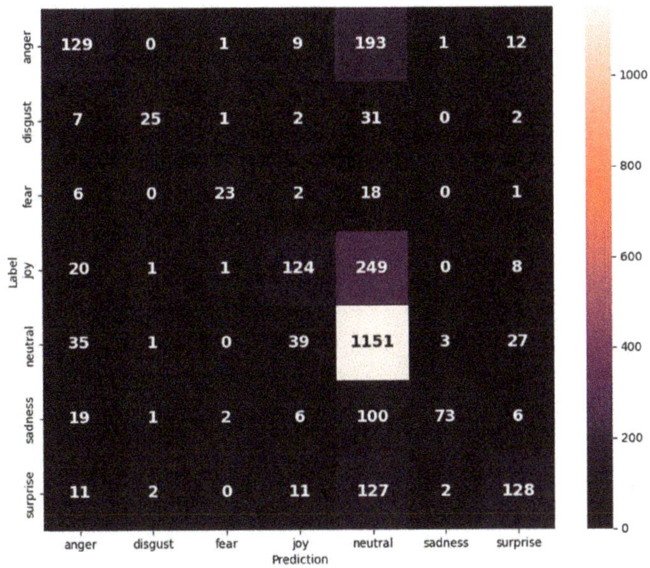


**Figure 10.** Confusion matrix of the MELD dataset.

**Table 7.** Classification report results for the MELD on the text and audio modalities in the function of precision, recall, and F1-score and accuracy of the model.

| Classes | Precision | Recall | F1-score | Accuracy |
|---|---|---|---|---|
| Anger | 0.37 | 0.57 | 0.45 | |
| Disgust | 0.37 | 0.83 | 0.51 | |
| Fear | 0.46 | 0.82 | 0.59 | |
| Joy | 0.31 | 0.64 | 0.42 | 0.6333 |
| Neutral | 0.92 | 0.62 | 0.74 | |
| Sadness | 0.35 | 0.92 | 0.51 | |
| Surprise | 0.46 | 0.70 | 0.55 | |

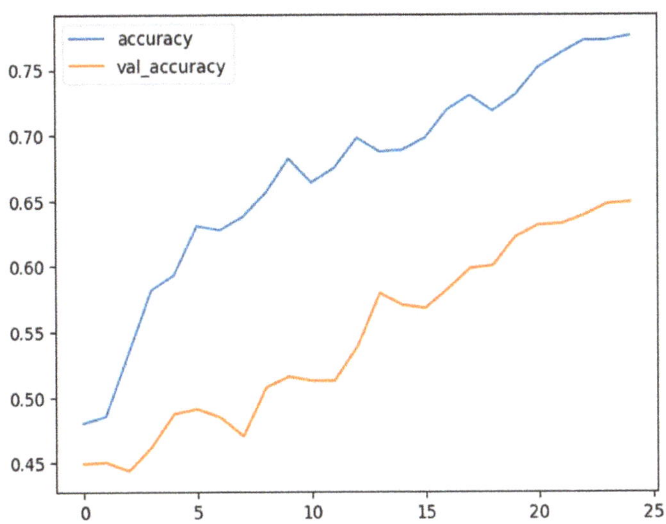


**Figure 11.** Training and validation accuracy versus epochs in MELD.

#### 4.3.3. Tweeteval Dataset

For the Tweeteval dataset, Figure 12 shows the confusion matrix for each task in the dataset. (a) Refers to emotion which includes four classes (anger, joy, optimism, and sadness); (b) hate which includes two classes (non-hate and hate); (c) irony also has two classes (non-irony and irony); (d) offensive includes two classes (non-offensive and offensive); (e) sentiment which includes three classes (negative, neutral, and positive); (f) stance also has three classes (none, against, and favor); and (g) emoji which includes twenty classes (red heart, smiling face with heart eyes, face with tears of joy, two hearts, fire, smiling face with smiling eyes, smiling face with sunglasses, sparkles, blue heart, face blowing kiss, camera, united states, sun, purple heart, winking face, hundred points, beaming face with smiling eyes, Christmas tree, camera with flash, and winking face with tongue).

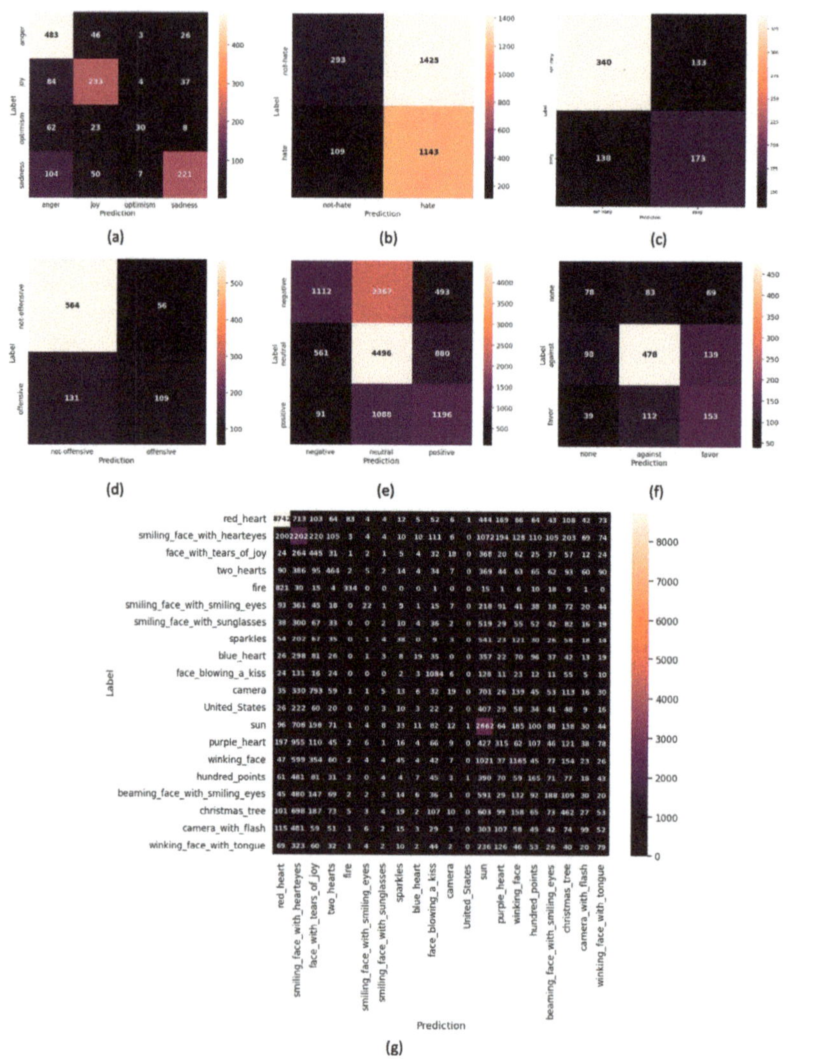


**Figure 12.** Confusion matrix for (**a**) emotion, (**b**) hate, (**c**) irony, (**d**) offensive, (**e**) sentiment, (**f**) stance, and (**g**) emoji in the Tweeteval dataset.

Table 8 shows the classification report for each task in the Tweeteval dataset. The offensive has the best average precision of 68%, 74% for average recall, and 70% for average F1 score. For accuracy, offensive task also has the best value of 78.26%. All results depend on text-only modality due to the data available in the Tweeteval dataset.

**Table 8.** Classification report results for the Tweeteval on the text and audio modalities in the function of precision, recall, and F1-score and accuracy of the model.

| Task | Average Precision | Average Recall | Average F1 Score | Accuracy |
|---|---|---|---|---|
| Emotion | 0.59 | 0.69 | 0.61 | 0.6806 |
| Hate | 0.54 | 0.59 | 0.44 | 0.4835 |
| Irony | 0.64 | 0.64 | 0.64 | 0.6543 |
| Offensive | 0.68 | 0.74 | 0.7 | 0.7826 |
| Sentiment | 0.51 | 0.56 | 0.51 | 0.5539 |
| Stance | 0.5 | 0.5 | 0.5 | 0.5677 |
| Emoji | 0.22 | 0.27 | 0.21 | 0.3701 |

In Figure 13, we visualize the training accuracy versus epochs of the Tweeteval dataset in each task. (a) Refers to emotion, (b) to hate, (c) to irony, (d) to offensive, (e) to sentiment, (f) to stance, and (g) to emoji.

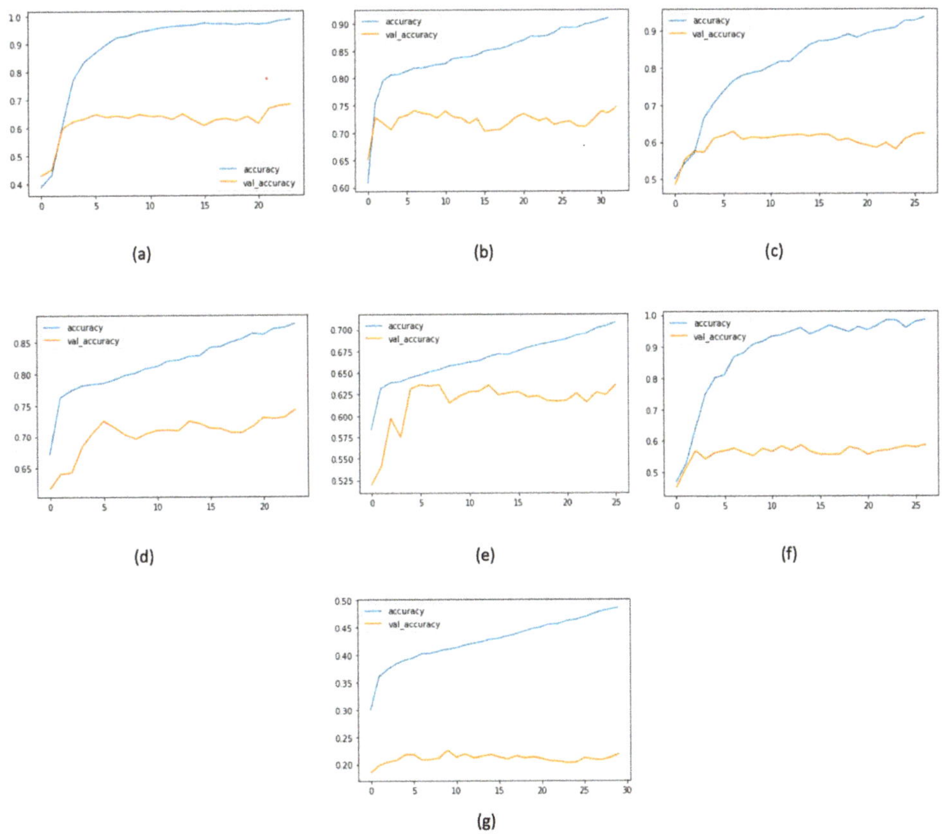


**Figure 13.** Training and validation accuracy versus epochs in (**a**) emotion, (**b**) hate, (**c**) irony, (**d**) offensive, (**e**) sentiment, (**f**) stance, and (**g**) emoji tasks in the Tweeteval dataset.

### 4.4. Comparison with State-of-the-Art Methods

We conducted an insightful investigation into the effectiveness of different feature representations within our model. We encountered a challenge in this regard due to the necessity of converting audio data into an image format suitable for training with the MobileNet model. This conversion was essential to align with the model architecture's requirements. As a result, we were unable to freeze the entire feature extraction process as it was integral to this conversion. As such, we examined the Mel spectrogram feature in isolation and compared it to the chromagram feature, both of which are commonly used representations for audio analysis. And presented in state-of-the-art comparison tables as Ours (Mel spectrogram) and Ours (chromagram). Our experimentation unveiled interesting findings. While both representations yielded promising results, the Mel spectrogram exhibited a slight advantage in terms of performance, signifying its greater discriminatory power in capturing essential acoustic characteristics. Intriguingly, when we merged these two features within our model, the outcomes were notably enhanced. This merging harnessed the complementary strengths of the Mel spectrogram and chromagram, resulting in a richer and more detailed feature set for training. Our findings highlight the significance of feature selection in optimizing model performance for audio-based tasks, emphasizing the benefits of leveraging a combination of feature representations to enhance the discriminative capacity of our model.

To further verify the competitiveness of MM-EMOR, it was compared to state-of-the-art methods. Our approach achieved higher accuracy, as shown in Table 9, for anger, happiness, neutral, and sadness classes in the IEMOCAP dataset. The increase in accuracy was up to 7%. Table 10 also shows that our approach achieved higher accuracy, up to 7%, for anger, happiness merged with excitement, neutral, and sadness classes. The consistent performance of MM-EMOR across different scenarios proves the strength and robustness of our multimodal emotion recognition system.

**Table 9.** Comparison of our proposed approach with state-of-the-art (anger, happiness, neutral, and sadness) classes on the IEMOCAP dataset.

| Methods | UA | WA |
|---|---|---|
| H.Xu et al. [12] | 0.709 | 0.725 |
| Guo et al. [10] | 0.725 | 0.719 |
| Zaidi et al. [20] | 0.756 | - |
| Ours (Chromagram) | 0.761 | 0.769 |
| Ours (Mel spectrogram) | 0.766 | 0.775 |
| Huddar et al. [14] | 0.766 | - |
| Wang et al. [37] | **0.77** | 0.765 |
| Ours | **0.771** | **0.779** |

**Table 10.** Comparison of our proposed approach with state-of-the-art (anger, happiness merged with excitement, neutral, and sadness) classes on the IEMOCAP dataset.

| Methods | UA | WA |
|---|---|---|
| S. Sahoo et al. [38] | 0.687 | - |
| H. Feng et al. [39] | 0.697 | 0.686 |
| S. Tripathi et al. [13] | 0.697 | - |
| Ours (Chromagram) | 0.744 | 0.753 |
| Kumar et al. [17] | 0.750 | 0.717 |
| Ours (Mel spectrogram) | 0.759 | 0.767 |
| Setyono et al. [40] | **0.76** | 0.76 |
| Ours | **0.762** | **0.771** |

We also compared our proposed approach with other state-of-the-art methods for the MELD dataset in Table 11. MM-EMOR offered higher accuracy with a difference ranging from 0.1% to 8.5%.

**Table 11.** Comparison of our proposed approach with state-of-the-art MELD dataset.

| Methods | Accuracy |
|---|---|
| Wang et al. [19] | 0.481 |
| Guo et al. [10] | 0.548 |
| Wang et al. [41] | 0.558 |
| Ours (Chromagram) | 0.588 |
| Ours (Mel spectrogram) | 0.592 |
| Lian et al. [42] | 0.62 |
| Ho et al. [43] | 0.632 |
| Ours | **0.633** |

For the Tweeteval dataset, we compared our model with other state-of-the-art methods. The results in Table 12 show that our approach achieved higher accuracy in emotion, emoji, and offensive tasks. The best UA improvement was obtained with the emoji task with an outstanding increase of 17.94%, followed by the offensive task scoring a 4.5% improvement. A similar performance was attained for the irony task. However, a performance gap exists for the hate, sentiment, and stance tasks. To this end, further tuning needs to be applied to enhance the performance.

**Table 12.** Comparison of our proposed approach with state-of-the-art Tweeteval dataset.

| Methods | Emotion | Hate | Irony | Offensive | Sentiment | Stance | Emoji |
|---|---|---|---|---|---|---|---|
| Li et al. [1] | 0.6770 | **0.5950** | **0.6667** | 0.7371 | **0.6143** | **0.6756** | 0.1907 |
| Ours | **0.6806** | 0.4835 | 0.6543 | **0.7826** | 0.5539 | 0.5677 | **0.3701** |

The state-of-the-art comparison shows that the IEMOCAP dataset achieves great results compared to other models, and Tweeteval too. The MELD dataset also achieved great performance compared to others but, in general, it achieved low accuracy because of the data collected from TV-shows which includes audience laughing and cinematic processing, so the data is not clear. In the case of Tweeteval, the observed accuracy is slightly lower, which can be attributed to the nature of social media data, known for its prevalence of human errors and incorrect text. This observation can be attributed to the inherent characteristics of the dataset at hand, which is exclusively comprised of text data and thus qualifies as an unimodal dataset. Unimodal datasets inherently possess a certain degree of simplicity in comparison to their multimodal counterparts, where the fusion of diverse data modalities can introduce added complexity.

A noteworthy finding concerns the length of the training process across the datasets. The training and feature extraction duration on the IEMOCAP dataset was 242 min, indicating the comprehensive nature of the model's learning process. On the MELD dataset, a similar pattern emerged, with the training phase lasting 325 min. Notably, the Tweeteval dataset deviated from this pattern, necessitating a 49-min training time. Furthermore, the execution time required to process a single instance is an important metric of efficiency. This metric was low at 19.9, 16.9, and 0.51 milliseconds for IEMOCAP, MELD, and Tweeteval, respectively, thus highlighting the computational efficiency that characterizes the MM-EMOR system's real-time functionality. To train our model, we used the capabilities of cloud computing infrastructure. The hardware requirements used in this cloud-based environment were Nvidia V100 GPU, which is well-known for its performance in deep learning applications, to expedite model training, and 52 gigabytes of RAM, which provided adequate memory resources to meet the complex needs of our training datasets. We used an 8-core CPU to support these computational operations.

## 5. Conclusions

In conclusion, the MM-EMOR system emerges as a promising forerunner of advanced emotion recognition capabilities in the context of multimodal data processing. The ingenious concatenation of preprocessed audio data, harnessed via the Mobilenet Convolutional Neural Network, with textual insights gleaned from the Roberta model is a key component of this system. This approach has been thoroughly examined and validated across three distinct benchmark datasets, each representing a distinct aspect of the emotion recognition landscape. The interactive emotional dyadic motion capture (IEMOCAP) dataset, the multimodal emotion lines dataset (MELD), and the Tweeteval dataset are among these. Notably, the last dataset only contains textual modality but includes seven experimental scenarios, showing the MM-EMOR system's versatility. MM-EMOR consistently outperformed its state-of-the-art counterparts by a significant margin across this broad spectrum of evaluation. In the case of the IEMOCAP dataset, MM-EMOR achieves improved accuracy, with gains ranging from 0.1% to an impressive 7%. Similarly, the MELD dataset sees an increase in accuracy ranging from 0.1% to 8.5%. The most impressive progress is seen in the Tweeteval dataset, where MM-EMOR achieves an astonishing 18% increase in accuracy. These findings highlight the approach's profound potential in the domain of social media analysis, implying that it should be expanded to include facial gesture recognition within videos, a path that promises to usher in a more comprehensive understanding of human emotions.

**Author Contributions:** Formal analysis, O.A. and K.M.F.; conceptualization, O.A. and K.M.F.; methodology, O.A., K.M.F. and A.A.E.; software, O.A.; validation, O.A.; investigation, O.A.; resources, O.A.; data curation, O.A. and K.M.F.; writing—original draft preparation, O.A.; writing—review and editing, K.M.F. and A.A.E.; visualization, O.A., K.M.F. and A.A.E.; supervision, K.M.F. and A.A.E.; project administration, K.M.F. and A.A.E. All authors have read and agreed to the published version of the manuscript.

**Funding:** This research received no external funding.

**Data Availability Statement:** We used the IEMOCAP dataset available at: https://www.kaggle.com/datasets/jamaliasultanajisha/iemocap-full accessed on 23 September 2022, MELD dataset available at: https://affective-meld.github.io/ accessed on 26 September 2022, and Tweeteval available at: https://github.com/cardiffnlp/tweeteval accessed on 8 December 2022.

**Conflicts of Interest:** The authors declare no conflict of interest.

## References

1. Li, J.; Mishra, S.; El-Kishky, A.; Mehta, S.; Kulkarni, V. NTULM: Enriching social media text representations with non-textual units. *arXiv* **2022**, arXiv:2210.16586.
2. Pablos, S.M.; García-Bermejo, J.G.; Zalama Casanova, E.; López, J. Dynamic facial emotion recognition oriented to HCI applications. *Interact. Comput.* **2015**, *27*, 99–119. [CrossRef]
3. Makiuchi, M.R.; Uto, K.; Shinoda, K. Multimodal emotion recognition with high-level speech and text features. In Proceedings of the 2021 IEEE Automatic Speech Recognition and Understanding Workshop (ASRU), Cartagena, Colombia, 13–17 December 2021; IEEE: Piscataway, NJ, USA; pp. 350–357.
4. Kandali, A.B.; Routray, A.; Basu, T.K. Emotion recognition from Assamese speeches using MFCC features and GMM classifier. In Proceedings of the TENCON 2008—2008 IEEE Region 10 Conference, Hyderabad, India, 19–21 November 2008; IEEE: Piscataway, NJ, USA; pp. 1–5.
5. Nwe, T.L.; Foo, S.W.; De Silva, L.C. Speech emotion recognition using hidden Markov models. *Speech Commun.* **2003**, *41*, 603–623. [CrossRef]
6. Graves, A.; Mohamed, A.R.; Hinton, G. Speech recognition with deep recurrent neural networks. In Proceedings of the 2013 IEEE International Conference on Acoustics, Speech and Signal Processing, Vancouver, BC, Canada, 26–31 May 2013; IEEE: Piscataway, NJ, USA; pp. 6645–6649.
7. Mao, Q.; Dong, M.; Huang, Z.; Zhan, Y. Learning salient features for speech emotion recognition using convolutional neural networks. *IEEE Trans. Multimed.* **2014**, *16*, 2203–2213. [CrossRef]
8. Breuel, T.M. High performance text recognition using a hybrid convolutional-lstm implementation. In Proceedings of the 2017 14th IAPR International Conference on Document Analysis and Recognition (ICDAR), Kyoto, Japan, 9–15 November 2017; IEEE: Piscataway, NJ, USA, 2017; Volume 1, pp. 11–16.
9. Jaderberg, M.; Simonyan, K.; Vedaldi, A.; Zisserman, A. Deep structured output learning for unconstrained text recognition. *arXiv* **2014**, arXiv:1412.5903.
10. Guo, L.; Wang, L.; Dang, J.; Fu, Y.; Liu, J.; Ding, S. Emotion Recognition with Multimodal Transformer Fusion Framework Based on Acoustic and Lexical Information. *IEEE MultiMedia* **2022**, *29*, 94–103. [CrossRef]

11. Guo, W.; Wang, J.; Wang, S. Deep multimodal representation learning: A survey. *IEEE Access* **2019**, *7*, 63373–63394. [CrossRef]
12. Xu, H.; Zhang, H.; Han, K.; Wang, Y.; Peng, Y.; Li, X. Learning alignment for multimodal emotion recognition from speech. In Proceedings of the 20th Annual Conference of the International Speech Communication Association (INTERSPEECH), Graz, Austria, 15–19 September 2019; pp. 3569–3573.
13. Tripathi, S.; Tripathi, S.; Beigi, H. Multimodal Emotion Recognition on IEMOCAP Dataset using Deep Learning. *arXiv* **2018**, arXiv:1804.05788.
14. Huddar, M.G.; Sannakki, S.S.; Rajpurohit, V.S. Attention-based multimodal contextual fusion for sentiment and emotion classification using bidirectional LSTM. *Multimed. Tools Appl.* **2021**, *80*, 13059–13076. [CrossRef]
15. Eyben, F.; Wollmer, M.; Schuller, B. Opensmile: The munich versatile and fast open-source audio feature extractor. In Proceedings of the 18th ACM International Conference on Multimedia, Firenze, Italy, 25–29 October 2010; pp. 1459–1462.
16. Busso, C.; Bulut, M.; Lee, C.C.; Kazemzadeh, A.; Mower, E.; Kim, S.; Chang, J.N.; Lee, S.; Narayanan, S.S. IEMOCAP: Interactive emotional dyadic motion capture database. *J. Lang. Resour. Eval.* **2008**, *42*, 335–359. [CrossRef]
17. Kumar, P.; Kaushik, V.; Raman, B. Towards the Explainability of Multimodal Speech Emotion Recognition. In Proceedings of the Interspeech, Brno, Czech Republic, 30 August–3 September 2021; pp. 1748–1752. [CrossRef]
18. Singh, P.; Srivastava, R.; Rana, K.P.S.; Kumar, V. A multimodal hierarchical approach to speech emotion recognition from audio and text. *Knowl. Based Syst.* **2021**, *229*, 107316. [CrossRef]
19. Wang, Y.; Gu, Y.; Yin, Y.; Han, Y.; Zhang, H.; Wang, S.; Li, C.; Quan, D. Multimodal transformer augmented fusion for speech emotion recognition. *Front. Neurorobotics* **2023**, *17*, 1181598. [CrossRef] [PubMed]
20. Zaidi, S.A.M.; Latif, S.; Qadi, J. Cross-Language Speech Emotion Recognition Using Multimodal Dual Attention Transformers. *arXiv* **2023**, arXiv:2306.13804.
21. Canal, F.Z.; Müller, T.R.; Matias, J.C.; Scotton, G.G.; de Sa Junior, A.R.; Pozzebon, E.; Sobieranski, A.C. A survey on facial emotion recognition techniques: A state-of-the-art literature review. *Inf. Sci.* **2022**, *582*, 593–617. [CrossRef]
22. Atrey, P.K.; Hossain, M.A.; El Saddik, A.; Kankanhalli, M.S. Multimodal fusion for multimedia analysis: A survey. *Multimed. Syst.* **2010**, *16*, 345–379. [CrossRef]
23. Huang, J.; Li, Y.; Tao, J.; Lian, Z.; Wen, Z.; Yang, M.; Yi, J. Continuous multimodal emotion prediction based on long short term memory recurrent neural network. In Proceedings of the 7th Annual Workshop on Audio/Visual Emotion Challenge, Mountain View, CA, USA, 23–27 October 2017; pp. 11–18.
24. Stappen, L.; Baird, A.; Christ, L.; Schumann, L.; Sertolli, B.; Messner, E.M.; Cambria, E.; Zhao, G.; Schuller, B.W. The MuSe 2021 multimodal sentiment analysis challenge: Sentiment, emotion, physiological-emotion, and stress. In Proceedings of the 2nd on Multimodal Sentiment Analysis Challenge, Virtual Event, 24 October 2021; pp. 5–14.
25. Liu, Y.; Ott, M.; Goyal, N.; Du, J.; Joshi, M.; Chen, D.; Levy, O.; Lewis, M.; Zettlemoyer, L.; Stoyanov, V. Roberta: A robustly optimized bert pretraining approach. *arXiv* **2019**, arXiv:1907.11692.
26. Devlin, J.; Chang, M.W.; Lee, K.; Toutanova, K. Bert: Pre-training of deep bidirectional transformers for language understanding. *arXiv* **2018**, arXiv:1810.04805.
27. Venkataramanan, K.; Rajamohan, H.R. Emotion recognition from speech. *arXiv* **2019**, arXiv:1912.10458.
28. Hao, M.; Cao, W.H.; Liu, Z.T.; Wu, M.; Xiao, P. Visual-audio emotion recognition based on multi-task and ensemble learning with multiple features. *Neurocomputing* **2020**, *391*, 42–51. [CrossRef]
29. McFee, B.; Raffel, C.; Liang, D.; Ellis, D.P.; McVicar, M.; Battenberg, E.; Nieto, O. librosa: Audio and music signal analysis in python. In Proceedings of the 14th Python in Science Conference, Austin, TX, USA, 6–12 July 2015; pp. 18–25.
30. Solovyev, R.A.; Vakhrushev, M.; Radionov, A.; Romanova, I.I.; Amerikanov, A.A.; Aliev, V.; Shvets, A.A. Deep learning approaches for understanding simple speech commands. In Proceedings of the 2020 IEEE 40th International Conference on Electronics and Nanotechnology (ELNANO), Kyiv, Ukraine, 22–24 April 2020; IEEE: Piscataway, NJ, USA; pp. 688–693.
31. Howard, A.G.; Zhu, M.; Chen, B.; Kalenichenko, D.; Wang, W.; Weyand, T.; Andreetto, M.; Adam, H. Mobilenets: Efficient convolutional neural networks for mobile vision applications. *arXiv* **2017**, arXiv:1704.04861.
32. Sifre, L. Rigid-Motion Scattering for Image Classification. Ph.D. Thesis, CMAP Ecole Polytechnique, Palaiseau, France, 2014.
33. Kingma, D.P.; Ba, J. Adam: A method for stochastic optimization. *arXiv* **2014**, arXiv:1412.6980.
34. Barbieri, F.; Camacho-Collados, J.; Neves, L.; Espinosa-Anke, L. Tweeteval: Unified benchmark and comparative evaluation for tweet classification. *arXiv* **2020**, arXiv:2010.12421.
35. Poria, S.; Hazarika, D.; Majumder, N.; Naik, G.; Cambria, E.; Mihalcea, R. Meld: A multimodal multi-party dataset for emotion recognition in conversations. *arXiv* **2018**, arXiv:1810.02508.
36. Plutchik, R. The Nature of Emotions. *J. Storage (JSTOR) Digit. Libr. Am. Sci. J.* **2001**, *89*, 344–350.
37. Wang, Y.; Shen, G.; Xu, Y.; Li, J.; Zhao, Z. Learning Mutual Correlation in Multimodal Transformer for Speech Emotion Recognition. In Proceedings of the Interspeech, Brno, Czech Republic, 30 August–3 September 2021; pp. 4518–4522. [CrossRef]
38. Sahoo, S.; Kumar, P.; Raman, B.; Roy, P.P. A Segment Level Approach to Speech Emotion Recognition using Transfer Learning. In Proceedings of the 5th Asian Conference on Pattern Recognition (ACPR), Auckland, New Zealand, 26–29 November 2019; pp. 435–448.
39. Feng, H.; Ueno, S.; Kawahara, T. End-to-end Speech Emotion Recognition Combined with Acoustic-to-Word ASR Model. In Proceedings of the Interspeech, Shanghai, China, 25–29 October 2020; pp. 501–505. [CrossRef]

40. Setyono, J.C.; Zahra, A. Data augmentation and enhancement for multimodal speech emotion recognition. *Bull. Electr. Eng. Inform.* **2023**, *12*, 3008–3015. [CrossRef]
41. Wang, N.; Cao, H.; Zhao, J.; Chen, R.; Yan, D.; Zhang, J. M2R2: Missing-Modality Robust emotion Recognition framework with iterative data augmentation. *IEEE Trans. Artif. Intell.* **2022**, *4*, 1305–1316. [CrossRef]
42. Lian, Z.; Liu, B.; Tao, J. CTNet: Conversational transformer network for emotion recognition. *IEEE/ACM Trans. Audio Speech Lang. Process.* **2021**, *29*, 985–1000. [CrossRef]
43. Ho, N.H.; Yang, H.J.; Kim, S.H.; Lee, G. Multimodal approach of speech emotion recognition using multi-level multihead fusion attention-based recurrent neural network. *IEEE Access* **2020**, *8*, 61672–61686. [CrossRef]

big data and
cognitive computing

MDPI

Article

# Social Trend Mining: Lead or Lag

**Hossein Hassani *, Nadejda Komendantova, Elena Rovenskaya and Mohammad Reza Yeganegi**

International Institute for Applied Systems Analysis (IIASA), Schloßpl. 1, 2361 Laxenburg, Austria; komendan@iiasa.ac.at (N.K.); rovenska@iiasa.ac.at (E.R.); yeganegi@iiasa.ac.at (M.R.Y.)
* Correspondence: hassani@iiasa.ac.at

**Abstract:** This research underscores the profound implications of Social Intelligence Mining, notably employing open access data and Google Search engine data for trend discernment. Utilizing advanced analytical methodologies, including wavelet coherence analysis and phase difference, hidden relationships and patterns within social data were revealed. These techniques furnish an enriched comprehension of social phenomena dynamics, bolstering decision-making processes. The study's versatility extends across myriad domains, offering insights into public sentiment and the foresight for strategic approaches. The findings suggest immense potential in Social Intelligence Mining to influence strategies, foster innovation, and add value across diverse sectors.

**Keywords:** social intelligence; data mining; open data; search engines; review; lead and lag

## 1. Introduction

The importance of social trend mining lies in its ability to provide stakeholders with a deep understanding of user behavior, preferences, and sentiments. By leveraging the vast amount of data available from the internet including search engines, businesses gain real-time insights into market dynamics, consumer trends, and competitive intelligence. These data also have a high potential to inform research in the public interest.

Big Data Analytics using available information provided by various search engines (for instance, Google) has opened a new golden opportunity [1,2]. In addition, such an approach can also be used to measure people's engagement, priority, and sentiment over time. For example, Google Trends [3] allows users to analyze the popularity of keywords and phrases on Google over time. It can be used for a variety of purposes, including analyzing public opinion, tracking the spread of information and news, and identifying trends in consumer behavior. Here are a few examples of studies that have used Google Trends data. For instance, Google Trends data has been used to predict the GDP growth in the United States [4]; Google Trends data was utilized to analyze tourism demand in various countries [5]; ref. [6] used Google Trends data to predict stock market returns in China; and [7] used Google Trends data to predict the spread of infectious diseases. These are just a few examples; there are many more studies and articles that have used Google Trends data in various ways (see, for example, [8–17]).

The landscape of online networks and interaction mediums has been significantly transformed by the advent of social media platforms. Unlike their predecessors, social media platforms exhibit distinctive characteristics such as openness, participatory dynamics, flexibility, robustness, and creativity. These platforms, akin to real-life social networks, establish virtual connections between individuals, giving rise to the small-world phenomena that characterize them (see, for instance, [18–25]. Recent statistics reveal that the average social media user maintains 8.4 active accounts and dedicates around 145 min daily to engaging across various social media platforms. Amidst this vibrant virtual environment, numerous challenges have emerged concerning the extraction and analysis of content generation, modification, and dissemination across diverse topics on social media. This paper draws on a recent review [26] and various other sources [27–33] to delve into the multifaceted challenges inherent in social media mining and analysis, shedding light on the complexities of understanding and navigating this dynamic landscape.

The novelty of this paper lies in its innovative approach to social trend mining, which leverages open access data and Google Search engine data to provide comprehensive insights into user behavior, and trend identification. By integrating time series analysis and advanced analytical methods, this paper offers a fresh perspective on understanding and harnessing social data for decision making. Furthermore, this paper introduces the novel application of wavelet coherence analysis and phase difference to uncover hidden relationships and patterns within social data, enhancing our ability to identify leading and lagging trends. This unique combination of methodologies and its adaptability to various domains make the paper a pioneering contribution to the field of social trend mining.

The ability to identify leading and lagging trends, perform trend analysis, and differentiate noise from meaningful events provides organizations with a powerful tool for informed decision making, proactive strategy formulation, and effective risk management.

The next section presents the methodology, which consists of two subsections: trend extraction based on indexed time series and coherence analysis between two time series. These methodologies are employed to explore the applicability of the proposed approach using real data extracted from Google Search.

The first subsection focuses on trend extraction based on indexed time series. By utilizing indexed time series data, trends can be identified and analyzed. This approach allows for the examination of temporal patterns and variations in search interest over time. The indexed time series data, obtained from Google Search data, serve as a valuable resource for trend analysis and provide insights into societal interests and preferences.

The second subsection delves into coherence analysis between two time series. This analysis explores the association between two time series, considering both time and frequencies. By examining the coherence between the two series, it becomes possible to understand the degree of similarity and correlation between them. This analysis provides valuable information about the interconnectedness and potential causal relationships between different social phenomena.

To evaluate the proposed approach, real data extracted from Google searches are utilized. These data represent actual search queries made by individuals, reflecting their interests and concerns. By employing this real-world data, the applicability and effectiveness of the methodology can be assessed, providing valuable insights into social intelligence mining.

## 2. Methodology and Methods

The methodology section of this paper briefly describes a trend extraction approach based on indexed time series and an approach to coherence analysis between two time series. By utilizing real data from Google Search, the proposed approach can be evaluated and its effectiveness in uncovering trends and associations can be assessed. These methodologies serve as powerful tools for exploring and analyzing social phenomena, offering valuable insights into the dynamics of our ever-evolving society.

### *2.1. Trend Analysis*

Search engine trend analysis has emerged as a valuable technique for gaining insights into user behavior, sentiment analysis, and trend identification. Among various search engines, Google Trends stands out as a prominent platform that provides a vast amount of data, which can be utilized for time series analysis and opinion mining of individuals.

By monitoring the frequency of searches related to specific keywords over time, researchers can obtain valuable insights into the popularity and dynamics of various topics. This data can be leveraged for time series analysis techniques, such as trend identification, seasonality detection, and forecasting. The data extracted from Google Trends data provides not only quantitative information but also a glimpse into public sentiment and opinions. By analyzing the content of search queries and associated search results, researchers can gain insights into the preferences, concerns, and sentiments of individuals. Opinion mining techniques, such as sentiment analysis and topic modeling, can be applied to extract meaningful insights from this data, enabling a deeper understanding of public opinion on specific subjects.

To facilitate trend analysis and comparison across different topics, the concept of creating indices based on selected keywords can be used. These indices capture the relative popularity or interest in specific topics over time. The formula for creating indices can be adjusted based on the desired characteristics and data normalization techniques. Two common types of indices are:

(a) Univariate Google Trend Index: This index represents the search interest for a single keyword or topic. It is calculated by normalizing the search volume or frequency for the chosen keyword over a specific time period. Normalization techniques, such as dividing by the maximum search volume or using z-scores, can be employed to standardize the data and make it comparable.

(b) Multivariate Google Trend Index: This index captures the comparative popularity of multiple keywords or topics. It involves selecting a set of keywords of interest and calculating the search volume or frequency for each keyword.

By utilizing these indices, researchers and practitioners can gain a comprehensive view of the relative popularity and trends associated with different keywords or topics over time. This enables them to identify emerging trends, track public sentiment, and compare the performance of various keywords or topics within their respective domains.

### 2.2. *Lead and Lag Analysis*

A wavelet transform is used to transform time series with complex periodic behavior to simplified signals, each of which has simple periodic behavior (with a single period). From a mathematical point of view, a wavelet transform is a generalization of Fourier transform. A Continuous Wavelet Transform, CWT, uses a mother wavelet function $\psi(\cdot)$ to transform a discrete-time time series $\{y_t\}_1^n$ to wavelet coefficients $W_\psi\{y\}(\tau, s)$, for the time localizing parameter $\tau$ and the scale parameter $s$.

#### 2.2.1. Univariate Case

The wavelet coefficients $W_\psi\{y\}(\tau, s)$ are defined as a convolution of time series $\{y_t\}_1^n$ with the localized mother wavelet $\psi(\cdot)$ (named child wavelet), localized in time and frequency space by $\tau$ and $s$ [34]:

$$W_\psi\{y\}(\tau, s) = \sum_{t=1}^{n} y_t \frac{1}{\sqrt{s}} \overline{\psi}\left(\frac{t-\tau}{s}\right),$$

where $\overline{\psi}(\cdot)$ is the complex conjugate of the mother wavelet $\psi(\cdot)$. The localization parameter $\tau$ exhibits periodic behavior over time, while the scale parameter $s$ localizes the periodic behavior in the frequency domain. When the scale parameter $s$ has larger values, this indicates long-term periodic behavior with low frequency. On the other hand, smaller values of the scale parameter $s$ reveal details in short-term periodic patterns with higher frequencies. One commonly used choice for the mother wavelet is the Morlet wavelet [33], which is formulated as follows:

$$\psi(t) = c_\omega \pi^{-\frac{1}{4}} \exp\left\{-\frac{t^2}{2}\right\}\left(e^{i\omega t} - \kappa_\omega\right),$$

where $\omega$ is the angular frequency, and $\kappa_\omega$ and $c_\omega$ are constants defined as:

$$c_\omega = \left(1 + e^{-\omega^2} - 2e^{-\frac{3}{4}\omega^2}\right)^{-\frac{1}{2}}, \kappa_\omega = e^{-\frac{1}{2}\omega^2}.$$

The $\omega = 6$ is a proper choice for the angular frequency since it makes the Morlet wavelet approximately analytic. Large absolute values of $W_\psi\{y\}(\tau, s)$ indicate powerful periodic patterns in time $\tau$ and period $s$. The wavelet coefficients can be used to construct the wavelet power spectrum of time series $\{y_t\}_1^n$:

$$Power_{\psi}\{y\}(\tau,s) = \frac{1}{s}\left|W_{\psi}\{y\}(\tau,s)\right|^2$$

The wavelet power spectrum, denoted as $Power_{\psi}\{y\}$, is a valuable tool for mapping periodic patterns in a given time series over time. To assess the significance of the wavelet power spectrum, it can be compared against the white noise spectrum using either the asymptotic chi-square statistic [34] or Monte Carlo simulation [35]. The Monte Carlo simulation approach is employed for evaluating the significance of the wavelet power spectrum.

2.2.2. Bivariate Case

Let us now consider the time series $\{x_t\}_1^n$ and $\{y_t\}_1^n$ as the bivariate case. A cross-wavelet transform can be used to investigate the relationship between $\{x_t\}_1^n$ and $\{y_t\}_1^n$ [34]:

$$W_{\psi}\{xy\}(\tau,s) = \frac{1}{s}W_{\psi}\{x\}(\tau,s)\,\overline{W}_{\psi}\{y\}(\tau,s),$$

where $\overline{W}$ denotes a complex conjugate and $W_{\psi}\{x\}(\tau,s)$ and $W_{\psi}\{y\}(\tau,s)$ are the wavelet coefficients in CWT of $\{x_t\}_1^n$ and $\{y_t\}_1^n$, respectively. The wavelet cross power spectrum, as modulus of wavelet coefficients, can be used to map the similarities between two time series' periodic behavior:

$$Power_{\psi}\{xy\}(\tau,s) = \left|W_{\psi}\{xy\}(\tau,s)\right|.$$

The $Power_{\psi}\{xy\}(\tau,s)$, like covariance, depends on the underlying time series' unit of measurement and may not properly interpret the degree of association between two series. Wavelet coherence between two time series $\{x_t\}_1^n$ and $\{y_t\}_1^n$ is defined as the local cross-correlation between the series, localized at time $\tau$ and scale $s$:

$$W_{\psi}\{xy\}(\tau,s) = \frac{\left|sW_{\psi}\{xy\}(\tau,s)\right|^2}{sPower_{\psi}\{x\}(\tau,s).sPower_{\psi}\{y\}(\tau,s)},$$

where prefix $s$ behind $W_{\psi}$ and $Power_{\psi}$ indicates smoothing is required. Similar to the power spectrum, the wavelet coherence between two series can also be examined using Monte Carlo simulation [36–39]. Monte Carlo simulation provides a robust approach for testing the significance of wavelet coherence and assessing the presence of coherent relationships between the two series under investigation.

The Continuous Wavelet Transform (CWT) reveals localized periodic patterns in a given time series $\{y_t\}_1^n$. The wavelet phase indicates the local displacement of the periodic behavior relative to the localization parameter τ, which is shifted across the time domain when τ is set as the origin. The wavelet phase is typically represented as an angle within the interval [−π, π].

$$Phase_{\psi}\{y\}(\tau,s) = \tan^{-1}\left(\frac{Im(W_{\psi}\{y\}(\tau,s))}{Re(W_{\psi}\{y\}(\tau,s))}\right),$$

where $Im(.)$ and $Re(.)$ are imaginary and real parts of the wavelet coefficient $W_{\psi}\{y\}(\tau,s)$.

Using the cross-wavelet coefficients, one can calculate the difference between wavelet phases from two time series (which is actually the difference between two phases):

$$Angle_{\psi}\{xy\}(\tau,s) = \tan^{-1}\left(\frac{Im(W_{\psi}\{xy\}(\tau,s))}{Re(W_{\psi}\{xy\}(\tau,s))}\right) = Phase_{\psi}\{x\}(\tau,s) - Phase_{\psi}\{y\}(\tau,s),$$

where $Angle_{\psi}\{xy\}(\tau,s)$ represents the phase difference between two time series $\{x_t\}_1^n$ and $\{y_t\}_1^n$. $Angle_{\psi}\{xy\}(\tau,s)$ can be used to determine which time series starts the periodic pattern first and which one is following, for a given time and frequency interval. Figure 1 shows the simplified interpretation of the phase difference between time series $\{x_t\}_1^n$ and $\{y_t\}_1^n$.

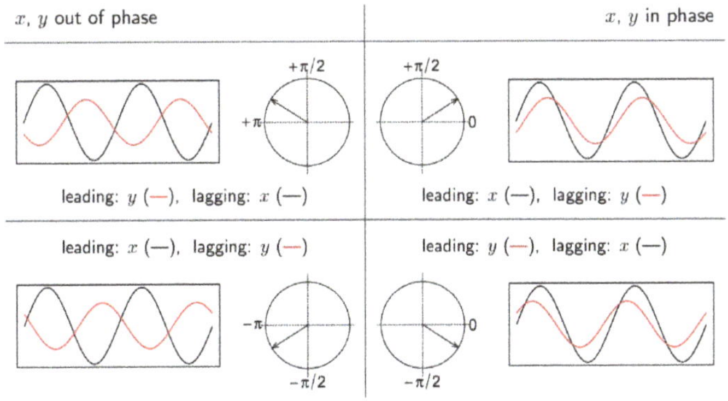


**Figure 1.** Interpretation of phase difference between signals x and y [32].

Various social phenomena can exhibit cyclical patterns [33]. By applying wavelet analysis to the social data, we can uncover the underlying periodic patterns in these phenomena. For example, analyzing people's interest in a subject through chat discussions or online searches can reveal if there are cycles where this subject becomes popular in society. Similarly, examining the number of participants in a social activity can expose the cycles of outbursts in that particular activity.

Furthermore, by utilizing a wavelet coherence analysis and phase difference, we can examine the correlation between two social phenomena, even if it was only for a brief period. This analysis can help determine which phenomenon had a leading role if such a relationship existed. The presence of coherency and phase difference between two social phenomena could indicate a causal relationship, where events from the leading series influence events in the lagging series. Alternatively, it could suggest that both series are influenced by another common social phenomenon. These findings serve as powerful tools for generating hypotheses about social events, trends, and their interrelations.

## 3. Applied Methodology: Real Data Implementation and Analysis

In this section, we focus on the application of social trend mining to several dimensions of human security including food, water, and energy security. These three dimensions of human security correspond to three UN Sustainable Development Goals (SDGs)—SDG 2, SDG 6, and SDG 7, respectively. Let us now provide a brief overview of three SDGs utilized in this paper.

Sustainable Development Goal 2 is about creating a world free of hunger by 2030. In 2020, between 720 million and 811 million persons worldwide were suffering from hunger, roughly 161 million more than in 2019. Also in 2020, a staggering 2.4 billion people, or above 30 percent of the world's population, were moderately or severely food insecure, lacking regular access to adequate food. The figure increased by nearly 320 million people in just one year. Globally, 149.2 million children under 5 years of age, or 22.0 percent, were suffering from stunting (low height for their age) in 2020, a decrease from 24.4 percent in 2015.

SDG 6 is about ensuring access to water and sanitization for all. Access to safe water, sanitation, and hygiene is the most basic human need for health and well-being. Billions of people will lack access to these basic services in 2030 unless progress quadruples. Demand for water is rising owing to rapid population growth, urbanization, and increasing water needs from the agriculture, industry, and energy sectors.

To reach universal access to drinking water, sanitation, and hygiene by 2030, the current rates of progress would need to increase fourfold. Achieving these targets would save 829,000 people annually, who die from diseases directly attributable to unsafe water, inadequate sanitation, and poor hygiene practices.

SDG 7 is about ensuring access to clean and affordable energy, which is key to the development of agriculture, business, communications, education, healthcare, and transportation. The lack of access to energy hinders economic and human development.

The latest data suggest that the world continues to advance towards sustainable energy targets. Nevertheless, the current pace of progress is insufficient to achieve Goal 7 by 2030. Huge disparities in access to modern sustainable energy persist.

It should be mentioned that significant challenges remain at the global level in terms of achieving these SDGs. This assessment is true for most of the world's major regions, while recent trends are mainly stagnating (in lower-income countries) or moderately increasing (in higher-income countries).

The actual or perceived lack of food, water, or energy security could be a source of social instability. In addition to the indicators describing the actual availability, accessibility, and affordability of food, water, and energy, such as prices and use, indicators describing people's perceptions provide important input for policy makers. Google Search data can inform such indicators which could be made available to policy makers almost in real time.

Figure 2, as an example, depicts the Google Search hits for the keywords "Food Security", "Energy Security", and "Water Security" over the past five years. These three concepts—food, energy, and water security—are important dimensions of human security. However, when comparing the search hits for the three subsets, it is evident that the search interest in food security is significantly higher than that for water security and energy security. The search interest in energy security is lower than the other two dimensions although only slightly lower than the search interest in water security.

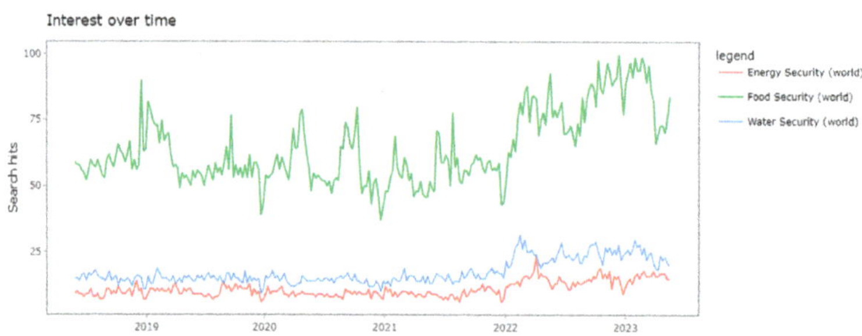


**Figure 2.** The Google Search hits for the keywords "Food Security", "Energy Security", and "Water Security" over the last 5 years.

Let us now explore the periodic behavior in the search interests for "Food Security" and "Water Security". Figure 3 displays the wavelet power spectrum for both series.

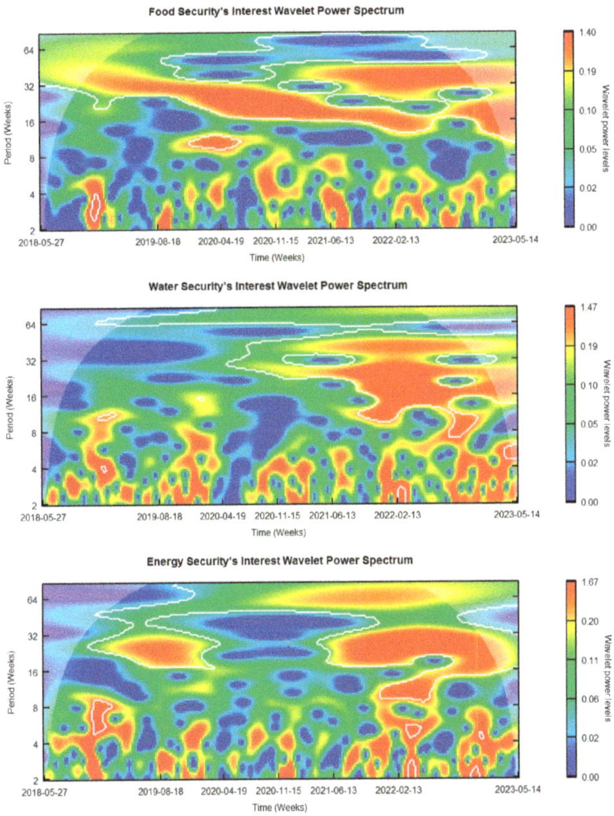


**Figure 3.** Wavelet power spectrum for "Food Security", "Water Security", and "Energy Security" Google Trends. White contours show significant powers at $\alpha = 0.1$ significance level.

As depicted in Figure 3 (top panel), the interest in "Food Security" has exhibited significant low and mid-frequency behaviors over the past five years. Note that low-frequency cycles are cycles with long periods, for example, longer than 32 weeks period in this data.

Mid-frequency cycles are cycles with mid-range periods, for example, around a 16-week period in this data. The significance test, conducted using Monte Carlo simulation with 5000 sample paths, confirms this observation. However, in recent years (i.e., after 19-Apr-2020), the power spectrum has shown an increase, with a focus on mid-range periods (around 16 and 32 weeks) and long-range periods (above 32 weeks). This indicates that the search frequency for the keyword "Food Security" has become more frequent over time.

In the middle panel of Figure 3, the power spectrum for "Water Security" displays significant patterns mostly concentrated on the right side of the timeline (approximately after 15-Nov-2020). The most powerful periodic patterns occur within the 16 to 32-week period range. In other words, approximately after 15-Nov-2020, there has been a periodic pattern in people's interest in the "Water Security" keyword. The length of each periodic pattern (the beginning of one surge of interest to the beginning of the next one) mostly includes periods below 16 to periods above 32 weeks.

As shown in the bottom panel, the significant "Energy Security" power spectrum also is mostly concentrated on low and mid frequencies (i.e., long periods in which the time from one surge of interest in "Energy Security" is between 32 to 64 weeks and mid-ranged periods which the time from one surge of interest in "Energy Security" is between 16 to 32 weeks). Periodic behavior of interest in the "Energy Security" keyword has become more powerful after 13-Jun-2021. Furthermore, the periodic behavior of interest in "Energy Security" includes shorter periods (higher frequency) as well. In other words, the power spectrum of interest in "Energy Security" shows that the interest in "Energy Security" has increased and the search for "Energy Security" has become more frequent.

These findings suggest that the interest in "Food Security", "Water Security", and "Energy Security" has increased in the last three years (after 14-Apr-2020), and searching for these keywords has become more frequent. Furthermore, it can be seen the increased interest in these keywords started with "Food Security" and, as the top panel shows an increased power spectrum at higher frequencies sooner. In other words, after 14-Apr-2020, the search for "Food Security" became more frequent, then the search for "Water Security", and, finally, "Energy Security".

Additionally, in recent years (especially after 13-Jun-2021) the periodic behavior of interest in these keywords demonstrates mid-range periodic patterns (with period lengths between 16 and 32 weeks.), which suggests that it takes 16 to 32 weeks (almost 4 to 8 months) form one surge of interest in these keywords to the next one.

In order to examine the relationship between interest in "Food Security" and "Water Security", the wavelet coherence analysis is applied to two series. The results are given in Figure 4.

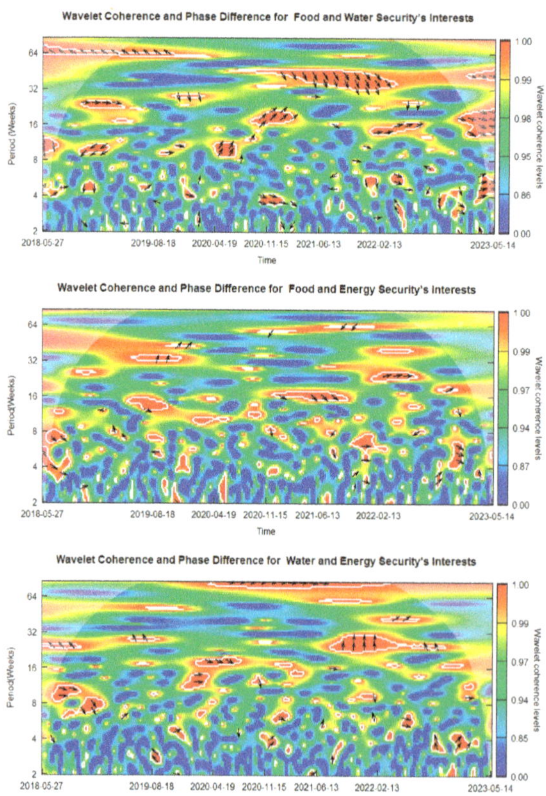


**Figure 4.** Wavelet coherence and phase difference angles (arrows) for "Food Security", "Water Security", and "Energy Security" interests. White contours show significant powers at $\alpha = 0.1$ significance level (90% confidence level). Phase difference angles are only presented for locations with significant coherence.

As previously mentioned, wavelet coherence and phase difference analysis are valuable tools for examining the relationship between two time series signals, such as food security and water security, in both the time and frequency domains. These techniques provide insights into the similarity, coherence, and phase relationship between the two signals at different scales or frequencies, over time.

Wavelet coherence quantifies the correlation between the two signals as a function of both time and frequency. It reveals the level of similarity or shared variability between the two time series across different frequency components. Higher coherence values indicate a stronger relationship, while lower coherence values suggest a weaker or non-existent relationship.

On the other hand, phase difference captures the phase lag or lead between the two signals. It indicates the relative timing or synchronization between the peaks and troughs of the two time series. A phase difference of zero denotes perfect synchronization, implying that the peaks and troughs of the two signals align precisely. Non-zero phase difference values indicate a lag or lead between the two signals, with the peaks and troughs occurring at different times.

By employing wavelet coherence and phase difference analysis, we can gain a deeper understanding of the relationship between food security and water security, uncovering their coherence and phase synchronization characteristics across different frequencies and over time.

In order to analyze the coherence between each two of these three time series, only the time intervals and periods in which the coherence of the two series is significant, and their power spectrum will be considered. Furthermore, in order to avoid overestimating the number of cyclical patterns, the results are presented for time intervals and periods whose length is not shorter than half of their period and at least 5 weeks apart.

The coherence analysis between "Food Security" and "Water Security" (top panel in Figure 4) reveals that the two series exhibit significant coherence mostly in mid-range periods, i.e., 15.45-week and 25.11-week periods, and long periods, i.e., 32-week, 38.1-week, 43.71-week, and 64-week periods (see Table A1 in Appendix A for more details). In other words, significant cyclical patterns are evident in both series and demonstrate a significant coherence between them, suggesting that they behave similarly, possibly with a time delay. It also can be seen that during the time, the frequencies in which two series have become coherent are increased (the length of periods is decreased). For instance, before the end of 2020, coherence between two series existed mostly at 43.71-week and 64-week periods ($\frac{1}{43.71} = 0.02287806$ and $\frac{1}{64} = 0.015625$ frequencies, respectively) while after June 2022, the coherence between the two series has occurred mostly in 25.11-week and 15.45-week periods ($\frac{1}{15.45} = 0.06472492$ and $\frac{1}{25.11} = 0.03982477$ frequencies, respectively).

The middle panel in Figure 4 shows significant coherency between "Food Security" and "Energy Security". Coupling these results with the power spectrum results (Figure 3) reveals that the two series exhibit significant coherence mostly in mid-range periods, i.e., 18.38-week and 25.11-week periods, and longer periods, i.e., 34.3-week and 39.4-week periods (see Table A2 for more details). This means there are significant cyclical patterns in both series with these periods in which they behave similarly in a time interval, possibly with a time delay. For instance, between 13-Feb-2022 and 3-Jul-2022, both series demonstrate a significant cyclical pattern in which it would take 25.11 weeks from one surge of interest in keywords to the next one, and two series have almost the same behavior possibly with a time delay.

Significant coherence between "Water Security" and "Energy Security" is presented in the bottom panel of Figure 4. Overlapping significant coherence areas (inside white contours) with "Water Security" and "Energy Security" power spectrums (Figure 3) shows that two series have significant coherence, mostly in mid-range periods (i.e., 17.15-week and 25.11-week periods) and very long period, i.e., 84.45-week period (see Table A3 for details).

The arrows displayed in above mentioned periods indicate the angular phase difference between the two series during each period. To convert these angular phase differences into a time unit (weeks), the following formulation can be utilized.

$$Phase.diff_{\psi}\{xy\}(\tau,s) = \frac{l_p}{2\pi} Angle_{\psi}\{xy\}(\tau,s),$$

where $Phase.diff_{\psi}\{xy\}(\tau,s)$ is the phase difference between two series, measured by time unit (which is "week" in our data), $l_p$ is the period length and $Angle_{\psi}\{xy\}(\tau,s)$ is angular phase difference measured in radians. For instance, in weekly data, the angular phase difference between two series for a 25.11-week period converts to phase difference in weeks as:

$$Phase.diff_{\psi}\{xy\}(\tau,s) = \frac{25.11}{2\pi} Angle_{\psi}\{xy\}(\tau,s)$$

Tables A1–A3 in Appendix A illustrate the phase difference and leading time series in each of the time intervals and periods discussed above. According to Tables A1–A3, all three time series have significant cyclical behavior with a 25.11-week period at some point over the time and each time series has significant coherence with another one. For instance, between 15-Apr-2022 and 3-Jul-2022, there is significant coherence between all pairs of time series, i.e., in the "Food Security"–"Water Security" pair, in the "Food Security"–"Energy Security" pair, and in "Water Security–Energy Security". In other words, in this short time interval, all three time series have a cyclical pattern in which the surge of interest in any one keyword to the next surge of interest in that keyword takes almost 25.11 weeks. Furthermore, the cyclical pattern is similar in all three time series, except for possible time delay. However, the cyclical pattern and coherence for each pair of time series may exceed this time interval differently. Phase difference analysis shows that between 15-Apr-2022 and 3-Jul-2022, in a cyclical pattern with a 25.11-week period, interest in "Food Security" leads both "Water Security" and "Energy Security" (with different phase difference values) and interest in "Energy Security" leads interest in "Water Security". These results imply that there is a cyclical pattern with a 25.11-week period length, between 15-Apr-2022 and 3-Jul-2022, in which the surge of interest in keywords first comes to "Food Security" and then "Energy Security" and, finally, "Water Security".

## 4. Search Engine: Comparative Analysis

Table 1 presents a comparative analysis of the trend extraction features offered by three prominent search engines: Google, Bing, and Yahoo. It serves as a valuable reference for users and decision makers seeking to understand the capabilities of these search engines in extracting and analyzing trending data.

In the table, various key features are assessed, including data accessibility, trend analysis tools, real-time data availability, geographic specificity, data visualization options, API support, and customization capabilities.

Google stands out with its extensive data accessibility, strong trend analysis tools, and real-time data availability, making it a robust choice for users interested in tracking and analyzing trends. Bing offers decent trend analysis capabilities and some customization options, making it a suitable alternative. Yahoo, on the other hand, offers limited data accessibility and trend analysis tools, making it less suited for in-depth trend extraction and analysis tasks.

Overall, this comparative analysis provides insights into the strengths and weaknesses of these search engines concerning trend extraction, enabling users to make informed choices based on their specific data analysis needs and preferences.

**Table 1.** Comparison of trend extraction features in popular search engines.

| Feature | Google | Bing | Yahoo |
|---|---|---|---|
| **Data Accessibility** | Extensive data availability | Good data accessibility | Limited data accessibility |
| **Trend Analysis** | Strong trend analysis tools | Decent trend analysis | Limited trend analysis |
| **Real-time Data** | Provides real-time data | Offers real-time data | Limited real-time data |
| **Geographic Specificity** | Offers precise location data | Provides location-based results | Limited location data |
| **Data Visualization** | Offers data visualization tools | Basic data visualization | Limited data visualization |
| **API Support** | Robust API for data access | API support available | Limited API support |
| **Customization** | Customizable search parameters | Some customization options | Limited customization |

## 5. Discussion

This paper delves into the concept of Social Intelligence Mining, highlighting the importance of leveraging open access data and Google Search engine data for trend analysis. This approach offers several notable advantages for both researchers and practitioners.

Firstly, the analysis of Google Search data as a time series provides a powerful tool for trend identification. By examining search queries over time, researchers can pinpoint emerging trends, recognize seasonality patterns, and even make predictions about future developments. This temporal perspective is invaluable for staying ahead in rapidly evolving fields and industries.

Furthermore, the application of opinion mining techniques to search queries offers a more profound understanding of public sentiment and preferences. This sentiment analysis adds a layer of nuance to the data, enabling decision makers to make informed choices regarding strategy formulation and risk management. In sum, Social Intelligence Mining, driven by open access data and Google Search engine data, equips organizations with comprehensive insights into user behavior, sentiment analysis, and trend identification.

This approach facilitates competitive advantages, informed decision making, and meaningful engagement with target audiences. As technology and data continue to evolve, the potential of Social Intelligence Mining for shaping strategies, driving innovation, and creating value across diverse domains remains substantial.

Additionally, the paper highlights the utility of wavelet coherence analysis and phase difference in uncovering hidden relationships and patterns within social data. These techniques offer a more profound understanding of the dynamics between social phenomena. By identifying leading and lagging trends, researchers and practitioners can make well-informed decisions based on a more comprehensive view of the data.

The specific case study presented in the paper on "Food Security", "Energy Security", and "Water Security" serves as an illustrative example. However, the methodology outlined can be applied to a wide range of domains and topics. Open access data sources like Google Trends offer valuable insights into public sentiment, emerging trends, and proactive strategy development.

Looking ahead, further research in this field should consider expanding the analysis to encompass additional relevant social phenomena. This expansion will allow for the exploration of more complex relationships and patterns. Additionally, incorporating data from social media and news sources can provide a more comprehensive understanding of social dynamics.

Addressing the challenges associated with data quality, privacy, and bias is also crucial for ensuring the reliability and validity of results in Social Intelligence Mining. As this field continues to evolve, these challenges must be carefully addressed to maintain the integrity of the research and its practical applications.

## 6. Conclusions

In conclusion, this paper underscores the pivotal role of Social Intelligence Mining, accentuating the utility of open access data and Google Search engine data for in-depth trend analysis. Harnessing these resources, coupled with sophisticated analytical methods, empowers organizations to secure a competitive advantage, make evidence-based decisions, and more effectively engage their target demographics.

Our findings spotlight the efficacy of wavelet coherence analysis and phase difference in elucidating concealed relationships and patterns within social datasets. Such techniques facilitate a more profound grasp of social phenomena dynamics, subsequently refining decision-making protocols.

However, this research is not without its limitations. The focus on a singular case study, albeit comprehensive, may not capture the entire spectrum of possibilities within Social Intelligence Mining. Looking forward, the versatility of the presented methodology suggests its applicability across diverse domains and subjects. Still, future research should venture into investigating intricate relationships and broaden its analytical scope to encapsulate various social phenomena. Integrating data from diverse platforms, such as social media and news outlets, will enrich the analysis. Addressing pressing concerns of data integrity, privacy, and potential biases will be paramount to buttress the dependability of subsequent Social Intelligence Mining endeavors.

To encapsulate, Social Intelligence Mining stands poised to redefine strategy formulation, spur innovation, and offer unparalleled value across sectors. Its sustained evolution augurs well for refining both decision-making paradigms and the comprehension of intricate social dynamics.

**Author Contributions:** H.H., N.K., E.R. and M.R.Y. conceptualied and designed the study and methodology. H.H. and M.R.Y. developed the software code and conducted formal data analysis. H.H. prepared the original draft. E.R., M.R.Y. and N.K. reviewed and edited the paper. All authors have read and agreed to the published version of the manuscript.

**Funding:** IIASA internal funding.

**Data Availability Statement:** Data are available upon request.

**Acknowledgments:** We extend our sincere gratitude to the referees and the editor, whose insightful comments and suggestions significantly contributed to the enhancement of our paper.

**Conflicts of Interest:** The authors declare no conflict of interest.

## Appendix A

**Table A1.** The phase difference and leading time series for "Food Security"–" Water Security" pair.

| Date | Coherence | Phase Diff. (Angular) | Phase Diff. (Temporal) | Leading Series | Date | Coherence | Phase Diff. (Angular) | Phase Diff. (Temporal) | Leading Series |
|---|---|---|---|---|---|---|---|---|---|
| **15.45-week period** | | | | | | | | | |
| 23-Jan-22 | 0.9958 | 0.0343 | 0.0844 | Food Security | 17-Apr-22 | 0.9966 | 0.2759 | 0.6786 | Food Security |
| 30-Jan-22 | 0.9963 | 0.0468 | 0.1151 | Food Security | 24-Apr-22 | 0.9966 | 0.3023 | 0.7436 | Food Security |
| 6-Feb-22 | 0.9966 | 0.0607 | 0.1493 | Food Security | 1-May-22 | 0.9966 | 0.3285 | 0.808 | Food Security |

**Table A1.** *Cont.*

| Date | Coherence | Phase Diff. (Angular) | Phase Diff. (Temporal) | Leading Series | Date | Coherence | Phase Diff. (Angular) | Phase Diff. (Temporal) | Leading Series |
|---|---|---|---|---|---|---|---|---|---|
| 13-Feb-22 | 0.9969 | 0.0761 | 0.1872 | Food Security | 8-May-22 | 0.9967 | 0.3543 | 0.8715 | Food Security |
| 20-Feb-22 | 0.997 | 0.093 | 0.2288 | Food Security | 15-May-22 | 0.9968 | 0.3793 | 0.933 | Food Security |
| 27-Feb-22 | 0.9971 | 0.1115 | 0.2743 | Food Security | 22-May-22 | 0.9969 | 0.4032 | 0.9918 | Food Security |
| 6-Mar-22 | 0.9971 | 0.1314 | 0.3232 | Food Security | 29-May-22 | 0.9971 | 0.4255 | 1.0466 | Food Security |
| 13-Mar-22 | 0.997 | 0.1528 | 0.3758 | Food Security | 5-Jun-22 | 0.9973 | 0.4458 | 1.0966 | Food Security |
| 20-Mar-22 | 0.997 | 0.1755 | 0.4317 | Food Security | 12-Jun-22 | 0.9974 | 0.4639 | 1.1411 | Food Security |
| 27-Mar-22 | 0.9969 | 0.1994 | 0.4905 | Food Security | 19-Jun-22 | 0.9975 | 0.4793 | 1.179 | Food Security |
| 3-Apr-22 | 0.9968 | 0.2242 | 0.5515 | Food Security | 26-Jun-22 | 0.9975 | 0.4916 | 1.2092 | Food Security |
| 10-Apr-22 | 0.9967 | 0.2498 | 0.6144 | Food Security | 3-Jul-22 | 0.9974 | 0.5004 | 1.2309 | Food Security |
| **25.11-week period** | | | | | | | | | |
| 15-May-22 | 0.9966 | −1.6031 | −6.4058 | Food Security | 26-Jun-22 | 0.9993 | −1.6985 | −6.787 | Food Security |
| 22-May-22 | 0.9971 | −1.6216 | −6.4797 | Food Security | 3-Jul-22 | 0.9994 | −1.7098 | −6.8321 | Food Security |
| 29-May-22 | 0.9977 | −1.6392 | −6.55 | Food Security | 10-Jul-22 | 0.9993 | −1.7194 | −6.8705 | Food Security |
| 5-Jun-22 | 0.9981 | −1.6558 | −6.6163 | Food Security | 17-Jul-22 | 0.999 | −1.7269 | −6.9004 | Food Security |
| 12-Jun-22 | 0.9986 | −1.6713 | −6.6783 | Food Security | 24-Jul-22 | 0.9983 | −1.7322 | −6.9216 | Food Security |
| 19-Jun-22 | 0.999 | −1.6856 | −6.7354 | Food Security | 31-Jul-22 | 0.9972 | −1.7347 | −6.9316 | Food Security |
| **32-week period** | | | | | | | | | |
| 24-Oct-21 | 0.9963 | −1.0047 | −5.1169 | Water Security | 19-Dec-21 | 0.9974 | −1.0521 | −5.3583 | Water Security |
| 31-Oct-21 | 0.9966 | −1.0097 | −5.1424 | Water Security | 26-Dec-21 | 0.9973 | −1.059 | −5.3934 | Water Security |
| 7-Nov-21 | 0.9968 | −1.015 | −5.1694 | Water Security | 2-Jan-22 | 0.9973 | −1.0661 | −5.4296 | Water Security |
| 14-Nov-21 | 0.9971 | −1.0206 | −5.1979 | Water Security | 9-Jan-22 | 0.9972 | −1.0733 | −5.4663 | Water Security |
| 21-Nov-21 | 0.9972 | −1.0265 | −5.2279 | Water Security | 16-Jan-22 | 0.997 | −1.0806 | −5.5035 | Water Security |
| 28-Nov-21 | 0.9973 | −1.0326 | −5.259 | Water Security | 23-Jan-22 | 0.9968 | −1.0881 | −5.5416 | Water Security |
| 5-Dec-21 | 0.9974 | −1.0389 | −5.2911 | Water Security | 30-Jan-22 | 0.9966 | −1.0957 | −5.5804 | Water Security |
| 12-Dec-21 | 0.9974 | −1.0454 | −5.3242 | Water Security | 6-Feb-22 | 0.9964 | −1.1034 | −5.6196 | Water Security |
| **38.1-week period** | | | | | | | | | |
| 11-Apr-21 | 0.9963 | −1.2271 | −7.432 | Water Security | 17-Oct-21 | 0.9979 | −1.0351 | −6.2692 | Water Security |

**Table A1.** *Cont.*

| Date | Coherence | Phase Diff. (Angular) | Phase Diff. (Temporal) | Leading Series | Date | Coherence | Phase Diff. (Angular) | Phase Diff. (Temporal) | Leading Series |
|---|---|---|---|---|---|---|---|---|---|
| **38.1-week period** | | | | | | | | | |
| 18-Apr-21 | 0.9965 | −1.2208 | −7.3939 | Water Security | 24-Oct-21 | 0.9979 | −1.0271 | −6.2207 | Water Security |
| 25-Apr-21 | 0.9966 | −1.2145 | −7.3557 | Water Security | 31-Oct-21 | 0.9979 | −1.0191 | −6.1723 | Water Security |
| 2-May-21 | 0.9967 | −1.2081 | −7.317 | Water Security | 7-Nov-21 | 0.9979 | −1.0109 | −6.1226 | Water Security |
| 9-May-21 | 0.9969 | −1.2017 | −7.2782 | Water Security | 14-Nov-21 | 0.9979 | −1.0027 | −6.0729 | Water Security |
| 16-May-21 | 0.997 | −1.1952 | −7.2388 | Water Security | 21-Nov-21 | 0.9978 | −0.9945 | −6.0233 | Water Security |
| 23-May-21 | 0.9971 | −1.1886 | −7.1989 | Water Security | 28-Nov-21 | 0.9978 | −0.9862 | −5.973 | Water Security |
| 30-May-21 | 0.9972 | −1.182 | −7.1589 | Water Security | 5-Dec-21 | 0.9978 | −0.9778 | −5.9221 | Water Security |
| 6-Jun-21 | 0.9972 | −1.1753 | −7.1183 | Water Security | 12-Dec-21 | 0.9978 | −0.9693 | −5.8706 | Water Security |
| 13-Jun-21 | 0.9973 | −1.1685 | −7.0771 | Water Security | 19-Dec-21 | 0.9977 | −0.9608 | −5.8192 | Water Security |
| 20-Jun-21 | 0.9974 | −1.1617 | −7.0359 | Water Security | 26-Dec-21 | 0.9977 | −0.9521 | −5.7665 | Water Security |
| 27-Jun-21 | 0.9975 | −1.1547 | −6.9935 | Water Security | 2-Jan-22 | 0.9976 | −0.9435 | −5.7144 | Water Security |
| 4-Jul-21 | 0.9975 | −1.1477 | −6.9511 | Water Security | 9-Jan-22 | 0.9976 | −0.9347 | −5.6611 | Water Security |
| 11-Jul-21 | 0.9976 | −1.1407 | −6.9087 | Water Security | 16-Jan-22 | 0.9975 | −0.9259 | −5.6078 | Water Security |
| 18-Jul-21 | 0.9976 | −1.1336 | −6.8657 | Water Security | 23-Jan-22 | 0.9975 | −0.917 | −5.5539 | Water Security |
| 25-Jul-21 | 0.9977 | −1.1264 | −6.8221 | Water Security | 30-Jan-22 | 0.9974 | −0.908 | −5.4994 | Water Security |
| 1-Aug-21 | 0.9977 | −1.1191 | −6.7779 | Water Security | 6-Feb-22 | 0.9973 | −0.8989 | −5.4443 | Water Security |
| 8-Aug-21 | 0.9978 | −1.1118 | −6.7337 | Water Security | 13-Feb-22 | 0.9973 | −0.8897 | −5.3885 | Water Security |
| 15-Aug-21 | 0.9978 | −1.1044 | −6.6889 | Water Security | 20-Feb-22 | 0.9972 | −0.8805 | −5.3328 | Water Security |
| 22-Aug-21 | 0.9978 | −1.097 | −6.6441 | Water Security | 27-Feb-22 | 0.9971 | −0.8712 | −5.2765 | Water Security |
| 29-Aug-21 | 0.9978 | −1.0895 | −6.5986 | Water Security | 6-Mar-22 | 0.9971 | −0.8618 | −5.2196 | Water Security |
| 5-Sep-21 | 0.9979 | −1.0819 | −6.5526 | Water Security | 13-Mar-22 | 0.997 | −0.8523 | −5.162 | Water Security |
| 12-Sep-21 | 0.9979 | −1.0742 | −6.506 | Water Security | 20-Mar-22 | 0.9969 | −0.8427 | −5.1039 | Water Security |
| 19-Sep-21 | 0.9979 | −1.0665 | −6.4593 | Water Security | 27-Mar-22 | 0.9968 | −0.833 | −5.0451 | Water Security |
| 26-Sep-21 | 0.9979 | −1.0588 | −6.4127 | Water Security | 3-Apr-22 | 0.9967 | −0.8232 | −4.9858 | Water Security |
| 3-Oct-21 | 0.9979 | −1.051 | −6.3655 | Water Security | 10-Apr-22 | 0.9967 | −0.8134 | −4.9264 | Water Security |
| 10-Oct-21 | 0.9979 | −1.0431 | −6.3176 | Water Security | | | | | |

**Table A1.** *Cont.*

| Date | Coherence | Phase Diff. (Angular) | Phase Diff. (Temporal) | Leading Series | Date | Coherence | Phase Diff. (Angular) | Phase Diff. (Temporal) | Leading Series |
|---|---|---|---|---|---|---|---|---|---|
| **43.71-week period** | | | | | | | | | |
| 4-Oct-20 | 0.996 | −0.8364 | −5.819 | Water Security | 23-May-21 | 0.9998 | −0.9986 | −6.9474 | Water Security |
| 11-Oct-20 | 0.9961 | −0.8437 | −5.8698 | Water Security | 30-May-21 | 0.9998 | −0.9982 | −6.9447 | Water Security |
| 18-Oct-20 | 0.9962 | −0.8512 | −5.922 | Water Security | 6-Jun-21 | 0.9997 | −0.9975 | −6.9398 | Water Security |
| 25-Oct-20 | 0.9963 | −0.8587 | −5.9741 | Water Security | 13-Jun-21 | 0.9997 | −0.9964 | −6.9321 | Water Security |
| 1-Nov-20 | 0.9964 | −0.8662 | −6.0263 | Water Security | 20-Jun-21 | 0.9996 | −0.995 | −6.9224 | Water Security |
| 8-Nov-20 | 0.9965 | −0.8738 | −6.0792 | Water Security | 27-Jun-21 | 0.9996 | −0.9932 | −6.9099 | Water Security |
| 15-Nov-20 | 0.9967 | −0.8813 | −6.1314 | Water Security | 4-Jul-21 | 0.9995 | −0.9911 | −6.8953 | Water Security |
| 22-Nov-20 | 0.9969 | −0.8887 | −6.1829 | Water Security | 11-Jul-21 | 0.9994 | −0.9885 | −6.8772 | Water Security |
| 29-Nov-20 | 0.997 | −0.8961 | −6.2343 | Water Security | 18-Jul-21 | 0.9993 | −0.9857 | −6.8577 | Water Security |
| 6-Dec-20 | 0.9972 | −0.9033 | −6.2844 | Water Security | 25-Jul-21 | 0.9993 | −0.9825 | −6.8354 | Water Security |
| 13-Dec-20 | 0.9974 | −0.9104 | −6.3338 | Water Security | 1-Aug-21 | 0.9992 | −0.9789 | −6.8104 | Water Security |
| 20-Dec-20 | 0.9976 | −0.9174 | −6.3825 | Water Security | 8-Aug-21 | 0.9991 | −0.975 | −6.7833 | Water Security |
| 27-Dec-20 | 0.9978 | −0.9241 | −6.4291 | Water Security | 15-Aug-21 | 0.9989 | −0.9708 | −6.754 | Water Security |
| 3-Jan-21 | 0.9979 | −0.9307 | −6.4751 | Water Security | 22-Aug-21 | 0.9988 | −0.9662 | −6.722 | Water Security |
| 10-Jan-21 | 0.9981 | −0.937 | −6.5189 | Water Security | 29-Aug-21 | 0.9987 | −0.9613 | −6.6879 | Water Security |
| 17-Jan-21 | 0.9983 | −0.9431 | −6.5613 | Water Security | 5-Sep-21 | 0.9986 | −0.9561 | −6.6518 | Water Security |
| 24-Jan-21 | 0.9985 | −0.949 | −6.6024 | Water Security | 12-Sep-21 | 0.9985 | −0.9506 | −6.6135 | Water Security |
| 31-Jan-21 | 0.9986 | −0.9545 | −6.6406 | Water Security | 19-Sep-21 | 0.9983 | −0.9448 | −6.5731 | Water Security |
| 7-Feb-21 | 0.9988 | −0.9598 | −6.6775 | Water Security | 26-Sep-21 | 0.9982 | −0.9386 | −6.53 | Water Security |
| 14-Feb-21 | 0.9989 | −0.9648 | −6.7123 | Water Security | 3-Oct-21 | 0.9981 | −0.9322 | −6.4855 | Water Security |
| 21-Feb-21 | 0.9991 | −0.9695 | −6.745 | Water Security | 10-Oct-21 | 0.9979 | −0.9255 | −6.4389 | Water Security |
| 28-Feb-21 | 0.9992 | −0.9738 | −6.7749 | Water Security | 17-Oct-21 | 0.9978 | −0.9185 | −6.3902 | Water Security |
| 7-Mar-21 | 0.9993 | −0.9778 | −6.8027 | Water Security | 24-Oct-21 | 0.9976 | −0.9113 | −6.3401 | Water Security |
| 14-Mar-21 | 0.9994 | −0.9815 | −6.8285 | Water Security | 31-Oct-21 | 0.9975 | −0.9038 | −6.2879 | Water Security |
| 21-Mar-21 | 0.9995 | −0.9849 | −6.8521 | Water Security | 7-Nov-21 | 0.9974 | −0.896 | −6.2336 | Water Security |
| 28-Mar-21 | 0.9996 | −0.9879 | −6.873 | Water Security | 14-Nov-21 | 0.9972 | −0.8879 | −6.1773 | Water Security |

**Table A1.** *Cont.*

| Date | Coherence | Phase Diff. (Angular) | Phase Diff. (Temporal) | Leading Series | Date | Coherence | Phase Diff. (Angular) | Phase Diff. (Temporal) | Leading Series |
|---|---|---|---|---|---|---|---|---|---|
| **43.71-week period** | | | | | | | | | |
| 4-Apr-21 | 0.9997 | −0.9905 | −6.8911 | Water Security | 21-Nov-21 | 0.9971 | −0.8796 | −6.1195 | Water Security |
| 11-Apr-21 | 0.9997 | −0.9927 | −6.9064 | Water Security | 28-Nov-21 | 0.9969 | −0.8711 | −6.0604 | Water Security |
| 18-Apr-21 | 0.9998 | −0.9946 | −6.9196 | Water Security | 5-Dec-21 | 0.9968 | −0.8624 | −5.9999 | Water Security |
| 25-Apr-21 | 0.9998 | −0.9962 | −6.9307 | Water Security | 12-Dec-21 | 0.9966 | −0.8534 | −5.9373 | Water Security |
| 2-May-21 | 0.9998 | −0.9973 | −6.9384 | Water Security | 19-Dec-21 | 0.9965 | −0.8442 | −5.8733 | Water Security |
| 9-May-21 | 0.9998 | −0.9981 | −6.944 | Water Security | 26-Dec-21 | 0.9963 | −0.8348 | −5.8079 | Water Security |
| 16-May-21 | 0.9998 | −0.9985 | −6.9468 | Water Security | 2-Jan-22 | 0.9962 | −0.8253 | −5.7418 | Water Security |
| **64-week period** | | | | | | | | | |
| 16-Dec-18 | 0.9983 | 2.6076 | 26.5608 | Water Security | 2-Jun-19 | 0.998 | 2.5975 | 26.4579 | Water Security |
| 23-Dec-18 | 0.9983 | 2.607 | 26.5547 | Water Security | 9-Jun-19 | 0.998 | 2.5975 | 26.4579 | Water Security |
| 30-Dec-18 | 0.9983 | 2.6065 | 26.5496 | Water Security | 16-Jun-19 | 0.998 | 2.5975 | 26.4579 | Water Security |
| 6-Jan-19 | 0.9983 | 2.6059 | 26.5435 | Water Security | 23-Jun-19 | 0.998 | 2.5976 | 26.4589 | Water Security |
| 13-Jan-19 | 0.9983 | 2.6054 | 26.5384 | Water Security | 30-Jun-19 | 0.9979 | 2.5978 | 26.461 | Water Security |
| 20-Jan-19 | 0.9982 | 2.6048 | 26.5323 | Water Security | 7-Jul-19 | 0.9979 | 2.598 | 26.463 | Water Security |
| 27-Jan-19 | 0.9982 | 2.6043 | 26.5272 | Water Security | 14-Jul-19 | 0.9979 | 2.5983 | 26.4661 | Water Security |
| 3-Feb-19 | 0.9982 | 2.6037 | 26.5211 | Water Security | 21-Jul-19 | 0.9979 | 2.5987 | 26.4701 | Water Security |
| 10-Feb-19 | 0.9982 | 2.6032 | 26.516 | Water Security | 28-Jul-19 | 0.9979 | 2.5992 | 26.4752 | Water Security |
| 17-Feb-19 | 0.9982 | 2.6027 | 26.5109 | Water Security | 4-Aug-19 | 0.9978 | 2.5997 | 26.4803 | Water Security |
| 24-Feb-19 | 0.9982 | 2.6022 | 26.5058 | Water Security | 11-Aug-19 | 0.9978 | 2.6004 | 26.4875 | Water Security |
| 3-Mar-19 | 0.9982 | 2.6017 | 26.5007 | Water Security | 18-Aug-19 | 0.9978 | 2.6011 | 26.4946 | Water Security |
| 10-Mar-19 | 0.9982 | 2.6012 | 26.4956 | Water Security | 25-Aug-19 | 0.9977 | 2.602 | 26.5038 | Water Security |
| 17-Mar-19 | 0.9981 | 2.6007 | 26.4905 | Water Security | 1-Sep-19 | 0.9977 | 2.6029 | 26.5129 | Water Security |
| 24-Mar-19 | 0.9981 | 2.6003 | 26.4864 | Water Security | 8-Sep-19 | 0.9977 | 2.604 | 26.5241 | Water Security |
| 31-Mar-19 | 0.9981 | 2.5998 | 26.4813 | Water Security | 15-Sep-19 | 0.9976 | 2.6052 | 26.5363 | Water Security |
| 7-Apr-19 | 0.9981 | 2.5994 | 26.4773 | Water Security | 22-Sep-19 | 0.9976 | 2.6065 | 26.5496 | Water Security |
| 14-Apr-19 | 0.9981 | 2.5991 | 26.4742 | Water Security | 29-Sep-19 | 0.9975 | 2.6079 | 26.5639 | Water Security |

**Table A1.** *Cont.*

| Date | Coherence | Phase Diff. (Angular) | Phase Diff. (Temporal) | Leading Series | Date | Coherence | Phase Diff. (Angular) | Phase Diff. (Temporal) | Leading Series |
|---|---|---|---|---|---|---|---|---|---|
| **64-week period** | | | | | | | | | |
| 21-Apr-19 | 0.9981 | 2.5987 | 26.4701 | Water Security | 6-Oct-19 | 0.9975 | 2.6094 | 26.5791 | Water Security |
| 28-Apr-19 | 0.9981 | 2.5984 | 26.4671 | Water Security | 13-Oct-19 | 0.9974 | 2.6111 | 26.5964 | Water Security |
| 5-May-19 | 0.9981 | 2.5981 | 26.464 | Water Security | 20-Oct-19 | 0.9973 | 2.6129 | 26.6148 | Water Security |
| 12-May-19 | 0.998 | 2.5979 | 26.462 | Water Security | 27-Oct-19 | 0.9972 | 2.6149 | 26.6352 | Water Security |
| 19-May-19 | 0.998 | 2.5977 | 26.46 | Water Security | 3-Nov-19 | 0.9972 | 2.617 | 26.6565 | Water Security |
| 26-May-19 | 0.998 | 2.5976 | 26.4589 | Water Security | | | | | |

**Table A2.** The phase difference and leading time series for "Food Security"–"Energy Security" pair.

| Date | Coherence | Phase Diff. (Angular) | Phase Diff. (Temporal) | Leading Series | Date | Coherence | Phase Diff. (Angular) | Phase Diff. (Temporal) | Leading Series |
|---|---|---|---|---|---|---|---|---|---|
| **18.38-week period** | | | | | | | | | |
| 9-Oct-22 | 0.9969 | 0.4942 | 1.4456 | Food Security | 20-Nov-22 | 0.9993 | 0.5193 | 1.519 | Food Security |
| 16-Oct-22 | 0.998 | 0.4976 | 1.4555 | Food Security | 27-Nov-22 | 0.999 | 0.5243 | 1.5336 | Food Security |
| 23-Oct-22 | 0.9988 | 0.5014 | 1.4667 | Food Security | 4-Dec-22 | 0.9986 | 0.5296 | 1.5492 | Food Security |
| 30-Oct-22 | 0.9992 | 0.5055 | 1.4787 | Food Security | 11-Dec-22 | 0.9981 | 0.5351 | 1.5652 | Food Security |
| 6-Nov-22 | 0.9994 | 0.5099 | 1.4915 | Food Security | 18-Dec-22 | 0.9976 | 0.5407 | 1.5816 | Food Security |
| 13-Nov-22 | 0.9994 | 0.5145 | 1.505 | Food Security | 25-Dec-22 | 0.9971 | 0.5466 | 1.5989 | Food Security |
| **25.11-week period** | | | | | | | | | |
| 13-Feb-22 | 0.9966 | 0.3524 | 1.4081 | Food Security | 1-May-22 | 0.9996 | 0.1925 | 0.7692 | Food Security |
| 20-Feb-22 | 0.9971 | 0.3329 | 1.3302 | Food Security | 8-May-22 | 0.9997 | 0.1838 | 0.7344 | Food Security |
| 27-Feb-22 | 0.9976 | 0.3144 | 1.2563 | Food Security | 15-May-22 | 0.9997 | 0.1763 | 0.7045 | Food Security |
| 6-Mar-22 | 0.9979 | 0.297 | 1.1868 | Food Security | 22-May-22 | 0.9996 | 0.1698 | 0.6785 | Food Security |
| 13-Mar-22 | 0.9983 | 0.2806 | 1.1212 | Food Security | 29-May-22 | 0.9994 | 0.1645 | 0.6573 | Food Security |
| 20-Mar-22 | 0.9986 | 0.2652 | 1.0597 | Food Security | 5-Jun-22 | 0.9992 | 0.1605 | 0.6413 | Food Security |
| 27-Mar-22 | 0.9988 | 0.2507 | 1.0018 | Food Security | 12-Jun-22 | 0.9989 | 0.1579 | 0.6309 | Food Security |
| 3-Apr-22 | 0.9991 | 0.2371 | 0.9474 | Food Security | 19-Jun-22 | 0.9985 | 0.1567 | 0.6262 | Food Security |
| 10-Apr-22 | 0.9993 | 0.2245 | 0.8971 | Food Security | 26-Jun-22 | 0.9979 | 0.1572 | 0.6281 | Food Security |
| 17-Apr-22 | 0.9994 | 0.2129 | 0.8507 | Food Security | 3-Jul-22 | 0.9971 | 0.1594 | 0.6369 | Food Security |

**Table A2.** *Cont.*

| Date | Coherence | Phase Diff. (Angular) | Phase Diff. (Temporal) | Leading Series | Date | Coherence | Phase Diff. (Angular) | Phase Diff. (Temporal) | Leading Series |
|---|---|---|---|---|---|---|---|---|---|
| **25.11-week period** | | | | | | | | | |
| 24-Apr-22 | 0.9995 | 0.2022 | 0.808 | Food Security | | | | | |
| **34.3-week period** | | | | | | | | | |
| 19-May-19 | 0.9962 | 1.2155 | 6.6348 | Food Security | 1-Sep-19 | 0.9994 | 1.2373 | 6.7538 | Food Security |
| 26-May-19 | 0.9966 | 1.214 | 6.6266 | Food Security | 8-Sep-19 | 0.9994 | 1.2425 | 6.7822 | Food Security |
| 2-Jun-19 | 0.9969 | 1.2129 | 6.6206 | Food Security | 15-Sep-19 | 0.9994 | 1.2482 | 6.8133 | Food Security |
| 9-Jun-19 | 0.9972 | 1.2121 | 6.6162 | Food Security | 22-Sep-19 | 0.9994 | 1.2545 | 6.8477 | Food Security |
| 16-Jun-19 | 0.9975 | 1.2118 | 6.6146 | Food Security | 29-Sep-19 | 0.9994 | 1.2614 | 6.8853 | Food Security |
| 23-Jun-19 | 0.9978 | 1.2118 | 6.6146 | Food Security | 6-Oct-19 | 0.9993 | 1.2687 | 6.9252 | Food Security |
| 30-Jun-19 | 0.9981 | 1.2123 | 6.6173 | Food Security | 13-Oct-19 | 0.9991 | 1.2767 | 6.9689 | Food Security |
| 7-Jul-19 | 0.9983 | 1.2132 | 6.6222 | Food Security | 20-Oct-19 | 0.9989 | 1.2852 | 7.0153 | Food Security |
| 14-Jul-19 | 0.9985 | 1.2146 | 6.6299 | Food Security | 27-Oct-19 | 0.9987 | 1.2943 | 7.0649 | Food Security |
| 21-Jul-19 | 0.9987 | 1.2164 | 6.6397 | Food Security | 3-Nov-19 | 0.9985 | 1.304 | 7.1179 | Food Security |
| 28-Jul-19 | 0.9989 | 1.2187 | 6.6523 | Food Security | 10-Nov-19 | 0.9982 | 1.3143 | 7.1741 | Food Security |
| 4-Aug-19 | 0.999 | 1.2214 | 6.667 | Food Security | 17-Nov-19 | 0.9978 | 1.3252 | 7.2336 | Food Security |
| 11-Aug-19 | 0.9992 | 1.2246 | 6.6845 | Food Security | 24-Nov-19 | 0.9974 | 1.3367 | 7.2964 | Food Security |
| 18-Aug-19 | 0.9993 | 1.2283 | 6.7047 | Food Security | 1-Dec-19 | 0.9969 | 1.3488 | 7.3624 | Food Security |
| 25-Aug-19 | 0.9994 | 1.2325 | 6.7276 | Food Security | 8-Dec-19 | 0.9964 | 1.3615 | 7.4317 | Food Security |
| **39.4-week period** | | | | | | | | | |
| 6-Feb-22 | 0.9967 | −0.2142 | −1.3431 | Energy Security | 24-Apr-22 | 0.9997 | −0.1654 | −1.0371 | Energy Security |
| 13-Feb-22 | 0.9972 | −0.2095 | −1.3136 | Energy Security | 1-May-22 | 0.9997 | −0.1615 | −1.0126 | Energy Security |
| 20-Feb-22 | 0.9978 | −0.2049 | −1.2848 | Energy Security | 8-May-22 | 0.9996 | −0.1577 | −0.9888 | Energy Security |
| 27-Feb-22 | 0.9982 | −0.2002 | −1.2553 | Energy Security | 15-May-22 | 0.9995 | −0.1541 | −0.9662 | Energy Security |
| 6-Mar-22 | 0.9986 | −0.1956 | −1.2264 | Energy Security | 22-May-22 | 0.9993 | −0.1506 | −0.9443 | Energy Security |
| 13-Mar-22 | 0.9989 | −0.1911 | −1.1982 | Energy Security | 29-May-22 | 0.9991 | −0.1473 | −0.9236 | Energy Security |
| 20-Mar-22 | 0.9992 | −0.1866 | −1.17 | Energy Security | 5-Jun-22 | 0.9988 | −0.1442 | −0.9042 | Energy Security |
| 27-Mar-22 | 0.9994 | −0.1822 | −1.1424 | Energy Security | 12-Jun-22 | 0.9985 | −0.1413 | −0.886 | Energy Security |
| 3-Apr-22 | 0.9995 | −0.1778 | −1.1148 | Energy Security | 19-Jun-22 | 0.9982 | −0.1386 | −0.869 | Energy Security |

**Table A2.** *Cont.*

| Date | Coherence | Phase Diff. (Angular) | Phase Diff. (Temporal) | Leading Series | Date | Coherence | Phase Diff. (Angular) | Phase Diff. (Temporal) | Leading Series |
|---|---|---|---|---|---|---|---|---|---|
| **39.4-week period** | | | | | | | | | |
| 10-Apr-22 | 0.9997 | −0.1736 | −1.0885 | Energy Security | 26-Jun-22 | 0.9978 | −0.1361 | −0.8534 | Energy Security |
| 17-Apr-22 | 0.9997 | −0.1694 | −1.0622 | Energy Security | 3-Jul-22 | 0.9974 | −0.1339 | −0.8396 | Energy Security |

**Table A3.** The phase difference and leading time series for "Water Security"–"Energy Security" pair.

| Date | Coherence | Phase Diff. (Angular) | Phase Diff. (Temporal) | Leading Series | Date | Coherence | Phase Diff. (Angular) | Phase Diff. (Temporal) | Leading Series |
|---|---|---|---|---|---|---|---|---|---|
| **17.14-week period** | | | | | | | | | |
| 30-Oct-22 | 0.9968 | 0.641 | 1.7494 | Water Security | 18-Dec-22 | 0.9995 | 0.714 | 1.9487 | Water Security |
| 6-Nov-22 | 0.998 | 0.6543 | 1.7857 | Water Security | 25-Dec-22 | 0.9992 | 0.7201 | 1.9653 | Water Security |
| 13-Nov-22 | 0.9988 | 0.6669 | 1.8201 | Water Security | 1-Jan-23 | 0.9988 | 0.7252 | 1.9793 | Water Security |
| 20-Nov-22 | 0.9994 | 0.6786 | 1.8521 | Water Security | 8-Jan-23 | 0.9983 | 0.7295 | 1.991 | Water Security |
| 27-Nov-22 | 0.9997 | 0.6891 | 1.8807 | Water Security | 15-Jan-23 | 0.9979 | 0.7329 | 2.0003 | Water Security |
| 4-Dec-22 | 0.9998 | 0.6986 | 1.9067 | Water Security | 22-Jan-23 | 0.9975 | 0.7357 | 2.0079 | Water Security |
| 11-Dec-22 | 0.9997 | 0.7068 | 1.929 | Water Security | | | | | |
| **25.11-week period** | | | | | | | | | |
| 8-Aug-21 | 0.9967 | 1.4767 | 5.9007 | Water Security | 27-Mar-22 | 0.9974 | 1.7032 | 6.8057 | Energy Security |
| 15-Aug-21 | 0.9972 | 1.4849 | 5.9334 | Water Security | 3-Apr-22 | 0.9972 | 1.7131 | 6.8453 | Energy Security |
| 22-Aug-21 | 0.9976 | 1.4928 | 5.965 | Water Security | 10-Apr-22 | 0.9971 | 1.7234 | 6.8865 | Energy Security |
| 29-Aug-21 | 0.998 | 1.5003 | 5.995 | Water Security | 17-Apr-22 | 0.997 | 1.734 | 6.9288 | Energy Security |
| 5-Sep-21 | 0.9983 | 1.5075 | 6.0237 | Water Security | 24-Apr-22 | 0.9968 | 1.7448 | 6.972 | Energy Security |
| 12-Sep-21 | 0.9985 | 1.5143 | 6.0509 | Water Security | 1-May-22 | 0.9967 | 1.756 | 7.0167 | Energy Security |
| 19-Sep-21 | 0.9987 | 1.5209 | 6.0773 | Water Security | 8-May-22 | 0.9966 | 1.7675 | 7.0627 | Energy Security |
| 26-Sep-21 | 0.9989 | 1.5272 | 6.1025 | Water Security | 15-May-22 | 0.9966 | 1.7793 | 7.1098 | Energy Security |
| 3-Oct-21 | 0.999 | 1.5334 | 6.1272 | Water Security | 22-May-22 | 0.9965 | 1.7914 | 7.1582 | Energy Security |
| 10-Oct-21 | 0.9991 | 1.5393 | 6.1508 | Water Security | 29-May-22 | 0.9965 | 1.8037 | 7.2073 | Energy Security |
| 17-Oct-21 | 0.9992 | 1.545 | 6.1736 | Water Security | 5-Jun-22 | 0.9965 | 1.8163 | 7.2577 | Energy Security |
| 24-Oct-21 | 0.9992 | 1.5507 | 6.1964 | Water Security | 12-Jun-22 | 0.9966 | 1.8292 | 7.3092 | Energy Security |
| 31-Oct-21 | 0.9992 | 1.5563 | 6.2187 | Water Security | 19-Jun-22 | 0.9967 | 1.8423 | 7.3616 | Energy Security |

**Table A3.** *Cont.*

| Date | Coherence | Phase Diff. (Angular) | Phase Diff. (Temporal) | Leading Series | Date | Coherence | Phase Diff. (Angular) | Phase Diff. (Temporal) | Leading Series |
|---|---|---|---|---|---|---|---|---|---|
| **25.11-week period** | | | | | | | | | |
| 7-Nov-21 | 0.9993 | 1.5618 | 6.2407 | Water Security | 26-Jun-22 | 0.9968 | 1.8557 | 7.4151 | Energy Security |
| 14-Nov-21 | 0.9993 | 1.5673 | 6.2627 | Water Security | 3-Jul-22 | 0.9969 | 1.8692 | 7.469 | Energy Security |
| 21-Nov-21 | 0.9992 | 1.5728 | 6.2847 | Energy Security | 10-Jul-22 | 0.9971 | 1.883 | 7.5242 | Energy Security |
| 28-Nov-21 | 0.9992 | 1.5783 | 6.3067 | Energy Security | 17-Jul-22 | 0.9973 | 1.897 | 7.5801 | Energy Security |
| 5-Dec-21 | 0.9992 | 1.584 | 6.3294 | Energy Security | 24-Jul-22 | 0.9975 | 1.9111 | 7.6365 | Energy Security |
| 12-Dec-21 | 0.9991 | 1.5897 | 6.3522 | Energy Security | 31-Jul-22 | 0.9977 | 1.9253 | 7.6932 | Energy Security |
| 19-Dec-21 | 0.9991 | 1.5956 | 6.3758 | Energy Security | 7-Aug-22 | 0.998 | 1.9397 | 7.7508 | Energy Security |
| 26-Dec-21 | 0.999 | 1.6016 | 6.3998 | Energy Security | 14-Aug-22 | 0.9982 | 1.9542 | 7.8087 | Energy Security |
| 2-Jan-22 | 0.9989 | 1.6079 | 6.4249 | Energy Security | 21-Aug-22 | 0.9985 | 1.9687 | 7.8666 | Energy Security |
| 9-Jan-22 | 0.9988 | 1.6143 | 6.4505 | Energy Security | 28-Aug-22 | 0.9987 | 1.9832 | 7.9246 | Energy Security |
| 16-Jan-22 | 0.9987 | 1.621 | 6.4773 | Energy Security | 4-Sep-22 | 0.9988 | 1.9977 | 7.9825 | Energy Security |
| 23-Jan-22 | 0.9986 | 1.6279 | 6.5049 | Energy Security | 11-Sep-22 | 0.999 | 2.0121 | 8.0401 | Energy Security |
| 30-Jan-22 | 0.9985 | 1.6351 | 6.5336 | Energy Security | 18-Sep-22 | 0.999 | 2.0264 | 8.0972 | Energy Security |
| 6-Feb-22 | 0.9984 | 1.6425 | 6.5632 | Energy Security | 25-Sep-22 | 0.999 | 2.0405 | 8.1535 | Energy Security |
| 13-Feb-22 | 0.9982 | 1.6503 | 6.5944 | Energy Security | 2-Oct-22 | 0.9989 | 2.0544 | 8.2091 | Energy Security |
| 20-Feb-22 | 0.9981 | 1.6583 | 6.6263 | Energy Security | 9-Oct-22 | 0.9988 | 2.068 | 8.2634 | Energy Security |
| 27-Feb-22 | 0.998 | 1.6666 | 6.6595 | Energy Security | 16-Oct-22 | 0.9985 | 2.0813 | 8.3166 | Energy Security |
| 6-Mar-22 | 0.9978 | 1.6753 | 6.6943 | Energy Security | 23-Oct-22 | 0.9982 | 2.0941 | 8.3677 | Energy Security |
| 13-Mar-22 | 0.9977 | 1.6843 | 6.7302 | Energy Security | 30-Oct-22 | 0.9978 | 2.1065 | 8.4173 | Energy Security |
| 20-Mar-22 | 0.9975 | 1.6936 | 6.7674 | Energy Security | 6-Nov-22 | 0.9973 | 2.1184 | 8.4648 | Energy Security |
| 12-Apr-20 | 0.9987 | −0.0654 | −0.879 | Energy Security | 11-Apr-21 | 0.9997 | −0.0594 | −0.7984 | Energy Security |
| 19-Apr-20 | 0.9989 | −0.0616 | −0.8279 | Energy Security | 18-Apr-21 | 0.9997 | −0.0616 | −0.8279 | Energy Security |
| 26-Apr-20 | 0.999 | −0.0581 | −0.7809 | Energy Security | 25-Apr-21 | 0.9997 | −0.0638 | −0.8575 | Energy Security |
| 3-May-20 | 0.9991 | −0.0548 | −0.7365 | Energy Security | 2-May-21 | 0.9996 | −0.0661 | −0.8884 | Energy Security |
| 10-May-20 | 0.9992 | −0.0517 | −0.6949 | Energy Security | 9-May-21 | 0.9996 | −0.0685 | −0.9207 | Energy Security |
| 17-May-20 | 0.9992 | −0.0488 | −0.6559 | Energy Security | 16-May-21 | 0.9996 | −0.0708 | −0.9516 | Energy Security |

**Table A3.** *Cont.*

| Date | Coherence | Phase Diff. (Angular) | Phase Diff. (Temporal) | Leading Series | Date | Coherence | Phase Diff. (Angular) | Phase Diff. (Temporal) | Leading Series |
|---|---|---|---|---|---|---|---|---|---|
| **25.11-week period** | | | | | | | | | |
| 24-May-20 | 0.9993 | −0.0461 | −0.6196 | Energy Security | 23-May-21 | 0.9996 | −0.0732 | −0.9838 | Energy Security |
| 31-May-20 | 0.9994 | −0.0436 | −0.586 | Energy Security | 30-May-21 | 0.9996 | −0.0757 | −1.0174 | Energy Security |
| 7-Jun-20 | 0.9995 | −0.0412 | −0.5537 | Energy Security | 6-Jun-21 | 0.9996 | −0.0782 | −1.051 | Energy Security |
| 14-Jun-20 | 0.9995 | −0.0391 | −0.5255 | Energy Security | 13-Jun-21 | 0.9995 | −0.0807 | −1.0846 | Energy Security |
| 21-Jun-20 | 0.9996 | −0.0371 | −0.4986 | Energy Security | 20-Jun-21 | 0.9995 | −0.0833 | −1.1196 | Energy Security |
| 28-Jun-20 | 0.9996 | −0.0353 | −0.4744 | Energy Security | 27-Jun-21 | 0.9995 | −0.0859 | −1.1545 | Energy Security |
| 5-Jul-20 | 0.9997 | −0.0336 | −0.4516 | Energy Security | 4-Jul-21 | 0.9995 | −0.0886 | −1.1908 | Energy Security |
| 12-Jul-20 | 0.9997 | −0.0321 | −0.4314 | Energy Security | 11-Jul-21 | 0.9995 | −0.0913 | −1.2271 | Energy Security |
| 19-Jul-20 | 0.9998 | −0.0308 | −0.414 | Energy Security | 18-Jul-21 | 0.9995 | −0.094 | −1.2634 | Energy Security |
| 26-Jul-20 | 0.9998 | −0.0296 | −0.3978 | Energy Security | 25-Jul-21 | 0.9994 | −0.0968 | −1.301 | Energy Security |
| 2-Aug-20 | 0.9998 | −0.0286 | −0.3844 | Energy Security | 1-Aug-21 | 0.9994 | −0.0996 | −1.3387 | Energy Security |
| 9-Aug-20 | 0.9999 | −0.0276 | −0.371 | Energy Security | 8-Aug-21 | 0.9994 | −0.1024 | −1.3763 | Energy Security |
| 16-Aug-20 | 0.9999 | −0.0269 | −0.3615 | Energy Security | 15-Aug-21 | 0.9994 | −0.1052 | −1.4139 | Energy Security |
| 23-Aug-20 | 0.9999 | −0.0262 | −0.3521 | Energy Security | 22-Aug-21 | 0.9994 | −0.1081 | −1.4529 | Energy Security |
| 30-Aug-20 | 0.9999 | −0.0257 | −0.3454 | Energy Security | 29-Aug-21 | 0.9994 | −0.111 | −1.4919 | Energy Security |
| 6-Sep-20 | 0.9999 | −0.0253 | −0.34 | Energy Security | 5-Sep-21 | 0.9993 | −0.114 | −1.5322 | Energy Security |
| 13-Sep-20 | 0.9999 | −0.0251 | −0.3374 | Energy Security | 12-Sep-21 | 0.9993 | −0.117 | −1.5725 | Energy Security |
| 20-Sep-20 | 1 | −0.0249 | −0.3347 | Energy Security | 19-Sep-21 | 0.9993 | −0.12 | −1.6128 | Energy Security |
| 27-Sep-20 | 1 | −0.0249 | −0.3347 | Energy Security | 26-Sep-21 | 0.9993 | −0.123 | −1.6532 | Energy Security |
| 4-Oct-20 | 1 | −0.025 | −0.336 | Energy Security | 3-Oct-21 | 0.9993 | −0.1261 | −1.6948 | Energy Security |
| 11-Oct-20 | 1 | −0.0252 | −0.3387 | Energy Security | 10-Oct-21 | 0.9993 | −0.1291 | −1.7352 | Energy Security |
| 18-Oct-20 | 1 | −0.0255 | −0.3427 | Energy Security | 17-Oct-21 | 0.9993 | −0.1322 | −1.7768 | Energy Security |
| 25-Oct-20 | 1 | −0.0259 | −0.3481 | Energy Security | 24-Oct-21 | 0.9992 | −0.1354 | −1.8198 | Energy Security |
| 1-Nov-20 | 1 | −0.0264 | −0.3548 | Energy Security | 31-Oct-21 | 0.9992 | −0.1385 | −1.8615 | Energy Security |
| 8-Nov-20 | 1 | −0.027 | −0.3629 | Energy Security | 7-Nov-21 | 0.9992 | −0.1417 | −1.9045 | Energy Security |
| 15-Nov-20 | 1 | −0.0276 | −0.371 | Energy Security | 14-Nov-21 | 0.9992 | −0.1449 | −1.9475 | Energy Security |

**Table A3.** *Cont.*

| Date | Coherence | Phase Diff. (Angular) | Phase Diff. (Temporal) | Leading Series | Date | Coherence | Phase Diff. (Angular) | Phase Diff. (Temporal) | Leading Series |
|---|---|---|---|---|---|---|---|---|---|
| **84.45-week period** | | | | | | | | | |
| 22-Nov-20 | 1 | −0.0284 | −0.3817 | Energy Security | 21-Nov-21 | 0.9992 | −0.1481 | −1.9905 | Energy Security |
| 29-Nov-20 | 0.9999 | −0.0293 | −0.3938 | Energy Security | 28-Nov-21 | 0.9992 | −0.1514 | −2.0349 | Energy Security |
| 6-Dec-20 | 0.9999 | −0.0302 | −0.4059 | Energy Security | 5-Dec-21 | 0.9992 | −0.1546 | −2.0779 | Energy Security |
| 13-Dec-20 | 0.9999 | −0.0313 | −0.4207 | Energy Security | 12-Dec-21 | 0.9991 | −0.1579 | −2.1222 | Energy Security |
| 20-Dec-20 | 0.9999 | −0.0324 | −0.4355 | Energy Security | 19-Dec-21 | 0.9991 | −0.1612 | −2.1666 | Energy Security |
| 27-Dec-20 | 0.9999 | −0.0336 | −0.4516 | Energy Security | 26-Dec-21 | 0.9991 | −0.1645 | −2.2109 | Energy Security |
| 3-Jan-21 | 0.9999 | −0.0348 | −0.4677 | Energy Security | 2-Jan-22 | 0.9991 | −0.1679 | −2.2566 | Energy Security |
| 10-Jan-21 | 0.9999 | −0.0362 | −0.4865 | Energy Security | 9-Jan-22 | 0.9991 | −0.1712 | −2.301 | Energy Security |
| 17-Jan-21 | 0.9999 | −0.0376 | −0.5054 | Energy Security | 16-Jan-22 | 0.9991 | −0.1746 | −2.3467 | Energy Security |
| 24-Jan-21 | 0.9999 | −0.039 | −0.5242 | Energy Security | 23-Jan-22 | 0.9991 | −0.178 | −2.3924 | Energy Security |
| 31-Jan-21 | 0.9998 | −0.0406 | −0.5457 | Energy Security | 30-Jan-22 | 0.9991 | −0.1814 | −2.4381 | Energy Security |
| 7-Feb-21 | 0.9998 | −0.0422 | −0.5672 | Energy Security | 6-Feb-22 | 0.999 | −0.1848 | −2.4838 | Energy Security |
| 14-Feb-21 | 0.9998 | −0.0439 | −0.59 | Energy Security | 13-Feb-22 | 0.999 | −0.1882 | −2.5295 | Energy Security |
| 21-Feb-21 | 0.9998 | −0.0456 | −0.6129 | Energy Security | 20-Feb-22 | 0.999 | −0.1917 | −2.5765 | Energy Security |
| 28-Feb-21 | 0.9998 | −0.0474 | −0.6371 | Energy Security | 27-Feb-22 | 0.999 | −0.1951 | −2.6222 | Energy Security |
| 7-Mar-21 | 0.9998 | −0.0493 | −0.6626 | Energy Security | 6-Mar-22 | 0.999 | −0.1986 | −2.6693 | Energy Security |
| 14-Mar-21 | 0.9998 | −0.0512 | −0.6881 | Energy Security | 13-Mar-22 | 0.999 | −0.2021 | −2.7163 | Energy Security |
| 21-Mar-21 | 0.9997 | −0.0532 | −0.715 | Energy Security | 20-Mar-22 | 0.999 | −0.2056 | −2.7633 | Energy Security |
| 28-Mar-21 | 0.9997 | −0.0552 | −0.7419 | Energy Security | 27-Mar-22 | 0.999 | −0.2091 | −2.8104 | Energy Security |
| 4-Apr-21 | 0.9997 | −0.0573 | −0.7701 | Energy Security | 3-Apr-22 | 0.999 | −0.2126 | −2.8574 | Energy Security |

## References


1. Hassani, H.; Huang, X.; MacFeely, S.; Entezarian, M.R. Big Data and the United Nations Sustainable Development Goals (UN SDGs) at a Glance. *Big Data Cogn. Comput.* **2021**, *5*, 28. [CrossRef]
2. Hassani, H.; Huang, X.; Silva, E. Big Data and Climate Change. *Big Data Cogn. Comput.* **2019**, *3*, 12. [CrossRef]
3. Google Trend. Available online: Https://trends.google.com/ (accessed on 15 September 2023).
4. McCracken, J.T. Google Trends as a leading indicator of the real economy. *J. Monet. Econ.* **2016**, *82*, 1–14.
5. Li, J.Y.; Wang, Q. Google Trends as a proxy for tourism demand: An empirical study. *Tour. Manag.* **2016**, *52*, 190–199.
6. Wang, Y.; Chen, X. Using Google Trends to predict stock market returns: Evidence from China. *J. Bus. Res.* **2018**, *89*, 193–201.
7. Polgreen, P.M.; Brandt, A.L.; Janelle, J.M.; Lederberg, E.J. Predicting the spread of infectious diseases using Google Trends data. *J. Infect. Dis.* **2008**, *198*, 962–967.
8. Choi, H.; Varian, H. Predicting the present with Google Trends. *Econ. Rec.* **2012**, *88*, 2–9. [CrossRef]

9. Carneiro, H.A.; Mylonakis, E.; Google Trends Team. Google Trends: A web-based tool for real-time surveillance of disease outbreaks. *Clin. Infect. Dis.* **2009**, *49*, 1557–1564. [CrossRef] [PubMed]
10. Eysenbach, G. Infodemiology: Tracking flu-related searches on the web for syndromic surveillance. *AMIA Annu. Symp. Proc.* **2006**, *2006*, 244–248.
11. Nuti, S.V.; Wayda, B.; Ranasinghe, I.; Wang, S.; Dreyer, R.P.; Chen, S.I.; Krumholz, H.M. The use of Google Trends in health care research: A systematic review. *PLoS ONE* **2014**, *9*, e109583. [CrossRef] [PubMed]
12. Yang, A.C.; Tsai, S.J.; Huang, N.E.; Peng, C.K. Association of Internet search trends with suicide death in Taipei City, Taiwan, 2004–2009. *J. Affect. Disord.* **2010**, *124*, 307–311. [CrossRef]
13. Ginsberg, J.; Mohebbi, M.H.; Patel, R.S.; Brammer, L.; Smolinski, M.S.; Brilliant, L. Detecting influenza epidemics using search engine query data. *Nature* **2009**, *457*, 1012–1014. [CrossRef]
14. Pervaiz, F.; Pervaiz, M.K. Investigating the role of Google search on tourism demand in Australia: An econometric approach. *J. Travel Res.* **2012**, *51*, 470–480.
15. Haim, M.; Graefe, A.; Vogt, C.A. Investigating the use of destination-related online media in travel planning. *J. Travel Res.* **2011**, *50*, 571–582.
16. Marquet, O.; Alberico, C.; Miranda-Moreno, L.F. Disentangling the effects of accessibility on mode choice using Google Maps: Evidence from Santiago de Chile. *Transp. Res. Part A Policy Pract.* **2016**, *94*, 450–465.
17. Wang, Q.; Chen, J. Internet search behavior and tourism destination choice: A case study of Hainan Island, China. *J. Destin. Mark. Manag.* **2019**, *11*, 41–50.
18. Killworth, P.D.; Bernard, H.R. The reversal small-world experiment. *Soc. Netw.* **1978**, *1*, 159–192. [CrossRef]
19. Killworth, P.D.; McCarty, C.; Bernard, H.R.; House, M. The accuracy of small world chains in social networks. *Soc. Netw.* **2006**, *28*, 85–96. [CrossRef]
20. Kleinberg, J. The small-world phenomenon: An algorithmic perspective. In Proceedings of the Thirty-Second Annual ACM Symposium on Theory of Computing, Portland, OR, USA, 21–23 May 2000; ACM: New York, NY, USA, 2000; pp. 163–170.
21. Kleinberg, J. Small-world phenomena and the dynamics of information. In *Advances in Neural Information Processing Systems*; NeurIPS: Vancouver, BC, Canada, 2002; pp. 431–438.
22. Kleinfeld, J. Could it be a big world after all? The six degrees of separation myth. *Society* **2002**, *12*, 5-2.
23. Korte, C.; Milgram, S. Acquaintance networks between racial groups: Application of the small world method. *J. Personal. Soc. Psychol.* **1970**, *15*, 101. [CrossRef]
24. Badi, I.; Elghoul, E.M. Using Grey-ARAS Approach to Investigate the Role of Social Media Platforms in Spreading Fake News During COVID-19 Pandemic. *J. Intell. Manag. Decis.* **2023**, *2*, 66–73. [CrossRef]
25. Dong, K.; Tse, Y.K. Examining Public Perceptions of UK Rail Strikes: A Text Analytics Approach Using Twitter Data. *Inf. Dyn. Appl.* **2023**, *2*, 101–114. [CrossRef]
26. Thakur, N. Social Media Mining and Analysis: A Brief Review of Recent Challenges. *Information* **2023**, *14*, 484. [CrossRef]
27. Gundecha, P.; Liu, H. Mining Social Media: A Brief Introduction. In *2012 TutORials in Operations Research*; INFORMS: Catonsville, MD, USA, 2012; pp. 1–17.
28. Bhattacharya, A. A Multi-Agent Model to Study the Effects of Crowdsourcing on the Spread of Misinformation in Social Networks. Master's Thesis, University of Cincinnati, Cincinnati, OH, USA, 2023. Available online: http://rave.ohiolink.edu/etdc/view?acc_num=ucin1684770124758418 (accessed on 14 August 2023).
29. Belle Wong, J.D. Top Social Media Statistics and Trends of 2023. Available online: https://www.forbes.com/advisor/business/social-media-statistics/ (accessed on 14 August 2023).
30. Jones, I.; Liu, H. Mining Social Media: Challenges and Opportunities. In Proceedings of the 2013 International Conference on Social Intelligence and Technology, State College, PA, USA, 8–10 May 2013; pp. 90–99.
31. Thakur, N. Sentiment Analysis and Text Analysis of the Public Discourse on Twitter about COVID-19 and MPox. *Big Data Cogn. Comput.* **2023**, *7*, 116. [CrossRef]
32. Thakur, N. A Large-Scale Dataset of Twitter Chatter about Online Learning during the Current COVID-19 Omicron Wave. *Data* **2022**, *7*, 109. [CrossRef]
33. Thakur, N.; Han, C.Y. A Multimodal Approach for Early Detection of Cognitive Impairment from Tweets. In *Human Interaction, Emerging Technologies and Future Systems V*; Springer International Publishing: Cham, Switzerland, 2022; pp. 11–19. ISBN 9783030855390.
34. Carmona, R.; Hwang, W.L.; Torresani, B. *Practical Time Frequency Analysis: Gabor and Wavelet Transforms with an Implementation in S*; Academic Press: San Diego, CA, USA, 1998.
35. Morlet, J.; Arens, G.; Fourgeau, E.; Giard, D. Wave propagation and sampling theory—Part I: Complex signal and scattering in multilayered media. *Geophysics* **1982**, *47*, 203–221. [CrossRef]
36. Torrence, C.; Compo, G.P. A practical guide to wavelet analysis. *Bull. Am. Meteorol. Soc.* **1998**, *79*, 61–78. [CrossRef]
37. Ge, Z. Significance tests for the wavelet power and the wavelet power spectrum. *Ann. Geophys.* **2007**, *25*, 2259–2269. [CrossRef]

38. Maraun, D.; Kurths, J. Cross wavelet analysis: Significance testing and pitfalls. *Nonlinear Process. Geophys.* **2004**, *11*, 505–514. [CrossRef]
39. Ge, Z. Significance tests for the wavelet cross spectrum and wavelet linear coherence. *Ann. Geophys.* **2008**, *26*, 3819–3829. [CrossRef]

big data and
cognitive computing

MDPI

*Article*

# Analyzing Trends in Digital Transformation Korean Social Media Data: A Semantic Network Analysis

**Jong-Hwi Song [1] and Byung-Suk Seo [2,*]**

[1] Division of Software, Yonsei University, Wonju City 26493, Republic of Korea; jh_song@yonsei.ac.kr
[2] Department of Computer Engineering, Sangji University, Wonju City 26339, Republic of Korea
* Correspondence: seobs@sangji.ac.kr

**Abstract:** This study explores the impact of digital transformation on Korean society by analyzing Korean social media data, focusing on the societal and economic effects triggered by advancements in digital technology. Utilizing text mining techniques and semantic network analysis, we extracted key terms and their relationships from online news and blogs, identifying major themes related to digital transformation. Our analysis, based on data collected from major Korean portals using various related search terms, provides deep insights into how digital evolution influences individuals, businesses, and government sectors. The findings offer a comprehensive view of the technological and social trends emerging from digital transformation, including its policy, economic, and educational implications. This research not only sheds light on the understanding and strategic approaches to digital transformation in Korea but also demonstrates the potential of social media data in analyzing the societal impact of technological advancements, offering valuable resources for future research in effectively navigating the era of digital change.

**Keywords:** digital transformation; online news; blog; big data; text mining; semantic network analysis; CONCOR analysis

## 1. Introduction

The digital era's boom has ushered in a paradigm shift known as the "digital transformation", which significantly impacts various societal sectors, including business, education, and governance [1,2]. This shift transcends mere technological upgrades, signifying a comprehensive overhaul of organizations, culture, and operations [1]. As our society becomes increasingly digitalized, with a rise in public and private services based on such technologies, the ability to adapt to variable situations has become crucial [2].

In the face of inevitable digital transformation across most industries, companies facing a challenging internal and external business environment, especially those with limited resources, have been compelled to seek breakthroughs via open innovation in corporate and business models [3,4]. Accelerated by pandemics or radical technological innovations, businesses have had to continuously evolve and adjust their innovation strategies to maintain their operations [5].

Additionally, digital transformation in business dismantles barriers among people, businesses, and objects, enabling the creation of new products and services and the genesis of new ventures [6]. It emphasizes the construction of new business models, processes, and software and systems that lead to increased revenue, competitive advantage, and higher efficiency [6].

In this context, social media becomes a vital tool for capturing public sentiment and social trends. Especially regarding the digital transformation, the role of these platforms in shaping public discourse is crucial. Considering Korea's high social media penetration rate in harmony with traditional values and technological advancements, it presents an opportunity to gauge the social and individual impacts of the digital transformation.

Within the context of social media, semantic network analysis holds particular importance. It enables mapping how ideas, trends, and sentiments interconnect and evolve over time within digital conversations. Researchers can gain insights into public opinion, emerging trends, and the cultural zeitgeist of specific communities or societies by examining these networks [6,7]. This approach not only enhances our understanding of the dynamics of digital communication but also offers practical applications across various domains, from marketing to political science, providing a lens through which to view the complex tapestry of human thought and expression in the digital realm [7–9].

This study applies semantic network analysis techniques to social media data from Korea, focusing on the keyword "digital transformation" to analyze the relationships between keywords and phrases within social media posts. This method reflects systematic and meta-analytical techniques previously employed, with an emphasis on distilling core ideas from extensive data. Moreover, the research integrates perspectives based on social media, particularly within the non-Western context of Korea, to provide insights into the concept of digital transformation.

## 2. Related Studies

### 2.1. Digital Transformation

Digital transformation involves using new digital technologies such as social media, mobile technologies, analytics, or embedded devices to enable key business improvements, including enhanced customer experience, streamlined operations, or new business models [10]. It represents the use of technology to fundamentally improve a company's performance or reach, encompassing changes related to the application of digital technologies in all aspects of human society [11,12].

Digital technologies can be seen as key assets for leveraging organizational innovation, considering their disruptive nature and inter-organizational and systemic effects [13]. To achieve successful digital transformation, changes must occur at various levels within an organization, including the exchange of resources and capabilities, adaptation of core businesses, restructuring of processes and structures, and practical implementation of a digital culture [14–17].

It is argued that digital transformation needs to capture both technology-centric and actor-centric perspectives [18]. For leveraging the technology-centric view, the literature on technological disruptions was included and merged with research on digital transformation [18]. Regarding the actor-centric perspective, intrinsic implications were derived from the field of entrepreneurship, which is seen as capable of adding valuable insights into action-driven innovation and renewal processes within the framework [18].

The rapid development of various digital technologies enables the transformation into digital service, thereby facilitating the accelerated growth of the service industry through digital transformation [19,20]. Considered foundational technologies for digitalization, IoT (Internet of Things), cloud computing, and Big Data analytics provide service firms with the capacity to develop customer-oriented business models [21,22]. Manufacturing companies are also shifting their primary focus to ecosystems that integrate products with services to maximize customer value [20].

The recent trend is evolving the world into a single competitive market through one platform [23]. Consequently, suppliers and buyers strive to secure a competitive advantage by offering more choices in an increasingly fierce market [23]. As a result, digital transformation becomes a key strategic force that can enable innovation for creating customer value [23].

### 2.2. Big Data and Semantic Network Analysis

The term "Big Data" has attracted considerable attention since the early 21st century, with various researchers attempting to establish a widely accepted definition. One of the most common definitions introduced the challenge of Big Data through the 3Vs: volume (large amounts of data), velocity (rapid data streams), and variety (heterogeneous content) [24]. Big Data has been defined as large volumes of structured or unstructured data, indicating that traditional data processing technologies struggle to manage and process it due to the data's complexity and volume [25].

From a corporate perspective, the same information is required across various aspects such as customers, their needs, competition, products, distribution channels, service providers, and laws, making Big Data analytics necessary for making informed decisions [26]. Mobile marketing and social media platforms can extend knowledge by incorporating detailed personal information such as geographical location, time, interests, and gender [26].

Data exist almost everywhere in business and everyday life, and their volume is continuously increasing [26]. With the growing amount of data, scalability issues have become apparent, leading to increased processing times [27,28]. However, combining traditional algorithms with Big Data technologies has played a role in mitigating these scalability issues [27,29].

Among various Big Data analytics techniques, semantic network analysis is a method that models semantic relationships represented by graphs with nodes and edges [30]. Semantic networks can be automatically extracted from unstructured text data and used as a medium for visual text analysis, incorporating information retrieval and text mining techniques to extract relationships within the text [31–33].

Compared to traditional methods of text data analysis, semantic network analysis allows for the objective and accurate understanding of the structural relationships between individual words and the overall context, with relatively less reflection of the researcher's subjective thoughts [34,35]. In a semantic network model, nodes represent semantic or lexical units, while edges denote the associations and similarities, co-occurrences, or intensity between them [36,37]. Representing relationships with graphs that have labeled nodes and edges enables the identification of semantic relationships, patterns, and similarities between words regarding a specific topic, making it easier to discover insights [34,38]. Therefore, semantic network analysis can be actively used to explore the qualitative aspects or intrinsic meanings of issues by focusing on relationships within online Big Data, such as news on portal sites or posts on social media.

Data from various social media posts, news, and blogs on internet portal sites have become major sources supplying the raw materials necessary for Big Data analysis. Therefore, this paper aims to identify public perceptions related to digital transformation and discover widely recognized trends using text mining techniques and semantic network analysis.

## 3. Method

In this study, we analyzed the key thoughts of Korean users on digital transformation using text mining techniques and semantic network analysis on Big Data collected from the internet. Text mining is the process of extracting meaningful information from unstructured text data, exploring core themes and trends from multiple perspectives. Furthermore, to understand the relationships between the extracted keywords, semantic network analysis was utilized. This paper outlines an analytical process to comprehend the semantics between words related to digital transformation in online news articles and blogs, based on the degree of their co-occurrence. The overall process is illustrated in Figure 1.

### 3.1. Data Collection

For the semantic network analysis conducted in this research, a search was performed on Korea's two major portals, Naver [38] and Daum [39], using the keyword "digital transformation" to collect data from online news and blogs. Based on this, 1236 online news articles from Naver News and 1137 blog posts from both Naver and Daum blogs were collected as the data for analysis.

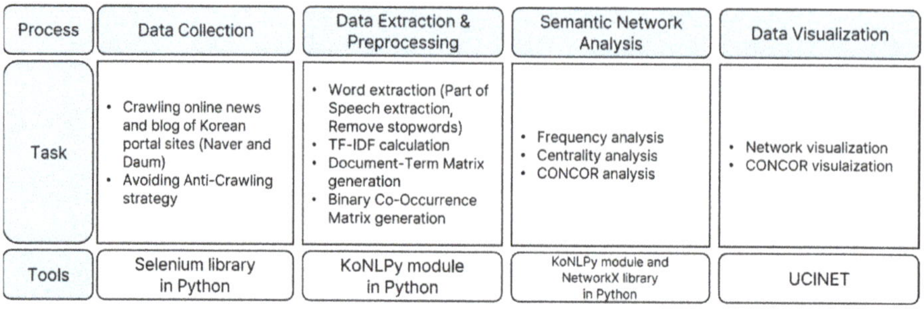


**Figure 1.** Data collection and analysis process for digital transformation.

Online news article texts were collected exclusively from Naver, as most Korean news articles can be accessed through it. However, given the occurrence of various news outlets providing identical articles, duplicates were removed from the collected news articles using cosine similarity on the texts. Since Naver and Daum blogs rarely contain posts with identical content on both platforms, no duplicate checks were conducted when collecting data from these blogs. Although the collection period was not specified, it was confirmed that over 90% of the data originated from within the last ten years.

During the collection process, it was observed that some websites had anti-crawling features. To circumvent these, the Selenium library, implemented in Python for automating web browser interactions, was utilized. The data thus collected were processed using the BeautifulSoup library and stored in the form of DataFrames using the Pandas library.

### *3.2. Data Extraction and Preprocessing*

The extraction and preprocessing of the data were performed using KoNLPy, a Python open-source library for natural language processing of the Korean language [40]. Utilizing KoNLPy, only nouns, verbs, and adjectives were selected as Korean unigrams, and stopwords, which are commonly used or insignificant words, were excluded to filter the data. The refined list of words was then used to calculate their TF-IDF (Term Frequency-Inverse Document Frequency) values, enabling the identification and weighting of the most relevant words within the dataset.

TF-IDF formally measures how the occurrence of a given word is concentrated in relatively fewer documents. It is calculated by multiplying two metrics: the word frequency in a document and the inverse document frequency of the word across a set of documents. This value is primarily used to gauge similarity within documents, in addition to assessing the relevance of a document in search queries and the importance of specific words in search results [41,42]. TF-IDF helps in highlighting words that are distinctive to certain documents, thereby facilitating more accurate and meaningful analysis of textual data.

After sorting the extracted TF-IDF word list in descending order, the top 50 words were selected as nodes for the semantic network analysis. During the selection of words, unrelated terms such as "person" and "society" were excluded, and semantically similar words were consolidated. For example, the frequency of "core" was combined with the frequency of "center", a similar term.

Based on the 50 selected words, a Document-Term Matrix (DTM) was created, representing the frequency of each word across various articles and blogs. DTM enables the quantification of the relationship between words and documents. Subsequently, a Co-Occurrence Matrix (COM) was constructed to represent the relationships of word co-occurrences across all documents. Due to the complexity of the analysis with the generated COM, all values were binarized by changing values higher than the median of all elements to 1 and those lower to 0, resulting in a binary matrix. This process simplifies dense values to 1 and 0, creating a looser relationship for network analysis. The semantic network analysis utilized this binary-structured keyword COM.

### 3.3. Semantic Network Analysis and Visualization

To discover the relationships among the top 50 words related to digital transformation, a semantic network analysis was conducted. This leverages the data mining techniques for unstructured Big Data analysis, a method distinct from social network analysis, which identifies the structural characteristics of social phenomena [43]. The co-occurrence relationships among the refined words within social media data were intuitively visualized using NetDraw 2.175, a network visualization software, with the previously created keyword COM [44].

To examine the connection structure of words related to digital transformation, the Python open-source package NetworkX [45] was utilized. Four types of network centrality metrics [46] were calculated using NetworkX for the keyword COM as follows:

1. Degree centrality, which calculates the number of nodes connected to a specific node, indicating the node's activity or popularity within the network;
2. Betweenness centrality, measuring a node's mediating role within the network, indicating its importance in facilitating information flow between other nodes;
3. Closeness centrality, calculating the inverse of the average distance to all other nodes, indicating how close a node is to all other nodes in the network, which can suggest its accessibility or centrality in the network's communication pathways;
4. Eigenvector centrality, a measure of a node's influence in the network, indicating not just how many connections a node has but also how important those connections are.

To identify mutually exclusive subgroups within the semantic network, a CONCOR (Convergence of Iterated Correlations) analysis was performed. CONCOR is based on structural equivalence, iteratively dividing nodes into subsets and then analyzing the Pearson correlation to identify groups with a certain level of similarity before forming clusters that include these groups [47]. This method is commonly used to find clusters of similar keywords and to identify the co-occurrence relationships between words across all possible terms [48]. UCINET 6.0 [49] was utilized to conduct the CONCOR analysis, and the results were visualized using NetDraw.

## 4. Results

### 4.1. The Frequencies of Keywords Related to Digital Transformation

The results of the word frequency analysis from online news articles and blogs, showing the top 50 words, are presented in Tables 1 and 2. The top five keywords from online news articles were "Education", "Innovation", "Corporation", "Information", and "Artificial Intelligence", highlighting a focus on how digital changes impact education, business innovation, and the integration of AI across sectors. Blogs, however, put "Artificial Intelligence", "Corporation", "Education", "Data", and "Innovation" at the forefront, indicating a stronger emphasis on the technical aspects of digital transformation, such as AI and data utilization, while still valuing education and innovation. This nuanced difference between online news articles and blogs suggests varying degrees of engagement with digital transformation themes across different platforms, but both recognize the importance of education and innovation in adapting to and capitalizing on digital advancements.

**Table 1.** Frequencies of 50 keywords related to digital transform in online news.

| Rank | Keyword | Freq. | Rank | Keyword | Freq. |
|---|---|---|---|---|---|
| 1 | Education | 2975 | 26 | Nation | 1137 |
| 2 | Innovation | 2643 | 27 | Strategy | 1068 |
| 3 | Corporation | 2437 | 28 | Cooperation | 1057 |
| 4 | Information | 2120 | 29 | New | 942 |
| 5 | Artificial Intelligence | 2093 | 30 | Smart | 937 |
| 6 | Project | 1955 | 31 | Human Resources | 852 |
| 7 | Data | 1942 | 32 | Operation | 831 |

**Table 1.** *Cont.*

| Rank | Keyword | Freq. | Rank | Keyword | Freq. |
|---|---|---|---|---|---|
| 8 | Future | 1830 | 33 | Citizens | 820 |
| 9 | Support | 1822 | 34 | Study | 791 |
| 10 | Government | 1816 | 35 | Student | 786 |
| 11 | Global | 1770 | 36 | Leading | 764 |
| 12 | Field | 1710 | 37 | Professor | 756 |
| 13 | Metaverse | 1644 | 38 | Competence | 745 |
| 14 | Policy | 1548 | 39 | Investment | 741 |
| 15 | Construction | 1502 | 40 | Training | 695 |
| 16 | Region | 1478 | 41 | Contents | 684 |
| 17 | Platform | 1460 | 42 | Institution | 668 |
| 18 | Promotion | 1458 | 43 | Tech | 657 |
| 19 | Era | 1444 | 44 | Bio | 646 |
| 20 | Center | 1394 | 45 | School | 639 |
| 21 | Service | 1363 | 46 | Finance | 544 |
| 22 | Economy | 1274 | 47 | Cloud | 543 |
| 23 | Development | 1267 | 48 | Infrastructure | 462 |
| 24 | Society | 1253 | 49 | Software | 439 |
| 25 | Plan | 1158 | 50 | Personal Information | 429 |

**Table 2.** Frequencies of 50 keywords related to digital transform in blogs.

| Rank | Keyword | Freq. | Rank | Keyword | Freq. |
|---|---|---|---|---|---|
| 1 | Artificial Intelligence | 2975 | 26 | New | 1137 |
| 2 | Corporation | 2643 | 27 | System | 1068 |
| 3 | Education | 2437 | 28 | Society | 1057 |
| 4 | Data | 2120 | 29 | Government | 942 |
| 5 | Innovation | 2093 | 30 | Market | 937 |
| 6 | Era | 1955 | 31 | Nation | 852 |
| 7 | Metaverse | 1942 | 32 | Cloud | 831 |
| 8 | Project | 1830 | 33 | Region | 820 |
| 9 | Service | 1822 | 34 | Study | 791 |
| 10 | Field | 1816 | 35 | Professor | 786 |
| 11 | Support | 1770 | 36 | Citizens | 764 |
| 12 | Future | 1710 | 37 | Corona | 756 |
| 13 | Global | 1644 | 38 | Online | 745 |
| 14 | Change | 1548 | 39 | Space | 741 |
| 15 | Information | 1502 | 40 | Human Resources | 695 |
| 16 | Development | 1478 | 41 | Personal Information | 684 |
| 17 | Platform | 1460 | 42 | Business | 668 |
| 18 | Construction | 1458 | 43 | Finance | 657 |
| 19 | Center | 1444 | 44 | Big Data | 646 |
| 20 | Strategy | 1394 | 45 | Leading | 639 |
| 21 | Economy | 1363 | 46 | Research | 544 |
| 22 | Smart | 1274 | 47 | Infrastructure | 543 |
| 23 | Promotion | 1267 | 48 | Industrial Revolution | 462 |
| 24 | Plan | 1253 | 49 | Software | 439 |
| 25 | Policy | 1158 | 50 | Science and Technology | 429 |

### *4.2. Analysis of Centralities of Keywords Related to Digital Transform*

In Section 4.1, we initially present the raw frequency results as shown in Tables 1 and 2. These results depict the unmodified occurrence of keywords across our dataset with common stopwords filtered out. Following this initial analysis, we apply the TF-IDF method to refine these frequencies, thereby highlighting words that hold unique significance within our corpus. The relationships and centrality of these keywords are then explored in greater depth using a Document-Term Matrix (DTM) and a binary-structured Co-Occurrence Matrix (COM).

Table 3 presents the results of the network centrality analysis from the keyword COM for online news articles. The keyword "Innovation" had the highest degree of association with other keywords, resulting in the highest degree centrality, followed by "Education", "Artificial Intelligence", "Support", and "Data". The order of betweenness centrality was high for "Data", "Education", "Innovation", "Support", and "Artificial Intelligence". Closeness centrality was high for "Innovation", "Education", "Artificial Intelligence", "Support", and "Data", in that order. Eigenvector centrality was high for "Innovation", "Artificial Intelligence", "Support", "Education", and "Corporation".

**Table 3.** Centralities of keywords related to digital transformation from news network.

| Rank | Keyword | Cd [1] | Keyword | Cb [2] | Keyword | Cc [3] | Keyword | Ce [4] |
|---|---|---|---|---|---|---|---|---|
| 1 | Innovation | 0.938776 | Data | 0.06154 | Innovation | 0.940408 | Innovation | 0.200111 |
| 2 | Education | 0.897959 | Education | 0.050155 | Education | 0.904239 | Artificial Intelligence | 0.195975 |
| 3 | Artificial Intelligence | 0.877551 | Innovation | 0.049585 | Artificial Intelligence | 0.887178 | Support | 0.194377 |
| 4 | Support | 0.877551 | Support | 0.035948 | Support | 0.887178 | Education | 0.192973 |
| 5 | Data | 0.857143 | Artificial Intelligence | 0.032176 | Data | 0.870748 | Corporation | 0.191475 |
| 6 | Corporation | 0.836735 | Corporation | 0.027202 | Corporation | 0.854917 | Data | 0.18877 |
| 7 | Global | 0.795918 | Global | 0.026213 | Global | 0.824919 | Global | 0.185881 |
| 8 | Project | 0.77551 | Government | 0.019029 | Project | 0.810697 | Future | 0.185531 |
| 9 | Future | 0.77551 | Future | 0.016818 | Future | 0.810697 | Project | 0.184892 |
| 10 | Government | 0.77551 | Project | 0.015426 | Government | 0.810697 | Field | 0.184361 |
| 11 | Information | 0.755102 | Information | 0.015374 | Information | 0.796956 | Government | 0.183366 |
| 12 | Field | 0.755102 | Metaverse | 0.015002 | Field | 0.796956 | Information | 0.18166 |
| 13 | Metaverse | 0.714286 | Policy | 0.013719 | Metaverse | 0.770826 | Promotion | 0.179538 |
| 14 | Policy | 0.714286 | Field | 0.01159 | Policy | 0.770826 | Policy | 0.176005 |
| 15 | Promotion | 0.714286 | Citizens | 0.011204 | Promotion | 0.770826 | Metaverse | 0.175198 |
| 16 | Construction | 0.693878 | Construction | 0.010303 | Construction | 0.758394 | Platform | 0.174759 |
| 17 | Platform | 0.693878 | Development | 0.009443 | Platform | 0.758394 | Center | 0.17228 |
| 18 | Service | 0.673469 | Platform | 0.008512 | Service | 0.746356 | Service | 0.172084 |
| 19 | Center | 0.653061 | Promotion | 0.007452 | Center | 0.734694 | Construction | 0.171971 |
| 20 | Development | 0.653061 | Service | 0.006708 | Development | 0.734694 | Nation | 0.170091 |
| 21 | Nation | 0.653061 | Society | 0.004792 | Nation | 0.734694 | Region | 0.166259 |
| 22 | Region | 0.632653 | Nation | 0.004159 | Region | 0.723391 | Society | 0.165359 |
| 23 | Society | 0.632653 | Region | 0.003406 | Society | 0.723391 | Economy | 0.164331 |
| 24 | Economy | 0.612245 | Center | 0.00286 | Economy | 0.71243 | Development | 0.163565 |
| 25 | Era | 0.591837 | Era | 0.001982 | Era | 0.701797 | Era | 0.159499 |
| 26 | Strategy | 0.571429 | Study | 0.001756 | Strategy | 0.691477 | Strategy | 0.156086 |
| 27 | Plan | 0.510204 | Economy | 0.001665 | Plan | 0.662259 | Plan | 0.141053 |
| 28 | Cooperation | 0.469388 | Human Resources | 0.001255 | Cooperation | 0.644115 | Cooperation | 0.13411 |
| 29 | New | 0.469388 | Strategy | 0.001233 | New | 0.644115 | New | 0.132824 |
| 30 | Citizens | 0.469388 | Leading | 0.001116 | Citizens | 0.644115 | Citizens | 0.127351 |
| 31 | Human Resources | 0.44898 | School | 0.001047 | Human Resources | 0.626939 | Human Resources | 0.122508 |
| 32 | Investment | 0.408163 | Student | 0.000984 | Investment | 0.61869 | Investment | 0.113464 |
| 33 | Smart | 0.367347 | Tech | 0.000834 | Smart | 0.602826 | Smart | 0.105321 |
| 34 | Operation | 0.326531 | Plan | 0.000646 | Operation | 0.587755 | Operation | 0.095075 |
| 35 | Bio | 0.326531 | Bio | 0.000354 | Bio | 0.587755 | Bio | 0.090109 |
| 36 | Leading | 0.265306 | Investment | 0.000338 | Leading | 0.566511 | Competence | 0.073346 |
| 37 | Study | 0.244898 | Training | $4.05 \times 10^{-5}$ | Study | 0.559767 | Leading | 0.071322 |
| 38 | Professor | 0.244898 | New | $3.87 \times 10^{-5}$ | Professor | 0.559767 | Contents | 0.07119 |
| 39 | Competence | 0.244898 | Cooperation | $3.15 \times 10^{-5}$ | Competence | 0.559767 | Professor | 0.070857 |
| 40 | Contents | 0.244898 | Smart | 0 | Contents | 0.553181 | Training | 0.064338 |

**Table 3.** *Cont.*

| Rank | Keyword | Cd [1] | Keyword | Cb [2] | Keyword | Cc [3] | Keyword | Ce [4] |
|---|---|---|---|---|---|---|---|---|
| 41 | Student | 0.22449 | Operation | 0 | Student | 0.546749 | Study | 0.058304 |
| 42 | Training | 0.22449 | Professor | 0 | Training | 0.546749 | Student | 0.052909 |
| 43 | School | 0.204082 | Competence | 0 | School | 0.540464 | Institution | 0.049241 |
| 44 | Tech | 0.183673 | Contents | 0 | Institution | 0.534323 | Cloud | 0.048228 |
| 45 | Institution | 0.163265 | Institution | 0 | Tech | 0.534323 | School | 0.044269 |
| 46 | Cloud | 0.163265 | Finance | 0 | Cloud | 0.534323 | Infrastructure | 0.042801 |
| 47 | Infrastructure | 0.142857 | Cloud | 0 | Infrastructure | 0.528319 | Tech | 0.040939 |
| 48 | Finance | 0.081633 | Infrastructure | 0 | Finance | 0.511091 | Finance | 0.02403 |
| 49 | Personal Information | 0.040816 | Software | 0 | Personal Information | 0.470204 | Personal Information | 0.010226 |
| 50 | Software | 0 | Personal Information | 0 | Software | 0 | Software | $1.45 \times 10^{-13}$ |

[1] Cd: degree centrality. [2] Cb: betweenness centrality. [3] Cc: closeness centrality. [4] Ce: eigenvector centrality.

In the analysis of network centrality from the keyword COM for blogs, as detailed in Table 4, "Artificial Intelligence" emerged as the most centrally connected term, exhibiting the highest degree of association with other keywords. This centrality was closely followed by the terms "Data", "Corporation", "Innovation", and "Service", in that order. Furthermore, "Artificial Intelligence" also led in betweenness centrality, suggesting its role as a pivotal bridge within the network. This pattern was similarly observed in closeness centrality, with "Artificial Intelligence", "Data", "Corporation", "Innovation", and "Service" ranking high, indicating their close connections within the network. Additionally, "Artificial Intelligence", "Data", "Corporation", "Service", and "Development" were found to have high eigenvector centrality, highlighting their influence across the network.

**Table 4.** Centralities of keywords related to digital transformation from blog network.

| Rank | Keyword | Cd [1] | Keyword | Cb [2] | Keyword | Cc [3] | Keyword | Ce [4] |
|---|---|---|---|---|---|---|---|---|
| 1 | Artificial Intelligence | 0.918367 | Artificial Intelligence | 0.092258 | Artificial Intelligence | 0.918802 | Artificial Intelligence | 0.193392 |
| 2 | Data | 0.877551 | Data | 0.041167 | Data | 0.881299 | Data | 0.192609 |
| 3 | Corporation | 0.857143 | Education | 0.031264 | Corporation | 0.863673 | Corporation | 0.191275 |
| 4 | Innovation | 0.816327 | Corporation | 0.028714 | Innovation | 0.830455 | Service | 0.18964 |
| 5 | Service | 0.816327 | Information | 0.018267 | Service | 0.830455 | Development | 0.189048 |
| 6 | Development | 0.816327 | Innovation | 0.017399 | Development | 0.830455 | Innovation | 0.18838 |
| 7 | Education | 0.795918 | Support | 0.017302 | Education | 0.814786 | Metaverse | 0.187632 |
| 8 | Metaverse | 0.795918 | Development | 0.014583 | Metaverse | 0.814786 | Construction | 0.186105 |
| 9 | Support | 0.795918 | Service | 0.013366 | Support | 0.814786 | Promotion | 0.186105 |
| 10 | Construction | 0.795918 | Construction | 0.012719 | Construction | 0.814786 | Support | 0.185267 |
| 11 | Promotion | 0.795918 | Promotion | 0.012719 | Promotion | 0.814786 | Project | 0.184539 |
| 12 | Project | 0.77551 | Metaverse | 0.010805 | Project | 0.799698 | Smart | 0.184539 |
| 13 | Smart | 0.77551 | Center | 0.009584 | Smart | 0.799698 | Education | 0.18388 |
| 14 | Field | 0.755102 | Project | 0.009436 | Field | 0.785158 | Field | 0.183307 |
| 15 | Center | 0.755102 | Smart | 0.009436 | Center | 0.785158 | Center | 0.18085 |
| 16 | Information | 0.734694 | Citizens | 0.008668 | Information | 0.771137 | Information | 0.175311 |
| 17 | Platform | 0.693878 | Strategy | 0.00794 | Platform | 0.744546 | Platform | 0.174807 |
| 18 | Strategy | 0.693878 | Field | 0.006863 | Strategy | 0.744546 | Strategy | 0.171625 |
| 19 | System | 0.693878 | Future | 0.005236 | System | 0.744546 | System | 0.171384 |
| 20 | Global | 0.653061 | System | 0.005218 | Global | 0.719728 | Global | 0.163418 |
| 21 | Future | 0.632653 | Global | 0.004499 | Future | 0.707929 | Plan | 0.159692 |
| 22 | Plan | 0.632653 | Plan | 0.003899 | Plan | 0.707929 | Future | 0.157281 |
| 23 | Era | 0.571429 | Platform | 0.003363 | Era | 0.674745 | Government | 0.151141 |
| 24 | Government | 0.571429 | Era | 0.002008 | Government | 0.674745 | Nation | 0.148339 |
| 25 | Citizens | 0.571429 | Government | 0.00083 | Citizens | 0.674745 | Era | 0.147639 |

**Table 4.** *Cont.*

| Rank | Keyword | Cd [1] | Keyword | Cb [2] | Keyword | Cc [3] | Keyword | Ce [4] |
|---|---|---|---|---|---|---|---|---|
| 26 | Nation | 0.55102 | Change | 0.000604 | Nation | 0.664364 | Citizens | 0.145619 |
| 27 | Policy | 0.530612 | Society | 0.000582 | Policy | 0.654298 | Policy | 0.143094 |
| 28 | Economy | 0.510204 | Economy | 0.000514 | Economy | 0.644532 | Economy | 0.137371 |
| 29 | Society | 0.489796 | Nation | 0.000448 | Society | 0.635054 | Cloud | 0.134165 |
| 30 | Cloud | 0.489796 | Policy | 0.000377 | Cloud | 0.635054 | Society | 0.130427 |
| 31 | Market | 0.469388 | Cloud | 0.000157 | Market | 0.62585 | Market | 0.129527 |
| 32 | Infrastructure | 0.44898 | Market | $7.91 \times 10^{-5}$ | Infrastructure | 0.61691 | Infrastructure | 0.124915 |
| 33 | Online | 0.428571 | New | 0 | Online | 0.608221 | Online | 0.120143 |
| 34 | Leading | 0.408163 | Region | 0 | Leading | 0.599773 | Leading | 0.114433 |
| 35 | Change | 0.387755 | Study | 0 | Change | 0.591557 | Region | 0.104145 |
| 36 | Region | 0.367347 | Professor | 0 | Region | 0.583563 | Change | 0.101043 |
| 37 | Study | 0.326531 | Corona | 0 | Study | 0.568206 | Study | 0.094029 |
| 38 | Big Data | 0.285714 | Online | 0 | Big Data | 0.553637 | Big Data | 0.082636 |
| 39 | New | 0.244898 | Space | 0 | New | 0.539796 | Finance | 0.070776 |
| 40 | Corona | 0.244898 | Human Resources | 0 | Corona | 0.539796 | Corona | 0.070504 |
| 41 | Finance | 0.244898 | Personal Information | 0 | Finance | 0.539796 | New | 0.065627 |
| 42 | Space | 0.183673 | Business | 0 | Space | 0.520285 | Space | 0.051164 |
| 43 | Business | 0.102041 | Finance | 0 | Business | 0.496364 | Business | 0.029595 |
| 44 | Human Resources | 0.081633 | Big Data | 0 | Human Resources | 0.48521 | Human Resources | 0.023802 |
| 45 | Personal Information | 0.061224 | Leading | 0 | Professor | 0.474546 | Personal Information | 0.016215 |
| 46 | Professor | 0.040816 | Research | 0 | Personal Information | 0.469388 | Professor | 0.011912 |
| 47 | Research | 0.020408 | Infrastructure | 0 | Research | 0.469388 | Research | 0.006106 |
| 48 | Industrial Revolution | 0 | Industrial Revolution | 0 | Industrial Revolution | 0 | Industrial Revolution | $3.72 \times 10^{-15}$ |
| 49 | Software | 0 | Software | 0 | Software | 0 | Software | $3.72 \times 10^{-15}$ |
| 50 | Science and Technology | 0 | Science and Technology | 0 | Science and Technology | 0 | Science and Technology | $3.72 \times 10^{-15}$ |

[1] Cd: degree centrality. [2] Cb: betweenness centrality. [3] Cc: closeness centrality. [4] Ce: eigenvector centrality.

### *4.3. CONCOR Analysis and Visualization*

A CONCOR analysis was conducted based on structural equivalence by analyzing the Pearson correlation from the keyword COM, resulting in clusters. Figure 2 presents the outcome of the CONCOR analysis performed on the digital transformation network generated from online news, identifying a total of seven clusters. The clusters are represented as [Word1, Word2, ...]. The cluster [Operation, Data, Bio, Development, Contents, Plan, Smart, Construction] can be interpreted as embodying the theme of technological advancement and strategic growth. The cluster [Support, Government, Corporation, Field, Project, Promotion] suggests that the government and various corporations collaborate to support innovative projects aimed at advancing key industrial sectors. The cluster [Training, Infrastructure, Cloud, Leading, Competence, Institution, Personal Information, Finance, Professor] represents the context of education, technology, and expertise development within an institutional framework. The [Software, Tech, Study, School, Student] cluster indicates an education or learning environment focused on technology and software. The cluster [Information, Innovation, Artificial Intelligence, Education, Future] encompasses future-oriented and technology-driven themes. The [Citizens, New, Human Resources, Investment, Region, Cooperation, Service, Metaverse] cluster can be seen as focusing on community and technological development within geographic or digital spaces. Lastly, the [Strategy, Platform, Society, Policy, Center, Global, Economy, Nation, Era] cluster can be interpreted as countries developing policies centered around digital platforms to drive economic growth or societal development in the global era.

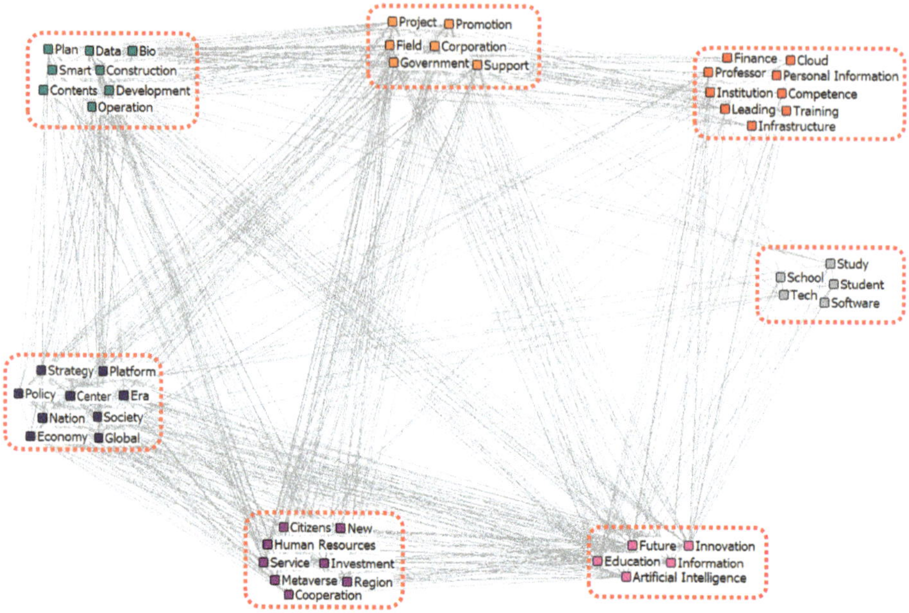


**Figure 2.** CONCOR analysis of news network of the digital transformation.

Figure 3 depicts the results of a CONCOR analysis on the digital transformation network generated from blogs, identifying a total of seven clusters. The cluster [Finance, Corona, Study, Space, Big Data, Change, Economy] can be interpreted as showing that the coronavirus pandemic has accelerated the digital transformation of finance and the economy, with research highlighting the importance of Big Data and digital spaces in driving change. The cluster [Personal Information, Citizens, Strategy, Information, Plan] suggests that themes of personal information, public engagement, and strategy are crucial regarding information. The cluster [Nation, Government, Era, Society, Policy, Future, Global] presents themes related to national and global governance, societal adaptation, and future-oriented policies in a digitally evolving world. The cluster [Development, Center, Education, System, Platform, Field] indicates that digital transformation is central to the development of the educational sector, platforms, and systems. The cluster [Region, Leading, Cloud, Infrastructure, Online, Market] points to a focus on regional development through cutting-edge cloud infrastructure and online marketplaces. The cluster [Project, Construction, Metaverse, Promotion, Corporation, Smart, Support, Service, Data, Innovation, Artificial Intelligence] represents a comprehensive approach to integrating advanced technologies into corporate projects and services. This can be interpreted as innovative projects initiated by corporations to focus on building smart services like the metaverse, supported by artificial intelligence and data analytics, to facilitate a new era of digital transformation and customer engagement. The cluster [New, Professor, Industrial Revolution, Human Resource, Software, Research, Science and Technology, Business] expresses the narrative of education and industrial evolution towards a technologically advanced future, emphasizing the collaborative role of academia and industry in pioneering R&D efforts using cutting-edge software and human resource innovation to underpin the digital transformation of businesses and society.

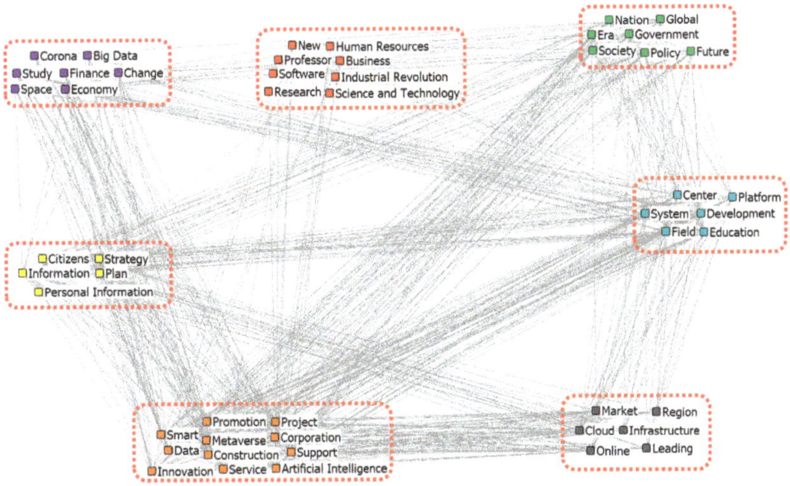


**Figure 3.** CONCOR analysis of blog network of digital transformation.

## 5. Discussion

The arrival of the digital transformation era brings technological advancements and consequential changes, fundamentally restructuring educational, occupational, and everyday life practices. The integration of digital technologies presents new opportunities and challenges across all age groups. This research utilizes text mining techniques to analyze social media data generated online on a large scale, aiming to understand the phenomena of digital transformation and its impacts. It delves into the effects of digital technology on various sectors such as society, economy, and education, seeking adaptation strategies and policy responses necessary for navigating the digital age. Also, this study has been conducted to elucidate the distinct patterns observed in both formal (articles) and informal (blogs) discourse on digital transformation within the Korean context. By analyzing these diverse sources of content, our intent is to provide a comprehensive view that enables readers from various countries to gain a nuanced understanding of how digital transformation is perceived and discussed in Korea.

Text mining related to digital transformation revealed the top five words with the highest frequency in online news articles as "Education", "Innovation", "Corporation", "Information", and "Artificial Intelligence" and in blogs as "Artificial Intelligence", "Corporation", "Education", "Data", and "Innovation". These results indicate the significant impact of technological advancement in various fields such as education, innovation, corporations, and artificial intelligence [50,51]. Additionally, artificial intelligence has been confirmed as one of the key elements in the era of digital transformation [52,53].

To refine our analysis, we employed the TF-IDF methodology, which assists in distinguishing significant keywords from those frequently appearing across different texts without substantial informational value.

An analysis of online news articles showed high centrality for "Innovation", "Education", "Artificial Intelligence", "Support", and "Data". These words are strongly interconnected as the main themes of digital transformation, highlighting the interaction between these themes in the context of technological progress in modern society, the evolution of education, data-driven decision-making processes, the expansion of artificial intelligence applications, and the importance of supporting systems in all these areas. A blog centrality analysis highlighted "Artificial Intelligence", "Data", "Corporation", "Innovation'", and "Service" as highly central. While "Education" and "Support" were emphasized as important themes in news articles, "Corporation" and "Service" showed greater centrality in blogs, suggesting that interests may vary depending on the community or platform discussing digital transformation. Blogs tend to focus more on in-depth analysis or opinions on personal or corporate experiences, products, and services, particularly highlighting corporate activities and service provision. In contrast, online news pays attention to broader topics like education and social support, dealing with the impact of digital transformation across society. The high centrality of "Innovation", "Data", and "Artificial Intelligence" in both mediums suggests that the advancement of digital technology is interconnected in areas such as innovation, data analysis, and artificial intelligence. The technological innovation enhances data-based decision-making processes, and the advancement of artificial intelligence enables new innovative solutions, driving change across various fields, as emphasized in other studies [54–56].

The CONCOR analysis identified seven clusters each in online news and blogs related to digital transformation, confirming the importance of technology and education in both mediums. The emphasis on technological advancements, particularly digital technologies like artificial intelligence, cloud computing, and Big Data, alongside education, is recognized as a leading force in driving the digital transformation. It suggests a focus on innovating educational systems to introduce new learning methods and competency development, potentially causing significant changes across the economy and society [57,58]. The acceleration of digitalization in economic activities due to the pandemic is a phenomenon already reported in various studies [59–61].

Online news tends to focus more on digital transformation support and the promotion of industry development through collaboration between governments and corporations. Words such as "Government", "Corporation", "Support", "Field", "Project", and "Promotion" highlight the strategic partnerships' vital role in supporting digital transformation and fostering innovative projects in specific industrial sectors. For small-scale service businesses, the digital transformation aims to expand competitive advantage, improve business outcomes, and achieve growth. The government's role in this context is identified as supporting the construction of digital platforms for small-scale service businesses, enabling mobile/digital payments, providing digital education, and building a digital collaboration ecosystem [62]. In contrast, blogs, with terms like "Personal Information", "Citizens", "Strategy", "Information", "Plan", and "Metaverse", reflect how individuals and small communities integrate and use digital technologies, especially innovative services like artificial intelligence and the metaverse, in daily life and business, indicating experiences and impacts on these practices.

Thus, online news tends to view digital transformation from the perspective of policy, economy, and national strategy, while blogs explore it from a standpoint closer to everyday life. This difference stems from each medium's purpose and target audience [63]. Online news aims to provide information to a broad readership, offering insights useful to policymakers and businesspeople, whereas blogs cater to personal interests, in-depth analysis, and detailed exploration of specific topics, providing customized content for the general public, particularly users and small communities interested in digital technologies [63].

Our analysis identifies several features unique to the Korean context, which significantly influence the discourse on digital transformation on Korean social media platforms. For instance, Korea's collectivist cultural norms shape the adoption of technologies that emphasize communal benefits and organizational harmony [64]. Additionally, the country's leading position in digital transformation fosters a progressive environment for discussing advanced digital infrastructure [65]. Economically, the interplay between large conglomerates and dynamic SMEs creates diverse viewpoints on how digital transformation can drive business growth and innovation [66]. These unique cultural, technological, and economic contexts provide a distinctive backdrop to Korea's digital transformation discourse, offering insights into the challenges and opportunities specific to this setting.

We found that global opinion polls often focus on general technological adaptation and digital readiness, whereas our analysis dives deeper into the specific themes and concerns prevalent in Korea. For example, global surveys like those conducted by the IFRC [67] highlight varied regional responses to digital transformation, with Korea emphasizing advanced analytics and system interoperability compared to other regions. Our findings, which underscore the high centrality of innovation and artificial intelligence in Korean discourse, align with these global trends but also reveal unique local priorities and cultural influences.

In conclusion, the findings illuminate the multifaceted impacts of digital transformation, offering diverse perspectives on technological changes, social, and economic transitions as manifested through online news and blogs. The real-time feedback and variety of user content on social media are valuable for policymakers, entrepreneurs, educators, and the general public to understand the advancements in digital technology and how these can be applied to their fields and lives. The insights and user engagement provided by social media data can lead to the development of innovative approaches and strategies that guide the digital transformation era, contributing to socially meaningful conversations about upcoming technological changes.

## 6. Conclusions

This study leveraged text mining techniques and a semantic network analysis to extract keywords and their associations from social media data, online news, and blog content related to digital transformation. Focusing on Korean language data, it intensively collected data from major Korean portal sites using "digital transformation" and related search terms, ensuring the selection of keywords and consistency of data by exclusively targeting content in Korean.

Despite some limitations, the analysis of Korean data collected from Korean portal sites offers insights into digital transformation, contributing to a comprehensive understanding of various aspects related to the advancement of digital technologies, social changes, and economic impacts. The insights derived from this study provide essential foundational data for in-depth analysis of the continuous development of digital technologies and their impacts on individuals, corporations, and society.

Furthermore, the results can serve as an important reference for strategic planning and policy development related to digital transformation. The data and analysis will offer valuable information to policymakers, entrepreneurs, and academic researchers in integrating digital technologies, seeking social adaptation strategies, and exploring economic sustainability.

To enhance the practical relevance of these findings, we plan to incorporate feedback from industry experts through structured interviews and align our results with documented case studies. It will bridge the gap between theoretical research and practical applications, ensuring that our insights are grounded in real-world experiences and contribute to the development of actionable and effective strategies in digital transformation.

**Author Contributions:** Conceptualization, J.-H.S. and B.-S.S.; methodology, J.-H.S.; software, J.-H.S.; validation, J.-H.S. and B.-S.S.; formal analysis, J.-H.S.; investigation, J.-H.S.; resources, J.-H.S. and B.-S.S.; data curation, J.-H.S. and B.-S.S.; writing—original draft preparation, J.-H.S. and B.-S.S.; writing—review and editing, J.-H.S. and B.-S.S.; visualization, J.-H.S. and B.-S.S.; supervision, B.-S.S.; project administration, B.-S.S.; funding acquisition, J.-H.S. and B.-S.S. All authors have read and agreed to the published version of the manuscript.

**Funding:** This research received no external funding.

**Institutional Review Board Statement:** Not applicable.

**Informed Consent Statement:** Not applicable.

**Data Availability Statement:** The data presented in this study are available from the corresponding author on request.

**Conflicts of Interest:** The authors declare no conflicts of interest.

## References


1. Reis, J.; Amorim, M.; Melão, N. Digital transformation: A literature review and guidelines for future research. *Trends Adv. Inf. Syst. Technol.* **2018**, *1*, 411–421.
2. Lima, J.V.V.; Santos, W.B.; Rodrigues, C.; Alencar, F. Digital Transformation in the Public Sector: Preliminary Results of a Tertiary Literature Review. In Proceedings of the 2023 18th Iberian Conference on Information Systems and Technologies (CISTI), Aveiro, Portugal, 20–23 June 2023; pp. 1–7. [CrossRef]
3. Vaska, S.; Massaro, M.; Bagarotto, E.M.; Mas, F.D. The digital transformation of business model innovation: A structured literature review. *Front. Psychol.* **2021**, *11*, 539363. [CrossRef] [PubMed]
4. Dopfer, M.; Fallahi, S.; Kirchberger, M.; Gassmann, O. Adapt and strive: How ventures under resource constraints create value through business model adaptations. *Creat. Innov. Manag.* **2017**, *26*, 233–246. [CrossRef]
5. Peñarroya-Farell, M.; Miralles, F. Business model dynamics from interaction with open innovation. *J. Open Innov. Technol. Mark. Complex.* **2021**, *7*, 81. [CrossRef]
6. Schwertner, K. Digital transformation of business. *Trakia J. Sci.* **2017**, *15*, 388–393. [CrossRef]
7. Fronzetti Colladon, A.; Grassi, S.; Ravazzolo, F.; Violante, F. Forecasting financial markets with semantic network analysis in the COVID-19 crisis. *J. Forecast.* **2023**, *42*, 1187–1204. [CrossRef]
8. Luo, C.; Chen, A.; Cui, B.; Liao, W. Exploring public perceptions of the COVID-19 vaccine online from a cultural perspective: Semantic network analysis of two social media platforms in the United States and China. *Telemat. Inform.* **2021**, *65*, 101712. [CrossRef] [PubMed]
9. Shi, W.; Fu, H.; Wang, P.; Chen, C.; Xiong, J. # Climatechange vs.# Globalwarming: Characterizing two competing climate discourses on Twitter with semantic network and temporal analyses. *Int. J. Environ. Res. Public Health* **2020**, *17*, 1062. [CrossRef] [PubMed]
10. Fitzgerald, M.; Kruschwitz, N.; Bonnet, D.; Welch, M. Embracing digital technology: A new strategic imperative. *MIT Sloan Manag. Rev.* **2014**, *55*, 1.
11. Westerman, G.; Calméjane, C.; Bonnet, D.; Ferraris, P.; McAfee, A. Digital Transformation: A roadmap for billion-dollar organizations. *MIT Cent. Digit. Bus. Capgemini Consult.* **2011**, *1*, 1–68.
12. Stolterman, E.; Fors, A.C. Information technology and the good life. *Inf. Syst. Res. Relev. Theory Inf. Pract.* **2004**, *143*, 687–692.
13. Besson, P.; Rowe, F. Strategizing information systems-enabled organizational transformation: A transdisciplinary review and new directions. *J. Strateg. Inf. Syst.* **2012**, *21*, 103–124. [CrossRef]
14. Karimi, J.; Walter, Z. The role of dynamic capabilities in responding to digital disruption: A factor-based study of the newspaper industry. *J. Manag. Inf. Syst.* **2015**, *32*, 39–81. [CrossRef]
15. Cha, K.J.; Hwang, T.; Gregor, S. An integrative model of IT-enabled organizational transformation: A multiple case study. *Manag. Decis.* **2015**, *53*, 1755–1770. [CrossRef]
16. Resca, A.; Za, S.; Spagnoletti, P. Digital platforms as sources for organizational and strategic transformation: A case study of the Midblue project. *J. Theor. Appl. Electron. Commer. Res.* **2013**, *8*, 71–84. [CrossRef]
17. Llopis, J.; Gonzalez, M.R.; Gasco, J.L. Transforming the firm for the digital era: An organizational effort towards an E-culture. *Hum. Syst. Manag.* **2004**, *23*, 213–225. [CrossRef]
18. Nadkarni, S.; Prügl, R. Digital transformation: A review, synthesis and opportunities for future research. *Manag. Rev. Q.* **2021**, *71*, 233–341. [CrossRef]
19. Gebauer, H.; Paiola, M.; Saccani, N.; Rapaccini, M. Digital servitization: Crossing the perspectives of digitization and servitization. *Ind. Mark. Manag.* **2021**, *93*, 382–388. [CrossRef]
20. Coreynen, W.; Matthyssens, P.; Vanderstraeten, J.; van Witteloostuijn, A. Unravelling the internal and external drivers of digital servitization: A dynamic capabilities and contingency perspective on firm strategy. *Ind. Mark. Manag.* **2020**, *89*, 265–277. [CrossRef]
21. Paiola, M.; Gebauer, H. Internet of things technologies, digital servitization and business model innovation in BtoB manufacturing firms. *Ind. Mark. Manag.* **2020**, *89*, 245–264. [CrossRef]
22. Frank, A.G.; Dalenogare, L.S.; Ayala, N.F. Industry 4.0 technologies: Implementation patterns in manufacturing companies. *Int. J. Prod. Econ.* **2019**, *210*, 15–26. [CrossRef]
23. Chin, H.S.; Marasini, D.P.; Lee, D. Digital transformation trends in service industries. *Serv. Bus.* **2023**, *17*, 11–36. [CrossRef]
24. Laney, D. 3D data management: Controlling data volume, velocity and variety. *META Group Res. Note* **2001**, *6*, 1.
25. Kostakis, P.; Kargas, A. Big-Data Management: A Driver for Digital Transformation? *Information* **2021**, *12*, 411. [CrossRef]
26. Bosilj, N.; Jurinjak, I. The role of knowledge management in mobile marketing. *J. Inf. Organ. Sci.* **2009**, *33*, 231–241.
27. Fayyaz, Z.; Ebrahimian, M.; Nawara, D.; Ibrahim, A.; Kashef, R. Recommendation systems: Algorithms, challenges, metrics, and business opportunities. *Appl. Sci.* **2020**, *10*, 7748. [CrossRef]
28. Almohsen, K.A.; Al-Jobori, H. Recommender systems in light of big data. *Int. J. Electr. Comput. Eng.* **2015**, *5*, 1553–1563. [CrossRef]
29. Verma, J.P.; Patel, B.; Patel, A. Big data analysis: Recommendation system with Hadoop framework. In Proceedings of the 2015 IEEE International Conference on Computational Intelligence & Communication Technology, Ghaziabad, India, 13–14 February 2015; IEEE: Piscataway Township, NJ, USA, 2015.
30. Drieger, P. Semantic network analysis as a method for visual text analytics. *Procedia-Soc. Behav. Sci.* **2013**, *79*, 4–17. [CrossRef]

31. Feldman, R.; Sanger, J. *The Text Mining Handbook: Advanced Approaches in Analyzing Unstructured Data*; Cambridge University Press: Cambridge, UK, 2007.
32. Risch, J.; Kao, A.; Poteet, S.R.; Wu, J. Text visualization for visual text analytics. In *Visual Data Mining: Theory, Techniques and Tools for Visual Analytics*; Springer: Berlin/Heidelberg, Germany, 2008; pp. 154–171.
33. Berry, M.W.; Kogan, J. (Eds.) *Text Mining: Applications and Theory*; John Wiley & Sons: Hoboken, NJ, USA, 2010.
34. Kim, E.J.; Kim, J.Y. Exploring the Online News Trends of the Metaverse in South Korea: A Data-Mining-Driven Semantic Network Analysis. *Sustainability* **2023**, *15*, 16279. [CrossRef]
35. Kang, G.J.; Ewing-Nelson, S.R.; Mackey, L.; Schlitt, J.T.; Marathe, A.; Abbas, K.M.; Swarup, S. Semantic network analysis of vaccine sentiment in online social media. *Vaccine* **2017**, *35*, 3621–3638. [CrossRef]
36. Christensen, A.P.; Kenett, Y.N. Semantic network analysis (SemNA): A tutorial on preprocessing, estimating, and analyzing semantic networks. *Psychol. Methods* **2021**, *28*, 860–879. [CrossRef] [PubMed]
37. Collins, A.M.; Loftus, E.F. A spreading-activation theory of semantic processing. *Psychol. Rev.* **1975**, *82*, 407. [CrossRef]
38. NAVER. Available online: https://www.naver.com (accessed on 15 February 2024).
39. DAUM. Available online: https://www.daum.net (accessed on 15 February 2024).
40. Park, E.L.; Cho, S. KoNLPy: Korean natural language processing in Python. In Proceedings of the 26th Annual Conference on Human & Cognitive Language Technology, Chuncheon, Korea, 10–11 October 2014; Volume 6.
41. Leskovec, J.; Rajaraman, A.; Ullman, J. Recommender systems. In *Mining of Massive Datasets*; Springer: Berlin/Heidelberg, Germany, 2011; p. 327.
42. De Boom, C.; Van Canneyt, S.; Bohez, S.; Demeester, T.; Dhoedt, B. Learning semantic similarity for very short texts. In Proceedings of the 2015 IEEE International Conference on Data Mining WORKSHOP (icdmw), Atlantic City, NJ, USA, 14–17 November 2015; IEEE: Piscataway Township, NJ, USA, 2015.
43. Hong, Y. How the discussion on a contested technology in Twitter changes: Semantic network analysis of tweets about cryptocurrency and blockchain technology. In Proceedings of the 22nd Biennial Conference of the International Telecommunications Society (ITS), Beyond the Boundaries: Challenges for Business, Policy and Society, Seoul, Republic of Korea, 24–27 June 2018.
44. Borgatti, S.P. *NetDraw Software for Network Visualization*; Analytic Technologies: Lexington, KY, USA, 2002.
45. Hagberg, A.; Swart, P.; Chult, D.S. *Exploring Network Structure, Dynamics, and Function Using NetworkX. No. LA-UR-08-05495*; LA-UR-08-5495; Los Alamos National Lab. (LANL): Los Alamos, NM, USA, 2008.
46. Tabassum, S.; Pereira, F.S.F.; Fernandes, S.; Gama, J. Social network analysis: An overview. *Wiley Interdiscip. Rev. Data Min. Knowl. Discov.* **2018**, *8*, e1256. [CrossRef]
47. Breiger, R.L.; Boorman, S.A.; Arabie, P. An algorithm for clustering relational data with applications to social network analysis and comparison with multidimensional scaling. *J. Math. Psychol.* **1975**, *12*, 328–383. [CrossRef]
48. Kim, N.R.; Hong, S.G. Text mining for the evaluation of public services: The case of a public bike-sharing system. *Serv. Bus.* **2020**, *14*, 315–331. [CrossRef]
49. Borgatti, S.P.; Everett, M.G.; Freeman, L.C. *Ucinet for Windows: Software for Social Network Analysis*; Analytic Technologies: Harvard, MA, USA, 2002; Volume 6, pp. 12–15.
50. Truong, T.-C.; Diep, Q.B. Technological Spotlights of Digital Transformation in Tertiary Education. *IEEE Access* **2023**, *11*, 40954–40966. [CrossRef]
51. Gao, D.; Yan, Z.; Zhou, X.; Mo, X. Smarter and prosperous: Digital transformation and enterprise performance. *Systems* **2023**, *11*, 329. [CrossRef]
52. Neethirajan, S. Artificial intelligence and sensor technologies in dairy livestock export: Charting a digital transformation. *Sensors* **2023**, *23*, 7045. [CrossRef] [PubMed]
53. Lei, Y.; Liang, Z.; Ruan, P. Evaluation on the impact of digital transformation on the economic resilience of the energy industry in the context of artificial intelligence. *Energy Rep.* **2023**, *9*, 785–792. [CrossRef]
54. Bahoo, S.; Cucculelli, M.; Qamar, D. Artificial intelligence and corporate innovation: A review and research agenda. *Technol. Forecast. Soc. Chang.* **2023**, *188*, 122264. [CrossRef]
55. Mariani, M.M.; Machado, I.; Magrelli, V.; Dwivedi, Y.K. Artificial intelligence in innovation research: A systematic review, conceptual framework, and future research directions. *Technovation* **2023**, *122*, 102623. [CrossRef]
56. Reim, W.; Åström, J.; Eriksson, O. Implementation of artificial intelligence (AI): A roadmap for business model innovation. *AI* **2020**, *1*, 11. [CrossRef]
57. Mukul, E.; Büyüközkan, G. Digital transformation in education: A systematic review of education 4.0. *Technol. Forecast. Soc. Chang.* **2023**, *194*, 122664. [CrossRef]
58. Benavides, L.M.C.; Arias, J.A.T.; Serna, M.D.A.; Bedoya, J.W.B.; Burgos, D. Digital transformation in higher education institutions: A systematic literature review. *Sensors* **2020**, *20*, 3291. [CrossRef] [PubMed]
59. Amankwah-Amoah, J.; Khan, Z.; Wood, G.; Knight, G. COVID-19 and digitalization: The great acceleration. *J. Bus. Res.* **2021**, *136*, 602–611. [CrossRef] [PubMed]
60. Kutnjak, A. Covid-19 accelerates digital transformation in industries: Challenges, issues, barriers and problems in transformation. *IEEE Access* **2021**, *9*, 79373–79388. [CrossRef]
61. Kraus, N.; Kraus, K. Digitalization of business processes of enterprises of the ecosystem of Industry 4.0: Virtual-real aspect of economic growth reserves. *WSEAS Trans. Bus. Econ.* **2021**, *18*, 569–580. [CrossRef]

62. Chen, C.-L.; Lin, Y.-C.; Chen, W.-H.; Chao, C.-F.; Pandia, H. Role of government to enhance digital transformation in small service business. *Sustainability* **2021**, *13*, 1028. [CrossRef]
63. Tereszkiewicz, A. "I'm not sure what that means yet, but we'll soon find out"—The discourse of newspaper live blogs. *Stud. Linguist. Univ. Iagell. Cracoviensis* **2014**, *131*, 299–319.
64. Yul Kwon, O. A cultural analysis of South Korea's economic prospects. *Glob. Econ. Rev.* **2005**, *34*, 213–231. [CrossRef]
65. Chung, C.-S.; Choi, H.; Cho, Y. Analysis of digital governance transition in South Korea: Focusing on the leadership of the president for government Innovation. *J. Open Innov. Technol. Mark. Complex.* **2022**, *8*, 2. [CrossRef]
66. Kim, D.H.; Kim, S.; Lee, J.S. The rise and fall of industrial clusters: Experience from the resilient transformation in South Korea. *Ann. Reg. Sci.* **2023**, *71*, 391–413. [CrossRef] [PubMed]
67. Digital Transformation Poll Results. Available online: https://solferinoacademy.com/digital-transformation-messages-from-poll/ (accessed on 4 May 2022).

*big data and cognitive computing*

MDPI

*Article*

# BERTGuard: Two-Tiered Multi-Domain Fake News Detection with Class Imbalance Mitigation

**Mohammad Q. Alnabhan * and Paula Branco**

School of Electrical Engineering and Computer Science, University of Ottawa, 800 King Edward Ave., Ottawa, ON K1N5N6, Canada; pbranco@uottawa.ca

* Correspondence: malna035@uottawa.ca

**Abstract:** In an era where misinformation and fake news undermine social well-being, this work provides a complete approach to multi-domain fake news detection. Multi-domain news refers to handling diverse content across various subject areas such as politics, health, research, crime, and social concerns. Recognizing the lack of systematic research in multi-domain fake news detection, we present a fundamental structure by combining datasets from several news domains. Our two-tiered detection approach, BERTGuard, starts with domain classification, which uses a BERT-based model trained on a combined multi-domain dataset to determine the domain of a given news piece. Following that, domain-specific BERT models evaluate the correctness of news inside each designated domain, assuring precision and reliability tailored to each domain's unique characteristics. Rigorous testing on previously encountered datasets from critical life areas such as politics, health, research, crime, and society proves the system's performance and generalizability. For addressing the class imbalance challenges inherent when combining datasets, our study rigorously evaluates the impact on detection accuracy and explores handling alternatives—random oversampling, random upsampling, and class weight adjustment. These criteria provide baselines for comparison, fortifying the detection system against the complexities of imbalanced datasets.

**Keywords:** domain classification; fake news; class imbalance; deep learning

## 1. Introduction

Social media has become a primary source of news that is readily accessible to people worldwide. Today, sharing news content on social media platforms is as simple as clicking a button, leading to a constant stream of news from various domains being uploaded daily [1]. Unfortunately, the ease of uploading news content coincides with the increased spread of fake news. This poses a major problem in terms of reliability amongst consumers and can have negative impacts, not only on individuals but also on social mobility, potentially inciting widespread social panic [2]. In 2020, it was reported that one in every five U.S. adults relied on social media as their primary source for news about the 2020 U.S. presidential election [3]. Therefore, detecting fake news is essential for assessing the truthfulness of the content being consumed. Weibo, previously known as Sina Weibo (http://www.weibo.com/, accessed on 25 June 2024), and X (formerly Twitter) (http://www.X.com/, accessed on 25 June 2024) are among the primary platforms used by social media users to obtain news. According to Weibo's annual report, during the year 2020, a total of 76,107 fake news posts were detected and cleared. Similarly, a total of 8,711,000 engagements with Facebook stories were detected regarding the 2016 U.S. presidential election campaign, in comparison with a total of 7,367,000 engagements for the top 20 analyzed election stories on 19 major news websites (https://www.buzzfeednews.com/article/craigsilverman/viral-fake-election-news-outperformed-real-news-on-facebook, accessed on 25 June 2024).

While human judgment remains effective for fake news detection, machine learning (ML) models are increasingly employed by researchers for rapid and efficient fake news detection [4,5]. News gathered from social media has been categorized by topic into multiple domains, including health, social, politics, and entertainment [6]. While fake news detection is important in all domains, some domains have extremely limited fake news [7]. Further, fake news from specific domains has an impact on society that is deemed more significant [6]. For instance, the dissemination of fake news within the political domain during the 2016 U.S. presidential election may have had a significant effect on the election outcome [8,9]. Another example that led to a worldwide social panic [2] is linked to the influx of various information related to the COVID-19 pandemic. The impact of spreading fake news via various social media platforms led to a reduction in pandemic prevention measures [10]. As such, fake news within specific domains like political or health domains tends to be more significant than other fake news domains because of its ability to delve deeper, spread widely and more rapidly, and reach a larger audience.

To address the issue revolving around detecting fake news, a wide range of machine learning detection models have been suggested. Typically, these approaches concentrate on single-domain fake news detection (SFND) in areas such as politics or health, where an algorithm is specifically trained to identify fake news within a particular domain. However, models trained in this manner often produce inadequate results when applied to other domains [11]. Although it has been found that news pieces from various domains can be interconnected, which can potentially enhance the detection of fake news for the target domain, the use of SFND is currently viewed as a disadvantage [1,7]. As such, multi-domain fake news detection has been used to address the limitations of single-domain fake news detection [12]. Multi-domain fake news detection allows for the utilization of various domains at once within a given dataset, leading to an improvement in the performance in all domains. However, this comes with great complexity, creating a flaw within multi-domain fake news detection such that the detection of fake news in some domains will be higher at the expense of other domains performing worse [13].

In addition, the majority of the available real-world domains' instances belong to the real news class (the majority or negative class), while a significantly smaller number corresponds to the opposite fake news class (the minority or positive class), which is the most important class [14,15]. This is known as the class imbalance problem, in which the dominant class has an advantage when training predictive models, causing them to disregard the minority class. The imbalance problem leads to inadequate classification performance, and most algorithms perform badly when datasets are substantially imbalanced [16]. The class imbalance problem also affects fake news detection, as there is typically a lower representation of fake news instances compared to true news instances in the available data, which makes the model predict the true class (true news) almost all the time [15].

A critical gap remains from previous works, which is the lack of the details needed to train models for each domain comprehensively. Moreover, existing solutions typically train models on one or two datasets at most, which limits the scope of their analyses. Recognizing the lack of systematic and comprehensive work in multi-domain fake news detection, this study presents BERTGuard, a holistic approach to multi-domain fake news detection. BERTGuard combines datasets from a diverse set of news domains to build the groundwork for a complex detection system that functions on two levels. The first level of our detection system performs domain classification, which is a critical phase in which a Bidirectional Encoder Representations from Transformers (BERT) model is trained on a merged multi-domain dataset. This initial classification attempts to determine the exact domain to which a given news story belongs. The second level uses domain-specific BERT models to predict the news inside each designated domain. This hierarchical structure ensures that the detection process is tailored to the unique characteristics of each domain, increasing the overall system's precision and reliability.

To assess the performance and generalizability of our technique, we rigorously tested the created detection system in five distinct domains: politics, health, science, crime, and social. We considered as baselines a single-level BERT model (Baseline 1) and the model proposed by [12] (Baseline 2). Finally, we assessed the impact of class imbalance on detection accuracy and explored three different handling approaches; random downsampling, random upsampling, and class weight adjustment.

**Contributions:** The key contributions of this paper are as follows:

- **Introduction of BERTGuard**: We propose BERTGuard, a novel two-tier solution for multi-domain fake news detection utilizing BERT-based models, which addresses limitations in existing methods by improving the semantic understanding of textual data.
- **State-of-the-art performance**: We carry out an extensive evaluation of our proposed approach by comparing it against existing alternative solutions and demonstrate that BERTGuard outperforms existing methods by achieving a 27% improvement in accuracy and a 29% increase in F1-score across multiple benchmark datasets, establishing a new state-of-the-art in fake news detection.
- **Class imbalance analysis**: We analyze the impact of three methods for dealing with the class imbalance inherent to these domains, providing valuable insights for future research and practical applications in fake news detection.

**Organization:** The structure of this paper is as follows. Section 2 provides the background and a review of the literature. In Section 3, we present our proposed multi-stage solution for fake news detection, BERTGuard. In Section 4, we outline the experimental methodology employed in our study, providing details on the datasets and the evaluation metrics utilized. Section 5 presents the experimental results and analyzes the findings. Lastly, Section 6 concludes our paper.

## 2. Background and Related Work

### 2.1. Fake News Detection

Fake news consists of false information presented in various formats, including articles, images, or videos, that are spread across social media platforms to mimic genuine news content [17]. Such fake news is deliberately designed to deceive and influence readers into accepting the presented content as true [18]. Fake news can be generated using various tools, including bots, trolls, and social robots [18]. Social robots take advantage of fake news and input from readers. These bots are first designed as computer algorithms to engage in online disputes and spread false information. As a result, fake news causes a problem because of its ease of dissemination while potentially covering a wide variety of topics and targeting an immense number of websites. This directly influences readers since it deals with topics including democracy, media, health, money, and social concerns, among others [19].

Recent research examined how artificial intelligence (AI) influences user processing of and reactions to fake news in GenAI contexts [20]. The study used a heuristic–systematic model and diagnosticity to develop a cognitive model for processing fake news. It discovered that people with high heuristic processing mechanisms are more adept at distinguishing false information due to improved positive diagnostic perception than those with low heuristic processing. Furthermore, users' perceived diagnosticity of fake news from GenAI may be anticipated using their heuristic systematic evaluations [20].

Deep learning (DL) is becoming more popular for detecting false news than other traditional ML approaches. Manually constructing features in traditional ML can be time-consuming and can contribute to biases [21]. However, DL needs a bigger dataset to train the model [21]. On the other side, DL has proven outstanding for the identification of fake news by automatically extracting useful information. Fake news detection methods that include convolutional neural networks (CNNs) [22], recurrent neural networks (RNNs) [23], BERT [24], and graph neural networks (GNNs) [25] have all been used to detect fake news.

The COVID-19 pandemic in 2020 highlighted the critical need for detecting fake news, as false information about the virus spread extensively on social media. This 'infodemic', as termed by the World Health Organization (WHO) [26], encouraged researchers to propose a CNN-based model utilizing word embedding to detect COVID-19-related fake news [27]. This architecture employed a grid search to optimize the model hyperparameters, enabling the proposed CNN model to achieve impressive results, including a 96.19% mean accuracy and a 95% mean F1-score, outperforming several state-of-the-art machine learning algorithms.

Nasir et al. [28] developed a hybrid model that integrates a CNN and long short-term memory (LSTM). The CNN extracts the features, while LSTM captures the long-term dependencies. The approach uses Global Vectors for Word Representation (GloVe), which are pre-trained word embeddings that represent words as vectors. This methodology surpassed seven established ML techniques in terms of performance: logistic regression, random forest, multinomial naïve Bayes, k-nearest neighbors, decision tree, CNNs, and RNNs.

Kaliyar et al. [29] developed a deep-learning network based on pre-trained GloVe word embeddings called FNDNet. Three convolutional pooling layers gathered characteristics from word embedding vectors, which were then classified using additional convolutional pooling and dense layers.

Saleh et al. [30] proposed an optimized DL approach, OPCNN-FAKE, based on CNN architecture. They assessed its performance against RNNs, traditional ML approaches, and LSTM. The model includes an embedding layer that generates embedding vectors, a dropout layer for improved regularization, a convolution-pooling layer for extracting and reducing features, a flattened layer that produces a one-dimensional vector, and an output layer that decides whether the input text is fake or real based on the previous layer's output.

Yang et al. [31] designed another CNN-based approach (TI-CNN) that detects fake news by integrating obvious and hidden features from text and picture data. The authors evaluated their approach to various models such as LSTM, GRU, and CNN. Similarly, Raj and Meel [32] developed a CNN approach that uses text and image data to categorize internet news. However, these methods may fail to capture long-term contextual information. Furthermore, the word embedding does not capture context-specific information inside the text.

Hashmi et al. [33] proposed a thorough approach to detecting fake news utilizing three publicly available datasets: WELFake, FakeNewsNet, and FakeNewsPrediction. They combined FastText word embeddings with a range of ML and DL methods, improving these algorithms through regularization and hyperparameter optimization to reduce overfitting and to boost model generalization. Significantly, a hybrid model that integrated CNNs and LSTMs and was enhanced with FastText embeddings surpassed other methods in terms of classification performance across all datasets, achieving accuracy and F1-score of 99%, 97%, and 99%, in WELFake, FakeNewsPrediction, and FakeNewsNet respectively.

Transformer-based models have also been explored for fake news detection. Bidirectional Encoder Representations from Transformers (BERT) is a DL model designed to generate text, analyze sentiment, and understand natural language. It leverages transformer encoders to perform various natural language processing (NLP) tasks [34]. BERT has been employed in previous studies to achieve superior results on various sentiment analysis tasks [35]. Despite being pre-trained on a large set of textual data, BERT needs to be modified to perform efficiently on a given task [36]. Several versions of BERT exist in the literature, with each tailored to specific domains or tasks. Some notable BERT versions include:

- **RoBERTa**: Robustly Optimized BERT Pretraining Approach (RoBERTa) is a variant of BERT designed to improve the training process. It was developed by extending the training duration, using larger datasets with longer sequences, and employing larger mini-batches. The researchers achieved significant performance enhancements by adjusting several hyperparameters of the original BERT model [37].
- **DistilBERT**: DistilBERT is a more compact, faster, and cost-effective version of BERT, which is derived from the original model with certain architectural features removed to enhance efficiency [38]. Similar to BERT, DistilBERT can be fine-tuned to achieve strong performance in various natural language processing tasks [39].

- **BERT-large-uncased**: This is a BERT model with 12 more transformer layers, 230 million more parameters, and a larger embedding dimension, allowing the model to encapsulate much more information than BERT-base-uncased [40].
- **BERT-cased**: This is a model trained with the same hyperparameters as first published by Google [39].
- **ALBERT**: ALBERT is short for 'A Lite' BERT and is a streamlined version of BERT designed to reduce memory usage and accelerate training. It employs two primary parameter-reduction techniques: dividing the embedding matrix into two smaller matrices and using shared layers grouped together [41]. ALBERT is recognized for its reduced number of parameters compared to BERT, which enhances its memory efficiency while still delivering strong performance. These variations highlight BERT's adaptability across various domains and its potential for tailored customization to specific tasks.

Table 1 summarizes the latest key advancements in fake news detection utilizing transformer models.

**Table 1.** The main transformer-based fake news detection models.

| Ref. | Year | Model | Covered Multi-Domain? |
|---|---|---|---|
| [11] | 2023 | BiLSTM, Hybrid CNN+RNN, CNN, C-LSTM, and BERT | No |
| [42] | 2023 | BERT+LSTM model for text content analysis<br>GAT for modeling social network features | No |
| [43] | 2023 | BERT | No |
| [44] | 2023 | BERT + LightGBM | No |
| [45] | 2023 | SBERT, RoBERTa, and mBERT | No |
| [46] | 2023 | Hybrid ensemble learning model: BERT for text classification tasks<br>Ensemble learning models, including Voting Regressor and Boosting Ensemble | No |
| [47] | 2023 | CT-BERT with BiGRU | No |
| [48] | 2023 | BERT, LSTM, BiLSTM, and CNN-BiLSTM | No |
| [49] | 2023 | DistilBERT and RoBERTa | No |
| [50] | 2023 | hybrid model combining LSTM and BERT with GloVe embeddings | No |
| [51] | 2023 | TF-IDF N-gram, BERT, GloVe, and CNN | No |
| [52] | 2022 | Arabic-BERT, ARBERT, and QaribBert | No |
| [53] | 2022 | BERT, XLNet, RoBERTa, and Longformer | No |
| [34] | 2022 | BERT with Reinforcement Learning | Yes<br>Gossip (Social) and Political |
| [54] | 2022 | BERT | No |
| [55] | 2022 | BART and RoBERTa | No |
| [12] | 2021 | FakeBERT: Combination of BERT and 1d-CNN | Yes<br>Social and Political |
| [56] | 2021 | ALBERT, BERT, RoBERTa, XLNet, and DistilBERT | No |
| [57] | 2021 | BERT | Yes<br>Political, Social, and Health |

A comparative study was carried out involving five transformer models: BERT, ALBERT, RoBERTa, XLNet, and DistilBERT [56]. The authors ran the comparison with different hyperparameter combinations and yielded equivalent results.

Another comparative study investigated cross-domain models' effectiveness and found that BERT-based models achieved the best detection accuracy [11].

Subsequently, a deep learning model combining BART and RoBERTa was developed to differentiate between true and fake news articles [55]. The embeddings from both BART and RoBERTa were first processed through LSTM and CNN architectures and were then concatenated and further processed through additional LSTM and CNN layers.

In addition, Kaliyar et al. [12] introduced FakeBERT, a deep learning model for fake news detection that leverages BERT in conjunction with 1D CNNs of various kernel sizes and filter configurations to enhance classification performance. BERT is used to generate word embeddings, which are subsequently processed through three parallel convolutional layers with different kernel sizes. The dataset used encompasses both social and political news, with a focus on the U.S. presidential election of 2016. This BERT-based approach achieved an impressive 98.9% detection accuracy, outperforming other ML techniques.

An attention-based transformer model has also been employed to detect fake news [58]. The authors compared their approach with a hybrid CNN model that integrates both text and metadata. The transformer model demonstrated superior accuracy compared to the hybrid CNN model. However, transformer-based methods often involve significant computational costs and require large amounts of training data.

Using single datasets for fake news detection has several limitations. First, they provide limited coverage: single datasets may focus on specific topics or domains, such as political statements or social media posts, leading to a lack of diversity in the types of fake news covered [59]. Second, there is dataset bias: datasets constructed only with specific types of news, such as political or e-commerce news, can lead to biased models that perform poorly when detecting news related to other topics, resulting in dataset bias [59,60]. Third, there is a lack of labeled data: the shortage of labeled data for training detection models impedes the development of effective fake news detection systems [54]. Fourth, there are problems with overfitting and generalizability: limited and imbalanced datasets can cause machine learning and deep learning models to overfit or underfit, affecting their ability to generalize and perform well [61,62].

To overcome these limitations, it is crucial to utilize a diverse set of publicly available evaluation datasets for fake news detection and incorporate multiple domain-specific datasets. Implementing cross-domain, cross-language, or cross-topic analyses provides a comprehensive approach by incorporating datasets across various domains or topics, thereby enhancing the detection process and improving the generalizability of fake news detection models [63].

Many studies on the automated detection of fake news have relied on datasets confined to a single domain, such as political, social, or health domains, for model training and evaluation. This focus on a single domain is driven by the performance decline observed in these machine and deep learning techniques when they encounter unseen data from other domains. Domain-specific features, particularly style-based attributes, can vary significantly across different domains [63]; consequently, features must be carefully selected to distinguish between fake and real news within the specific domain being examined.

The current state-of-the-art highlights the need for further research across various domains. As a result, comprehensive cross-domain techniques for fake news detection are essential, despite some previous studies attempting to tackle this issue using cross-domain data, such as the recent work done by Alnabhan and Branco [11].

Other works exist that address the multiple-domain problem in detecting fake news. Han et al. [64] proposed a continuous learning approach for domain-agnostic fake news detection that utilized a graph neural network to sequentially learn across multiple domains. However, this method has two drawbacks: (1) it assumes that new domains will arrive in a specific sequence, and (2) it presumes that these domains are already known, which is not the case in real-world data streams. In contrast, Cardoso et al. [65] introduced a method that retains knowledge across different domains by leveraging a robust, optimized BERT model to select informative instances for manual annotation from a large, unlabeled dataset. Consequently, earlier studies explored integrating information from multiple domains to develop a cross-domain fake news detection model. For example, Castelo et al. [66] trained a model on the Celebrity dataset and tested it on the US-Election2016 dataset to determine the generalizability of their approach.

Huang et al. [57] proposed a novel framework for detecting fake news in new domains called DAFD: Domain Adaptation Framework for Fake News Detection. The framework combines domain adaptation and adversarial training strategies to align the data distribution of the source and target domains and enhance the model's generalization and robustness. The framework consists of a pre-training phase, where the data distribution alignment is performed, and a fine-tuning phase, where adversarial examples are generated to further improve the model's performance. Experiments conducted on real datasets, including PolitiFact, GossipCop, and COVID show that DAFD outperforms state-of-the-art methods for detecting fake news in new domains with a small amount of labeled data. The framework's components were analyzed, showing that both domain adaptation and adversarial training are crucial for improving detection performance.

Mosallanezhad et al. [34] proposed a domain-adaptive model called Reinforced Adaptive Learning Fake News Detection (REAL-FND). REAL-FND leverages generalized and domain-independent features to differentiate between fake and true news. This approach is based on prior findings that suggest domain-invariant features can improve the robustness and generalizability of fake news detection techniques. For example, it has been noted that fake news publishers frequently use clickbait writing styles to capture the attention of targeted audiences, highlighting a domain-invariant characteristic. Moreover, patterns derived from social contexts, such as a user's comment disputing a news article or interactions between users and identified fake news disseminators, offer critical supplementary information for classifying fake news within a specific domain. In REAL-FND, the approach departs from the conventional method of using adversarial learning to train cross-domain models. Instead, it employs a reinforcement learning (RL) component to transform the learned representation from the source domain to the target domain. Unlike other RL-based methods that modify model parameters, in REAL-FND, the RL agent adjusts the learned representations to obscure domain-specific features while preserving domain-invariant components. This method offers greater flexibility than adversarial training, as the RL agent can directly optimize the confidence values of any classifier without the need for a differentiable objective function.

### 2.2. Fake News Datasets

Researchers identify the primary challenge in fake news detection as the scarcity of large-scale datasets and the absence of a comprehensive benchmark dataset with reliable ground truth labels [67]. Furthermore, there are not many datasets with different labels, sizes, and application domains that can be found online for the detection of fake news. Certain datasets are sourced exclusively from political statements, while others come from postings on social media and even news articles. This diversity presents a significant obstacle in the field of fake news.

Additionally, acquiring datasets for academic research is challenging due to privacy restrictions on extracting data from online sources. One solution is to purchase data from these platforms or crowdsourcing websites. Another approach is to utilize existing datasets from the literature that align with the study's requirements, such as ISOT [68], PHEME [69], Liar [70], GossipCop [67], FakeNews (https://www.kaggle.com/competitions/fake-news/data, accessed on: 12 May 2024), Fake-OR-Real (https://github.com/joolsa/fake_real_news_dataset, accessed on: 12 May 2024), Snopes (http://fakenews.research.sfu.ca/, accessed on: 12 May 2024), COVID-19 FakeNews (https://data.mendeley.com/datasets/zwfdmp5syg/1, accessed on: 12 May 2024), COVIFN (https://ieee-dataport.org/documents/covifn-fake-news-covid19, accessed on: 12 May 2024), Politifact (https://www.kaggle.com/datasets/rmisra/politifact-fact-check-dataset, accessed on: 12 May 2024), Climate [71], and COVID-Claims (https://ieee-dataport.org/open-access/covid-19-fake-news-infodemic-research-dataset-covid19-fnir-dataset, accessed on: 12 May 2024).

### 2.3. Dealing with Class Imbalance

The class imbalance problem has received much attention [72]. However, when observing the particular domain of fake news detection, we verify that this issue still has received very little attention. Existing research has shown the prevalence of the imbalance problem as cases of fake news proliferate, sometimes surpassing those of true news [73]. This imbalance complicates the development of accurate and trustworthy fake news detection methods. Overfitting of neural networks due to class imbalance is a key challenge, highlighting the need for investigation and improvements in this domain [61].

While some studies have explored specific elements of fake news detection models, including dataset division, features, and classifiers, there has been an insufficient examination of the limitations of datasets and features and their impact on detection models, particularly regarding class imbalance [15,61]. Furthermore, the precision of detection models is significantly unsatisfactory, with a low rate of detection and a lengthy detection processing duration [59,63,73].

In addition, research has been performed to minimize domain biases and increase the accuracy of cross-domain fake news detection, which is connected to the difficulty of imbalanced data across different domains [60]. In addition, a comprehensive review highlights the constraints of current fake news detection models due to imbalanced and limited datasets, highlighting the necessity for comprehensive cross-domain techniques to address these difficulties [59].

Furthermore, learning techniques such as resampling methods (e.g., oversampling and undersampling), data augmentation, and cost-sensitive learning have been utilized to balance the class distribution and increase the accuracy of fake news detection models [74]. These approaches have yielded promising results in addressing the class imbalance problem in fake news detection.

Keya et al. [75] created 'AugFake-BERT' to control imbalances by data augmentation to boost the effectiveness of fake news categorization, specifically addressing the influence of imbalanced datasets on detection models. Similarly, [76] used back-translation as data augmentation, applying pre-trained word embeddings (Word2Vec, GloVe, and fastText) in CNN, bidirectional LSTM, and ResNet models for fake news detection.

Mouratidis et al. [77] addressed the class imbalance issue by implementing SMOTE (Synthetic Minority Over-sampling Technique) to oversample the minority class. SMOTE is a widely used method for generating synthetic samples to effectively reduce the class imbalance problem in machine learning models. This approach resulted in an improved accuracy and F1-score, improving from 95% and 99% to 98% and 100%, respectively.

Additionally, other methods such as the focal loss function and specific learning techniques have demonstrated the ability to achieve high accuracy and satisfactory recall, further mitigating the effects of class imbalance [78].

The amount of research studies conducted highlights ongoing efforts to tackle the imbalance in fake news detection.

## 3. BERTGuard: Two-Tiered Multi-Domain Fake News Detection with Class Imbalance Mitigation

In this section, we outline our BERTGuard approach for fake news detection, which utilizes a two-stage detection method based on BERT, and we explore the impact of domain-specific classification. The first stage of our solution entails classifying news domains with BERT to capture the nuances associated with various sources of information. Then, in the second stage, we use domain-specific BERT models to determine the validity of news within their respective areas. Figure 1 presents the BERTGuard detection approach.

We chose to base our approach on BERT due to its proven ability to enhance fake news detection by capturing complex linguistic patterns and contextual nuances. Numerous studies have validated BERT's effectiveness in this domain [11,22,79–82]. In addition, BERT's contextualized word representations enable it to capture the nuanced meaning and context of words and phrases within news articles, which is crucial for distinguishing between real and fake news. Moreover, BERT handles the issue of missing information in fake news detection through its advanced language understanding capabilities and the ability to capture contextual information. By analyzing the language used in news articles and comparing it to a database of known fake news, BERT can identify patterns and inconsistencies that suggest a news article may be fake, even in the presence of missing information [12]. Additionally, BERT's fine-tuning capability allows it to adapt to the nuances of the task and the dataset, which can help mitigate the impact of missing information on the overall fake news detection accuracy [12].

As depicted in Figure 1, the initial stage of the pipeline employs a BERT-based multi-class classification model. This model ingests diverse text formats, such as news articles, and analyzes them to determine and assign a relevant domain.

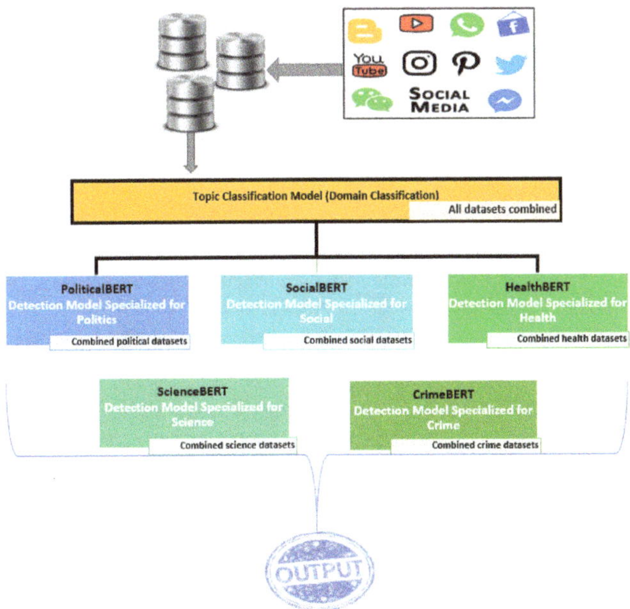


**Figure 1.** BERTGuard fake news detection architecture.

Following initial testing, we discovered that the base version of DistilBERT, a distilled variant of BERT, outperformed other BERT models and was notably faster in both the training and testing phases. It was therefore decided to use DistilBERT for classification.

We trained this model on all the training datasets allocated for this purpose. These training datasets were chosen from five different news domains. We merged the datasets FA-KES, COVID-FN, COVID-FNIR, FakeNews, ISOT, LIAR, Climate, and GossipCop into a single dataset for training DistilBERT. The data were preprocessed for classification by encoding the categories into integer values (crime: 0, health: 1, politics: 2, science: 3, and social media: 4).

The DistilBERT model was therefore set up to have five labels, one for each category. The training, validation, and testing datasets were encoded with the DistilBERT tokenizer and converted into PyTorch tensors. The motivation for domain classification arises from the fact that a news article often belongs to multiple domains or topics, i.e., news articles may overlap across different topics. In addition, the domain classification model was tuned by adjusting the number of epochs, and several iterations of a given epoch were performed to assess the performance at a given epoch.

Following the domain classification stage, the pipeline directs the input data to a domain-specific detection model. We conduct an initial testing phase to determine the most suitable BERT for each domain. The models in this stage are trained on datasets with data belonging to that specific domain, making them specialized at detecting fake content within that domain. The model outputs a boolean value denoting whether the input is detected as fake or not. Fine-tuning of the models was done to maximize the F1-scores. The fine-tuning that was done on all of the models above for each category includes epochs, optimizer, and learning rate.

Five domain-specific models were trained on each of the five domains considered (crime, health, politics, science, and social media). The pre-trained model selection consists of the tokenizer, the sequence classifier, and the model itself. Below are all the options used:

1. Tokenizer: BertTokenizer, classifier: BertForSequenceClassification, and model: **BERT (base uncased)**;
2. Tokenizer: BertTokenizer, classifier: BertForSequenceClassification, and model: **BERT (base cased)**;
3. Tokenizer: DistilBertTokenizer, classifier: DistilBertForSequenceClassification, and model: **DistilBERT (base uncased, finetuned SST-2 English)**;
4. Tokenizer: DistilBertTokenizer, classifier: DistilBertForSequenceClassification, and model: **DistilBERT (base uncased)**;
5. Tokenizer: RobertaTokenizer, classifier: RobertaForSequenceClassification, and model: **RoBERTa (base)**;
6. Tokenizer: AlbertTokenizer, classifier: AlbertForSequenceClassification, and model: **ALBERT (base v2)**.

Overall, BERTGuard incorporates an initial BERT model for domain classification, followed by five specialized BERT-based models tailored to the detected domain.

In addition, BERTGuard addresses the critical issue of class imbalance in fake news detection. Our approach includes balancing the training datasets through oversampling the minority class, undersampling the majority class, and adjusting class weights.

## 4. Materials and Methods

In our study, we developed BERTGuard: a two-stage fake news detection approach utilizing BERT and focusing on domain-specific classification. The first stage involves classifying news domains with BERT to capture the nuances of various information sources. In the second stage, domain-specific BERT models are used to assess the validity of news within their respective domains. To establish baselines for comparison, we developed a model that disregards news domains to serve as the first baseline. We also used a well-known model using a base version of BERT called FakeBERT [12] as another baseline. Various versions of BERT were utilized after carefully considering the performance of and the time required by these models. This dynamic selection process enhances the adaptability and effectiveness of our solution. Based on these initial experiments, we chose the following BERT versions for the following news domains:

- Crime: distilbert-base-uncased-finetuned-sst-2-english;
- Health: distilbert-base-uncased;
- Politics: bert-base-uncased;
- Science: roberta-base;
- Social media: distilbert-base-uncased.

For the first stage of our BERTGuard, we trained the DistilBERT model on a merged dataset comprising the FA-KES, COVID-FN, COVID-FNIR, FakeNews, LIAR, Climate, and GossipCop datasets. The dataset was preprocessed by encoding categories into integer values (crime: 0, health: 1, politics: 2, science: 3, and social media: 4). The DistilBERT model was configured with five labels: one for each category. In the second stage of BERTGuard, each domain's model was trained on the training datasets from the respective domain. For evaluation, we compared the performance of BERTGuard against the baselines.

Our BERTGuard also attempts to mitigate the class imbalance since this is a critical issue in the fake news detection context. These attempts include balancing the training datasets by oversampling the minority class, undersampling the majority class, and adjusting the class weights.

Oversampling entails boosting the number of instances in the minority class by randomly replicating them. This technique helps mitigate bias towards the majority class, enhancing the model's ability to learn from minority class examples.

Conversely, undersampling involves decreasing the number of instances in the majority class by randomly discarding some samples. This approach prevents the model from being dominated by the majority class, allowing it to concentrate more on learning from the minority class.

In addition, adjusting class weights is another technique used to address class imbalance. The idea is to assign higher weights to the minority class (e.g., fake news) and lower weights to the majority class (e.g., true news) during training to ensure that the model pays more attention to the minority class. The common approach to determining the optimal class weights is manual tuning. It is to manually tune the class weights based on domain knowledge or prior experience with the dataset. The choice of class weights is often a balance between overfitting and underfitting. Assigning excessively high weights to the minority class can lead to overfitting, where the model becomes overly biased towards classifying instances as the minority class, potentially reducing its overall accuracy and effectiveness. Conversely, setting the weights too low can result in underfitting, where the model does not adequately learn from the minority class, leading to poor performance in those examples. This required experimenting with different weight values and evaluating the model's performance on a validation set.

Regarding the main baseline model, we developed a BERT (base uncased) model for one-stage fake news detection (Baseline 1) without paying attention to news domains. The implementation is based on what we did in the domain classification model, with the following changes: the label column will be either '1' or '0', corresponding to 'fake' or 'not fake', and this will be the target for the detection model.

The other baseline (Baseline 2) that we used from the literature, FakeBERT [12], was trained on the same dataset used to train our main baseline.

### 4.1. Data Collection

We chose the most widely used and publicly available datasets related to the following fake news domains shown in Table 2.

**Table 2.** Characteristics of the datasets used in our experiments.

| Dataset | Domain | True | Fake |
|---|---|---|---|
| FA-KES | Crime | 426 | 378 |
| Snopes | Crime | 195 | 120 |
| COVID-FNIR | Health | 3795 | 3793 |
| COVID-Claims | Health | 1591 | 1230 |
| COVID-FN | Health | 2061 | 1058 |
| Pheme | Politics | 5089 | 1335 |
| Liar | Politics | 7176 | 5669 |
| ISOT | Politics | 23,481 | 21,417 |
| Politifact | Politics | 11,760 | 9392 |
| FakeNews | Politics | 10,413 | 10,387 |
| Climate | Science | 654 | 253 |
| ISOT-small (Science) | Science | 1000 | 1000 |
| GossipCop | Social | 16,817 | 5323 |
| ISOT-small (Social) | Social | 1000 | 1000 |

Regarding the training data and the preprocessing tasks, the model in the domain classification stage (the first stage of BERTGuard) is trained on a dataset composed of FA-KES, COVID-FN, COVID-FNIR, FakeNews, ISOT, LIAR, Climate, and GossipCop datasets.

These datasets represent the five news domains we selected. The data are processed into 'text' and 'label' columns. The 'text' column contains all the news article content (title and text) used for detection. The 'label' column will be either '1' or '0', corresponding to 'fake' or 'not fake', which serves as the target for the detection model.

Combining these datasets to train the domain classification model required careful consideration as some datasets include more textual columns such as 'title' and 'text'. We concatenated these fields into a single 'text' column containing all relevant content from the news articles. All other features were stored in a JSON dictionary and kept as supplementary metadata. By retaining the additional features in a metadata field, they can be readily accessed if needed for future refinements or other aspects of the project.

The datasets (news articles) we used were already labeled by their respective sources. We did not perform additional labeling. Each article was assigned a label in the 'label' column, where '1' indicates 'fake' and '0' indicates 'not fake'. Some datasets had more than two possible truth values, so articles that were labeled half-true or mostly-true were labeled as 0 (True), while articles labeled barely-true or pants-fire were labeled 1 (Fake). This label was used as the target variable for our detection models.

For domain labeling and categorization, we added a new 'domain' column to each dataset, indicating the specific category (domain) of each article based on its content and source. The domain label is consistent across all news articles within the same dataset. The domains were categorized into five primary areas: Politics, Health, Crime, Social, and Science. These categories were encoded into integer values for consistency and ease of processing as follows: Crime: 0, Health: 1, Politics: 2, Science: 3, and Social: 4. We followed a systematic approach, splitting the data into training, validation, and testing sets with a ratio of 7:1:2. We also included essential preprocessing steps and tracked the training duration to maintain consistency and optimize effectiveness. We tuned the models by adjusting the number of epochs and running multiple training iterations at a given epoch.

To achieve a unified format for all the collected datasets, comprehensive preprocessing was performed. This standardization ensures minimal additional processing in later stages and maintains data consistency across the project. The following steps detail the main preprocessing tasks:

1. Label standardization: The label values were standardized to ensure uniformity. The datasets that we utilized in this study were already labeled by their sources. In this project, '1' was assigned to indicate 'fake' news, and '0' indicates 'not fake' news. Some datasets originally had multiple classes representing varying degrees of 'fakeness'. These were binarized to a fake/not fake format based on criteria that balanced the dataset as effectively as possible.
2. Text processing:
    - Cleaning: raw data were cleaned to remove irrelevant content such as HTML tags, scripts, special characters, and advertisements.
    - Standardization: standardized formats were applied, such as converting dates to a consistent format and removing duplicates.
    - Normalization: text normalization included converting text to lowercase, removing punctuation, and expanding contractions.
    - Tokenization: the text was tokenized to break it down into individual words or tokens.
    - Stop word removal: common stop words were removed to focus on the most meaningful words.
    - Lemmatization: words were reduced to their base or root forms through lemmatization, aiding in the normalization of text data.
3. Handling missing data: entries with missing titles, bodies, or descriptions were removed to ensure the integrity of the dataset.
4. Multi-language data: for datasets containing multi-language data, articles were filtered to include only those written in English, ensuring uniformity in text processing and model training.

5. Metadata management: all other features not directly used in the primary analysis were stored in a JSON dictionary as supplementary metadata. This approach allows these features to be easily utilized if needed for further refinement of the results.

Regarding the detection stage (second stage of BERTGuard), each domain-specific model was trained using the training datasets that represent that domain. For example, we used the FakeNews, ISOT, and LIAR datasets to train the political BERT model. The remaining political datasets (Pheme and Politifact) were used later for final complete detection system testing.

Data were handled by loading them as a whole category to be used to train and test models. This process included detecting all feather files, loading them as dataframes, then concatenating all of the dataframes together to create a combined dataframe. This was then split using the train_validation_test_split from sklearn, and the subcategories were used accordingly.

### 4.2. Evaluation Metrics

The system's performance was evaluated using various metrics, as detailed in Table 3. These metrics were chosen to provide a thorough assessment of the model's effectiveness at detecting fake news. We included accuracy as well as several metrics suitable for imbalanced domains. This allowed us to obtain a better understanding of the performance of the models in an imbalanced setting [83].

**Table 3.** Performance evaluation metrics.

| Metric | Formula | Evaluation Focus |
|---|---|---|
| Accuracy | $\frac{tp+tn}{tp+fp+tn+fn}$ | The proportion of correctly predicted positive and negative instances out of the total instances. |
| Precision | $\frac{tp}{tp+fp}$ | Measures the positive patterns that are correctly predicted from the total predicted patterns in a positive class. |
| Recall | $\frac{tp}{tp+tn}$ | The proportion of actual positive instances that were correctly predicted by the model. |
| F1-Score | $\frac{2*pre*rec}{pre+rec}$ | The harmonic mean of precision and recall, balancing both metrics. |
| Geometric-mean (G-Mean) | $\sqrt{tp * tn}$ | The geometric mean of recall for each class, focusing on the balance between classification performance across classes. |

**Note:** *tp*—true positive; *tn*—true negative; *fp*—false positive; *fn*—false negative.

## 5. Results and Discussion

This section presents the results of our experimental evaluation. We begin by presenting the ad hoc testing results for selecting the best model for each case and domain. Afterward, we observe the performance of the baseline models considered and BERTGuard. Then we explore the effect of handling the imbalance problem by adjusting the class weights and applying upsampling and downsampling.

### 5.1. Model Selection

In this section, we present the initial ad hoc testing for picking the model for classification and detection. This is not meant to be an exhaustive test but rather a quick comparison between the models when deciding the best course of action.

To select the domain classification model, we used a balanced dataset, with 850 samples per category in the training/validation data and 160 samples/category in the testing data. The testing/validation split was 80% to 20%. Table 4 shows the accuracy for each model used.

**Table 4.** Domain classification model results.

| Model | Average Accuracy | Training Time |
|---|---|---|
| albert-base-v2 | 95.8% | 724.1373889 |
| bert-base-uncased | 92.9% | 1933.898181 |
| distilbert-base-uncased | 96.8% | 96.21049619 |
| distilbert-base-uncased-finetuned-sst-2-english | 91.8% | 220.6201028 |
| roberta-base | 98.6% | 1545.978703 |

While roberta-base achieved a slightly higher average accuracy (98.6%) compared to distilbert-base-uncased (96.8%), the training time for distilbert-base-uncased was significantly shorter. The distilbert-base-uncased model trained in 96.21 s, whereas the roberta-base model required 1545.98 s.

This substantial difference in training time demonstrates that distilbert-base-uncased is much more efficient, making it a more practical choice, especially when computational resources and time are constrained. Moreover, the marginal difference in accuracy (1.8%) does not justify the significantly higher training time and resource consumption of the roberta-base model. Therefore, we opted for the distilbert-base-uncased model to achieve a balance between high accuracy and efficient resource utilization.

For the domain-specialized detection model, another quick testing task took place for each news domain. Table 5 presents the average accuracies of different models tested for each domain.

**Table 5.** The average accuracies of various models on news domains.

| Model | Crime | Health | Politics | Science | Social |
|---|---|---|---|---|---|
| albert-base-v2 | 61.7% | 84.8% | 67.8% | 75.3% | 84.2% |
| bert-base-uncased | 61.3% | 87.5% | **87.8%** | 74.7% | 84.7% |
| distilbert-base-uncased | 60.1% | **88.3%** | 68.5% | 77.7% | **95.6%** |
| distilbert-base-uncased-finetuned-sst-2-english | **68.8%** | 86.2% | 61.3% | 71.6% | 84.5% |
| roberta-base | 56.9% | 87.6% | 71.8% | **80.2%** | 88.7% |
| bert-base-cased | 61.9% | 87.1% | 69.6% | 76.4% | 84.7% |

For deciding the best model for each news domain, we prioritized the highest average accuracy for each specific domain:

1. Crime: The distilbert-base-uncased-finetuned-sst-2-english model performed the best, with an average accuracy of 68.8%. This suggests that the finetuned version of DistilBERT is better at capturing the nuances in the crime domain.
2. Health: The distilbert-base-uncased model achieved the highest accuracy of 88.3%, making it the most suitable for health-related news detection.
3. Politics: The bert-base-uncased model had the highest accuracy at 87.8%, indicating its effectiveness in detecting political news.
4. Science: The roberta-base model excelled in the science domain, with an accuracy of 80.2%, suggesting it handles the specific language and context of science-related news more effectively.
5. Social: The distilbert-base-uncased model stood out, with a remarkable accuracy of 95.6%, making it the best choice for the social domain.

These selections balance accuracy across various domains, ensuring each model is optimized for the specific characteristics of the content it will encounter. While the roberta-base model showed strong performance in certain areas, the distilbert-base-uncased variants, including the finetuned version, offered competitive accuracies with the added benefit of being more resource-efficient, as noted in previous experiments.

### 5.2. Single-Stage Fake News Detection Results

In this section, we present the results of testing the first baseline model (Baseline 1) that we used in our work. This model is a BERT (base uncased) model trained on the merged datasets from the domains considered and tested on the set of unseen datasets. Domain categories were not taken into consideration when training this model. Table 6 presents the precision, recall, accuracy, F1-score, and G-mean when testing the Baseline 1 model, which is the main baseline model we developed.

**Table 6.** Single-stage fake news detection results—Baseline 1.

| Testing Dataset | Precision | Recall | F1-Score | Accuracy | G-Mean |
|---|---|---|---|---|---|
| Snope | 0.583011583 | 0.774358974 | 0.665198238 | 0.517460317 | 0.671907919 |
| COVID-Claims | 0.224425887 | 0.135135135 | 0.168693605 | 0.246979389 | 0.174148852 |
| Pheme | 0.313032887 | 0.192509363 | 0.238404453 | 0.744396015 | 0.245482712 |
| Politifact | 0.486368313 | 0.18228263 | 0.265180199 | 0.49520736 | 0.297752406 |
| Climate | 0.349206349 | 0.434782609 | 0.387323944 | 0.61631753 | 0.389652213 |
| GossipCop | 0.378611058 | 0.743565658 | 0.501743044 | 0.644941283 | 0.530586638 |
| ISOT-small | 0.581487556 | 0.789125069 | 0.669578551 | 0.556974212 | 0.677396788 |

The evaluation of Baseline 1 reveals the drawbacks of a non-domain-specific approach. Across several datasets from various domains, distinct patterns emerge. Starting with the accuracy and recall measures, the model's performance varies significantly between datasets. Notably, in the 'Snope' dataset, the model shows higher precision but lower recall, demonstrating a tendency to properly identify true news while missing a significant proportion of fake news cases. In contrast, in the 'GossipCop' dataset, the model achieves a greater recall, indicating a stronger capacity to detect fake news but with a trade-off in precision. On the other hand, Specialized datasets such as 'COVID-Claims' pose distinct difficulties for the baseline model, as evident in the reduced precision, recall, and F1-score. This underscores the importance of tailoring models to specific content domains, particularly those with distinct characteristics such as health-related information.

The baseline model has modest precision and recall in the 'Politifact' dataset, indicating some adaptation across platforms. However, the findings highlight the need for a more complex, domain-aware strategy for detecting fake and accurate news across several platforms. However, an unanticipated anomaly appears in the 'Pheme' dataset, where the model achieves significantly higher precision and accuracy than other performance measures. This anomaly necessitates more analyses to identify potential biases or unique dataset factors that could have influenced the model's performance. Understanding such anomalies is critical for improving the model's adaptability to different datasets.

The discovered disparities highlight the limitations of creating a one-size-fits-all fake news detection methodology. While the baseline approach shows promise in some domains, its performance limits emphasize the need for domain-specific modifications.

### 5.3. Multi-Domain Fake News Approach

In this section, we present the results of testing our BERTGuard, a multi-tier detection system, by using unseen datasets that we kept for this purpose. We compared Baseline 1, the existing Baseline 2 (FakeBERT) solution, and our proposed BERTGuard.

Table 7 shows the overall results of these three solutions averaged across all the testing datasets.

The results demonstrate that our domain-specific strategy significantly improves detection performance by leveraging domain-specific information. Moving into more detailed results of our experiments, Table 8 presents the precision, recall, accuracy, F1-score, and the G-mean when testing the proposed BERTGuard for each testing dataset.

**Table 7.** Overall results of Baseline 1, Baseline 2 (FakeBERT), and BERTGuard.

| Architecture | Precision | Recall | F1-Score | Accuracy |
|---|---|---|---|---|
| BERTGuard (Domain-Aware) | **82%** | **68%** | **70%** | **82%** |
| Baseline 1 (No Domains) | 42% | 46% | 41% | 55% |
| Baseline 2 (FakeBERT) | 32% | 34% | 31% | 38% |

These results reflect the average testing values for various unseen datasets.

**Table 8.** The proposed BERTGuard testing results.

| Testing Dataset | Precision | Recall | F1-Score | Accuracy | G-Mean |
|---|---|---|---|---|---|
| Snopes | 0.837719298 | 0.979487179 | 0.903073286 | 0.86984127 | 0.905834043 |
| COVID-Claims | 0.956869994 | 0.976115651 | 0.966397013 | 0.961620469 | 0.966444916 |
| Pheme | 0.847560976 | 0.624719101 | 0.71927555 | 0.89866127 | 0.727658939 |
| Politifact | 0.800490597 | 0.083248299 | 0.150812601 | 0.478772693 | 0.258146239 |
| Climate | 0.534798535 | 0.577075099 | 0.55513308 | 0.742006615 | 0.555534803 |
| GossipCop | 0.835781991 | 0.66259628 | 0.739180551 | 0.887579042 | 0.744168017 |
| ISOT-small | 0.898968008 | 0.868 | 0.883240223 | 0.8835 | 0.883362106 |

BERTGuard had significant improvements across multiple datasets that reflect multiple news domains, as evidenced by the precision, recall, F1-score, accuracy, and G-mean metrics.

The findings from the 'Snopes' and 'COVID-Claims' datasets demonstrate the success of our domain-specific approach in these crucial domains. With precision reaching 83% and recall over 97% in 'Snopes' and precision and recall both exceeding 95% in 'COVID-Claims', the model exhibits an impressive ability to properly classify both true and fake news instances, demonstrating its durability in areas where accuracy is critical.

In the 'Pheme' dataset, the model performs superbly, with a precision of 84.76%, suggesting a high proportion of accurately identified fake news events. However, the recall is rather low at 62.47%, indicating possible difficulties in collecting all instances of fake news in this particular domain. This intricate balance requires further investigation to fully comprehend the complexities of information transmission within the 'Pheme' dataset.

The model encounters significant obstacles in the 'Politifact' and 'Climate' domains. 'Politifact' has a lower precision of 80.05% and a recall of 8.32%, indicating a risk of false positives. The 'Climate' domain, despite obtaining moderate precision and recall, reveals complexity in differentiating between authentic and fake news in the science-related domain. In contrast, the model performs well on the 'GossipCop' and 'ISOT-small' datasets, with excellent precision and recall values. In 'ISOT-small', the model achieves precision and recall levels that exceed 89%, demonstrating its adaptability and effectiveness in dealing with a wide range of information across domains.

Across all datasets, our BERTGuard maintains a balanced F1-score, demonstrating its capacity to perform consistently across domains. Taking domains into consideration improved the detection performance compared to the baseline strategies, which did not consider news domains in their detection approach. Figure 2 shows the F1-score comparison between both baselines and the proposed BERTGuard solution with domain-specific knowledge.

Finally, the domain-specific technique appears to be a promising step toward a system that can detect fake news with greater adaptability and accuracy. The nuanced performance across domains emphasizes the significance of designing models to fit the complexities of unique information sets and domains.

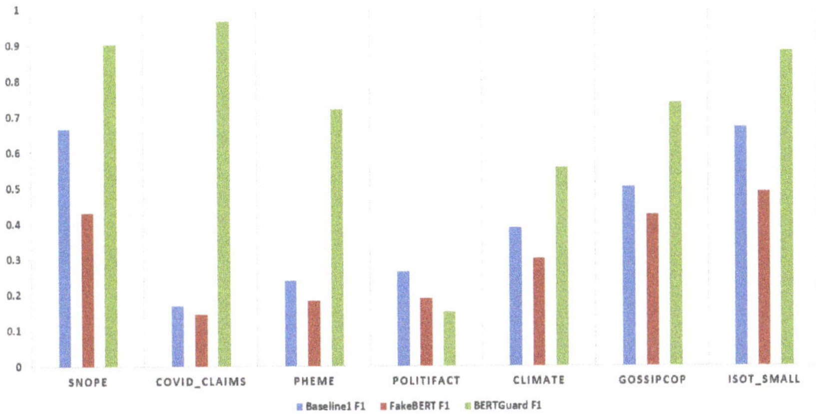


**Figure 2.** F1-scores for both baselines against BERTGuard.

### *5.4. Overall Impact of Class Imbalance Handling Strategies*

This section details the findings from the experiments aimed at assessing the effectiveness of various strategies for addressing class imbalance, including random upsampling, random downsampling, and adjustments to class weights. We tested this impact on Baseline 1, FakeBERT (Baseline 2), and our proposed BERTGuard. We trained each model on balanced datasets by applying each imbalance strategy individually. The performance obtained was compared against the model's performance without handling the class imbalance. The results shown in Table 9 show the results of this experiment averaged across all datasets tested. We observe that modifying class weights consistently outperforms the other strategies, indicating its effectiveness in mitigating the class imbalance in the context of fake news detection.

**Table 9.** The impact of class imbalance handling in BERTGuard.

| Strategy | Precision | Recall | F1-score | Accuracy |
|---|---|---|---|---|
| | | **BERTGuard** | | |
| No balancing | 82% | 68% | 70% | 82% |
| Random oversampling | 81% | 69% | 71% | 82% |
| Random undersampling | 79% | 61% | 65% | 68% |
| Adjusting class weights | **87%** | **68%** | **73%** | **78%** |
| | | **Baseline 1** | | |
| No balancing | 42% | 46% | 41% | 55% |
| Random oversampling | 46% | 48% | 45% | 52% |
| Random undersampling | 36% | 41% | 37% | 46% |
| Adjusting class weights | **62%** | **51%** | **55%** | **56%** |
| | | **FakeBERT (Baseline 2)** | | |
| No balancing | 32% | 34% | 31% | 38% |
| Random oversampling | 46% | 48% | 45% | 52% |
| Random undersampling | 36% | 41% | 37% | 46% |
| Adjusting class weights | **62%** | **51%** | **55%** | **56%** |

**Note:** These results reflect the average testing values for testing using various unseen datasets to include Snope, COVID-Claims, Pheme, Politifact, Climate, GossipCop, and ISOT-Science.

These findings underscore the importance of treating class imbalances in fake news detection, with class weighting being the most effective strategy among those tested, while downsampling did not consistently improve the model's performance across all testing datasets.

### 5.5. Impact of Balancing Datasets by Adjusting Weights

The effort to address class imbalance through adjusting weights in the dataset has yielded distinctive outcomes for the baselines and the proposed domain-specific fake news detection approaches. Figures 3 and 4 show the effect of handling the class imbalance issue by adjusting the class weights on both the main baseline (Baseline 1) we developed and the BERTGuard approaches.

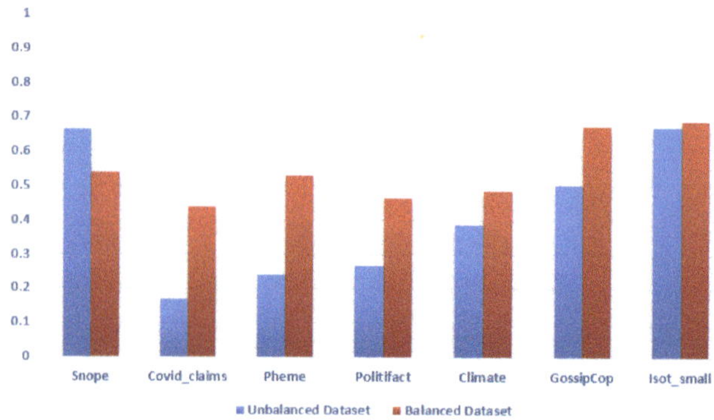


**Figure 3.** The effect of handling the imbalance on the main baseline (Baseline 1) F1-scores by adjusting the class weights.

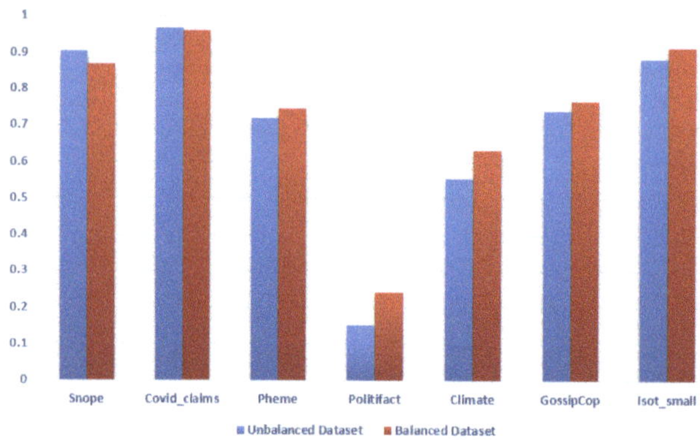


**Figure 4.** The effect of handling the imbalance on the BERTGuard F1-scores by adjusting the class weights.

The class weight adjustment approach results in significant improvements in some domains. Notably, 'Snopes' and 'COVID-Claims' show high precision and recall rates, demonstrating the model's expertise in these critical domains. The targeted changes in class weights show the potential efficacy of domain-specific techniques for resolving class imbalance. Significant advances in 'Pheme' and 'GossipCop' were demonstrated by gaining higher precisions, recalls, and F1-scores. The model's adaptability in capturing complex information transmission within these domains demonstrates the advantages of altering the class weight in the detection task as well as the possibility for domain-specific techniques to efficiently handle class imbalance over the baseline approach with no domains.

Despite efforts to address the imbalance, difficulties exist in some domains, as seen in the 'Politifact' and 'Climate' datasets. The models struggle to achieve a harmonious balance between precision and recall in several domains, implying that domain-specific complications may necessitate more specialized tactics than simple class weight modifications. Figure 5 presents a comparison between the baselines and the BERTGuard F1-scores after handling the class imbalance by adjusting the class weights.

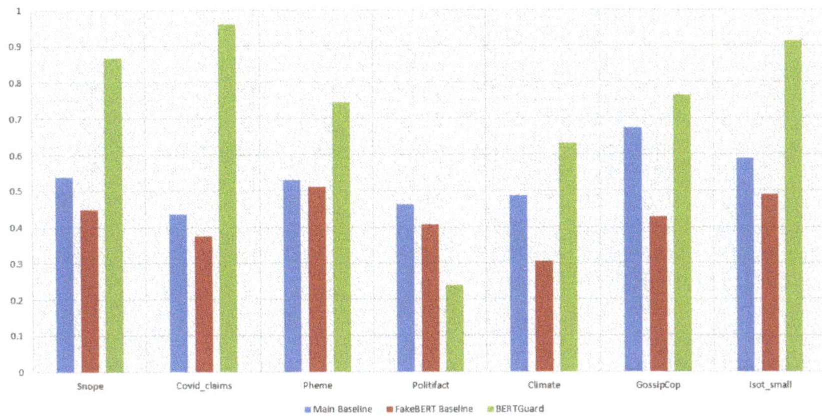


**Figure 5.** F1-scores of both baselines and BERTGuard after treating the imbalance by adjusting class weights.

In conclusion, the findings highlight the approach's versatility across a variety of datasets while also noting ongoing problems in specific domains. Balancing datasets by altering class weights is a complicated process. The domain-specific methodology demonstrates promising advances in the majority of the domains, emphasizing the importance of continuing refinement and adaptation. The findings also highlight the dynamic nature of fake news identification, emphasizing the need for specific imbalance handling solutions to navigate the intricacies of various domains.

### *5.6. Impact of Balancing Datasets by Upsampling*

In this section, we investigate handling class imbalance by using the random oversampling technique. Although this technique shows enhancements over unbalanced cases, upsampling performed more poorly compared to adjusting the class weights. Figure 6 shows the effect of the random oversampling technique on our proposed BERTGuard approach.

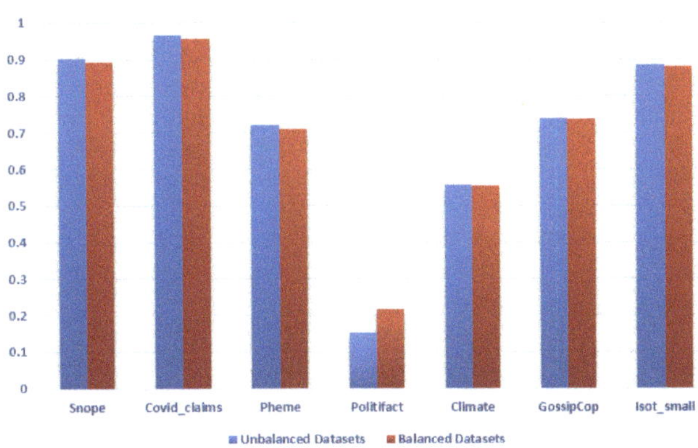


**Figure 6.** The effect of random oversampling on BERTGuard F1-scores.

The F1-scores improved notably across multiple datasets as compared to the baseline. The F1-scores topped 89% in 'Snope' and 'COVID-Claims', showing a significant improvement in both precision and recall. The approach achieved balanced precision and recall in both 'Gossipcop' and 'ISOT-small', indicating an effective trade-off between false positives and false negatives in both the baseline and domain-specific approaches.

While some datasets' F1-scores improved when compared to the imbalanced baseline, others declined. Notably, 'COVID-Claims' and 'Pheme' had lower F1-scores for the baseline approach, showing the limitations of the baseline model even in a balanced setting. In the domain-specific approach, several datasets, such as 'Politifact' and 'Climate', nevertheless faced difficulty, with lower F1-scores. This indicates that upsampling alone might not fully address the complexities in certain domains, and other techniques may perform better, such as adjusting the class weights.

In conclusion, while both upsampling and class weight adjustment strategies help to improve fake news detection in imbalanced datasets, the class weight adjustment approach displayed superior adaptability and competitive or superior F1-scores across many datasets, as presented in Figure 7.

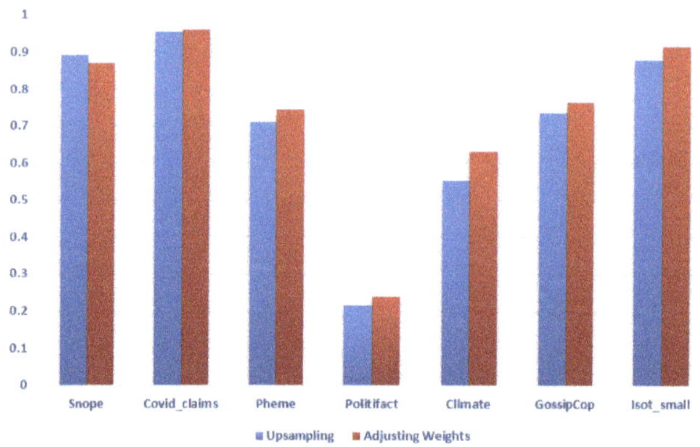


**Figure 7.** F1-score comparison between upsampling and class weight adjustments on BERTGuard.

### 5.7. Impact of Balancing Datasets by Downsampling

In this section, we investigate handling the class imbalance by using the downsampling technique. Although this technique has shown enhancement over the unbalanced case in different classification tasks, it did not help in our context.

Balancing the class distribution in the used dataset using random undersampling had a noticeable effect on the performance of our fake news detection approach. Looking at the F1-scores before and after treating the class imbalance using random undersampling, we can see that for most of the datasets, the F1-scores decreased slightly with training the models on balanced datasets by randomly undersampling the majority class. This could be due to the loss of information that occurs when randomly removing instances from the majority class during undersampling. However, it is important to note that while the F1-scores decreased, they are still relatively high, indicating that our model still performs well even after balancing the dataset. Figure 8 presents the F1-score results for the proposed BERTGuard after applying the random undersampling technique for handling class imbalance.

Regarding the baselines, downsampling produced mixed results. While there were advances in 'Pheme' and 'Politifact', other datasets such as 'COVID-Claims' and 'Snope' had difficulties, leading to decreased precision. Furthermore, 'GossipCop' and 'ISOT-small' revealed a trade-off between precision and recall, highlighting the difficulty of attaining balanced performance through downsampling alone. Figure 9 presents the F1-scores for the Baseline 1 (single-stage) detection model after applying the downsampling technique for handling the imbalance.

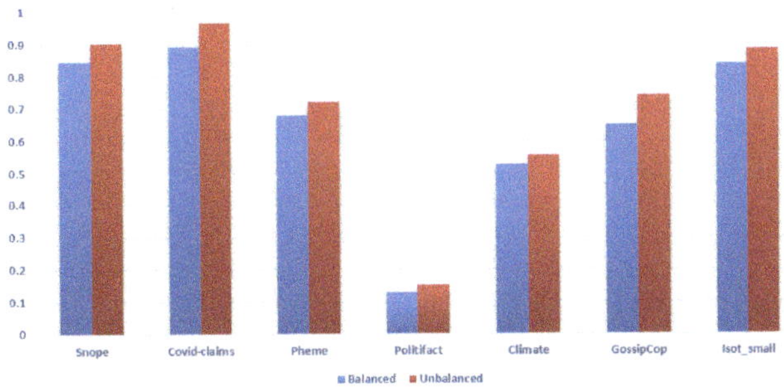


**Figure 8.** The effect of random undersampling on BERTGuard F1-scores.

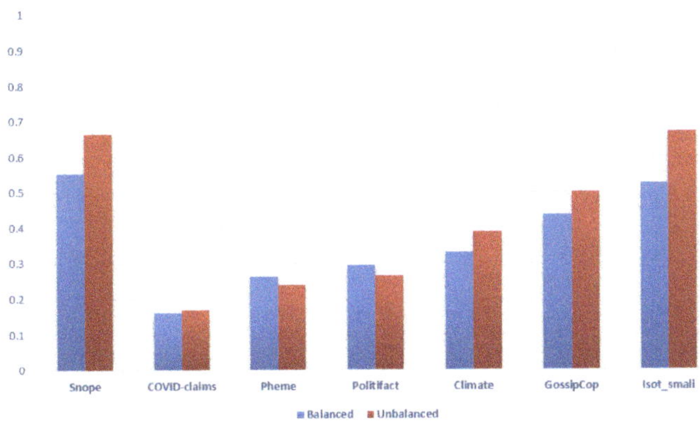


**Figure 9.** The effect of random undersampling on Baseline 1 F1-scores.

The performance of downsampling was context-dependent, with certain datasets favouring this approach over others. This underscores the importance of understanding the intricacies of each dataset to tailor the strategy accordingly. It performed well in some cases compared to the non-balanced case but had worse performance than the other imbalance handling techniques. Figure 10 compares the F1-score results when utilizing all three class imbalance handling techniques in the proposed domain-specific approach.

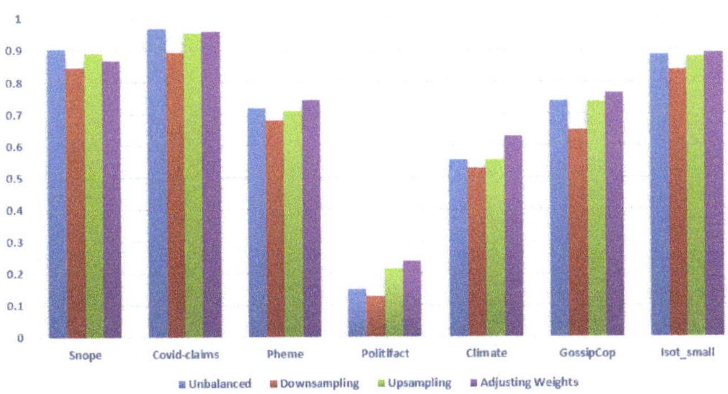


**Figure 10.** The effect of using upsampling, downsampling, and class weight adjustment vs. no handling technique on BERTGuard.

Another observation is that the F1-scores for some datasets, such as Climate, remained relatively stable after balancing, while others, like Politifact and Climate, saw a more significant decrease. This variation could be due to the nature of the datasets and the distribution of the classes within them.

Downsampling was not effective for improving the precision–recall balance and was not helpful in most other datasets. Upsampling, while beneficial in some domains, has drawbacks, including susceptibility to information nuances, and it handled more datasets than the downsampling technique. Class weight modification emerges as a consistently adaptive method for preserving and improving detection accuracy across diverse datasets and domains. The domain-specific character of issues, as seen in 'Politifact', necessitates nuanced studies and future model changes.

## 6. Conclusions

In this paper, we propose BERTGuard: a thorough, domain-specific strategy within the multi-domain fake news detection framework. Our methodology builds upon the unique characteristics that exist in various news domains. BERTGuard, a two-tiered fake news detection approach, includes domain categorization and domain-specific analyses using various BERT models. Our proposed approach demonstrates enhanced adaptability and precision across diverse information environments by leveraging multiple datasets from various domains simultaneously. To the best of our knowledge, this setting has not been explored previously. Our research also takes a methodical approach to the challenging obstacle of class imbalance in multi-domain fake news detection. We determine the best-performing strategy by rigorously evaluating handling strategies such as random oversampling, random undersampling, and class weight adjustment. This strengthens the detection system against the difficulties of imbalanced datasets. Our BERTGuard approach outperformed state-of-the-art solutions for fake news detection in a multi-domain setting.

Our findings provide a solid framework for comprehensive and deep fake news detection approaches and provides useful insights for practitioners looking for effective solutions to class imbalances in this critical domain. Future research should focus on refining the domain-specific methodology, exploring new information domains, and investigating developing solutions for dealing with increasing issues in the dynamic landscape of fake news. Future studies should also use transfer learning to build stronger models capable of detecting fake news patterns across other domains. In addition, future work will involve exploring various anomaly detection methods to address unanticipated anomalies and class imbalances more effectively. By integrating these techniques, we aim to refine our model's ability to manage diverse and imbalanced datasets, thereby improving detection accuracy and reliability across different domains.

**Author Contributions:** Conceptualization, M.Q.A. and P.B.; Data curation, M.Q.A.; Formal analysis, M.Q.A.; Investigation, M.Q.A.; Methodology, M.Q.A.; Project administration, M.Q.A. and P.B.; Resources, M.Q.A.; Software, M.Q.A.; Supervision, P.B.; Validation, M.Q.A. and P.B.; Visualization, M.Q.A.; Writing—original draft, M.Q.A.; Writing—review and editing, P.B. All authors have read and agreed to the published version of the manuscript.

**Funding:** This research received no external funding.

**Institutional Review Board Statement:** Not applicable.

**Informed Consent Statement:** Not applicable.

**Data Availability Statement:** The datasets used in this article are publicly available at the following links: https://github.com/joolsa/fake_real_news_dataset, accessed on: 12 May 2024; https://www.kaggle.com/competitions/fake-news/data, accessed on: 12 May 2024; http://fakenews.research.sfu.ca/, accessed on: 12 May 2024; https://data.mendeley.com/datasets/zwfdmp5syg/1, accessed on: 12 May 2024; https://ieee-dataport.org/documents/covifn-fake-news-covid19, accessed on: 12 May 2024; https://www.kaggle.com/datasets/rmisra/politifact-fact-check-dataset, accessed on: 12 May 2024; https://ieee-dataport.org/open-access/covid-19-fake-news-infodemic-research-dataset-covid19-fnir-datase, accessed on: 12 May 2024.

**Conflicts of Interest:** The authors declare no conflicts of interest.

## References

1. Silva, A.; Luo, L.; Karunasekera, S.; Leckie, C. Embracing domain differences in fake news: Cross-domain fake news detection using multi-modal data. In Proceedings of the AAAI Conference on Artificial Intelligence, Virtual, 2–9 February 2021; Volume 35, pp. 557–565.
2. Chen, Q. Coronavirus Rumors Trigger Irrational Behaviors among Chinese netizens. 2020. Available online: https://www.globaltimes.cn/content/1178157.shtml (accessed on 12 May 2024)
3. Sharma, K.; Qian, F.; Jiang, H.; Ruchansky, N.; Zhang, M.; Liu, Y. Combating fake news: A survey on identification and mitigation techniques. *Acm Trans. Intell. Syst. Technol. (TIST)* **2019**, *10*, 1–42.
4. Schuster, T.; Schuster, R.; Shah, D.J.; Barzilay, R. The limitations of stylometry for detecting machine-generated fake news. *Comput. Linguist.* **2020**, *46*, 499–510.
5. Shabani, S.; Sokhn, M. Hybrid machine-crowd approach for fake news detection. In Proceedings of the 2018 IEEE 4th International Conference on Collaboration and Internet Computing (CIC), Philadelphia, PA, USA, 18–20 October 2018; pp. 299–306.
6. Nan, Q.; Wang, D.; Zhu, Y.; Sheng, Q.; Shi, Y.; Cao, J.; Li, J. Improving Fake News Detection of Influential Domain via Domain- and Instance-Level Transfer. In Proceedings of the 29th International Conference on Computational Linguistics, Gyeongju, Republic of Korea, 12–17 October 2022; pp. 2834–2848.
7. Nan, Q.; Cao, J.; Zhu, Y.; Wang, Y.; Li, J. MDFEND: Multi-domain fake news detection. In Proceedings of the 30th ACM International Conference on Information & Knowledge Management, Virtual Event, QLD, Australia, 1–5 November 2021; pp. 3343–3347.
8. Allcott, H.; Gentzkow, M. Social media and fake news in the 2016 election. *J. Econ. Perspect.* **2017**, *31*, 211–236.
9. Vosoughi, S.; Roy, D.; Aral, S. The spread of true and false news online. *Science* **2018**, *359*, 1146–1151.
10. Bursztyn, L.; Rao, A.; Roth, C.P.; Yanagizawa-Drott, D.H. *Misinformation during a Pandemic*; Technical Report; National Bureau of Economic Research: Cambridge, MA, USA, 2020.
11. Alnabhan, M.Q.; Branco, P. Evaluating Deep Learning for Cross-Domains Fake News Detection. In Proceedings of the International Symposium on Foundations and Practice of Security, Bordeaux, France, 11–13 December 2023; Springer: Cham, Switzerland, 2024; pp. 40–51.
12. Kaliyar, R.K.; Goswami, A.; Narang, P. FakeBERT: Fake news detection in social media with a BERT-based deep learning approach. *Multimed. Tools Appl.* **2021**, *80*, 11765–11788.
13. Tang, H.; Liu, J.; Zhao, M.; Gong, X. Progressive layered extraction (ple): A novel multi-task learning (mtl) model for personalized recommendations. In Proceedings of the 14th ACM Conference on Recommender Systems, Rio de Janeiro, Brazil, 22–26 September 2020; pp. 269–278.
14. Seiffert, C.; Khoshgoftaar, T.M.; Van Hulse, J.; Napolitano, A. A comparative study of data sampling and cost sensitive learning. In Proceedings of the 2008 IEEE International Conference on Data Mining Workshops, Pisa, Italy, 15–19 December 2008; pp. 46–52.
15. Alnabhan, M.Q.; Branco, P. Fake News Detection Using Deep Learning: A Systematic Literature Review. *IEEE Access* **2024**, Volume 12, 1. [CrossRef]
16. Longadge, R.; Dongre, S. Class imbalance problem in data mining review. *arXiv* **2013**, arXiv:1305.1707.
17. Alenezi, M.N.; Alqenaei, Z.M. Machine learning in detecting COVID-19 misinformation on twitter. *Future Internet* **2021**, *13*, 244.
18. Moravec, P.; Kim, A.; Dennis, A. Flagging fake news: System 1 vs. System 2. In Proceedings of the 39th International Conference on Information Systems, San Francisco CA, USA, 13–16 December 2018.
19. Khweiled, R.; Jazzar, M.; Eleyan, D. Cybercrimes during COVID-19 pandemic. *Int. J. Inf. Eng. Electron. Bus.* **2021**, *13*, 1–10.
20. Shin, D.; Koerber, A.; Lim, J.S. Impact of misinformation from generative AI on user information processing: How people understand misinformation from generative AI. *New Media Soc.* **2024**, 14614448241234040. [CrossRef]
21. Qawasmeh, E.; Tawalbeh, M.; Abdullah, M. Automatic identification of fake news using deep learning. In Proceedings of the 2019 Sixth International Conference on Social Networks Analysis, Management and Security (SNAMS), Granada, Spain, 22–25 October 2019; pp. 383–388.
22. Kozik, R.; Kula, S.; Choraś, M.; Woźniak, M. Technical solution to counter potential crime: Text analysis to detect fake news and disinformation. *J. Comput. Sci.* **2022**, *60*, 101576.
23. Deepak, S.; Chitturi, B. Deep neural approach to Fake-News identification. *Procedia Comput. Sci.* **2020**, *167*, 2236–2243.
24. Sharma, S.; Saraswat, M.; Dubey, A.K. Fake News Detection Using Deep Learning. In Proceedings of the Knowledge Graphs and Semantic Web: Third Iberoamerican Conference and Second Indo-American Conference, KGSWC 2021, Kingsville, TX, USA, 22–24 November 2021; Springer: Berlin/Heidelberg, Germany, 2021; pp. 249–259.
25. Pilkevych, I.; Fedorchuk, D.; Naumchak, O.; Romanchuk, M. Fake news detection in the framework of decision-making system through graph neural network. In Proceedings of the 2021 IEEE 4th International Conference on Advanced Information and Communication Technologies (AICT), Lviv, Ukraine, 21–25 September 2021; pp. 153–157.
26. Manene, S. Mitigating misinformation about the COVID-19 infodemic on social media: A conceptual framework. *Jàmbá J. Disaster Risk Stud.* **2023**, *15*, 1416. [CrossRef]

27. Akhter, M.; Hossain, S.M.M.; Nigar, R.S.; Paul, S.; Kamal, K.M.A.; Sen, A.; Sarker, I.H. COVID-19 Fake News Detection using Deep Learning Model. *Ann. Data Sci.* **2024**, 1–32. [CrossRef]
28. Nasir, J.A.; Khan, O.S.; Varlamis, I. Fake news detection: A hybrid CNN-RNN based deep learning approach. *Int. J. Inf. Manag. Data Insights* **2021**, *1*, 100007.
29. Kaliyar, R.K.; Goswami, A.; Narang, P.; Sinha, S. FNDNet—A deep convolutional neural network for fake news detection. *Cogn. Syst. Res.* **2020**, *61*, 32–44.
30. Saleh, H.; Alharbi, A.; Alsamhi, S.H. OPCNN-FAKE: Optimized convolutional neural network for fake news detection. *IEEE Access* **2021**, *9*, 129471–129489.
31. Yang, Y.; Zheng, L.; Zhang, J.; Cui, Q.; Li, Z.; Yu, P.S. TI-CNN: Convolutional neural networks for fake news detection. *arXiv* **2018**, arXiv:1806.00749.
32. Raj, C.; Meel, P. ConvNet frameworks for multi-modal fake news detection. *Appl. Intell.* **2021**, *51*, 8132–8148.
33. Hashmi, E.; Yayilgan, S.Y.; Yamin, M.M.; Ali, S.; Abomhara, M. Advancing fake news detection: Hybrid deep learning with fasttext and explainable AI. *IEEE Access* **2024**, *12*, 44462–44480.
34. Mosallanezhad, A.; Karami, M.; Shu, K.; Mancenido, M.V.; Liu, H. Domain adaptive fake news detection via reinforcement learning. In Proceedings of the ACM Web Conference 2022, Lyon, France, 25–29 April 2022; pp. 3632–3640.
35. Li, X.; Fu, X.; Xu, G.; Yang, Y.; Wang, J.; Jin, L.; Liu, Q.; Xiang, T. Enhancing BERT representation with context-aware embedding for aspect-based sentiment analysis. *IEEE Access* **2020**, *8*, 46868–46876.
36. Xu, H.; Liu, B.; Shu, L.; Yu, P.S. BERT Post-Training for Review Reading Comprehension and Aspect-based Sentiment Analysis. *arXiv* **2019**, arXiv:1904.02232.
37. Kumar, B. *BERT Variants and Their Differences;* Technical report; 360DigiTMG: Hyderabad, India,2023.
38. Sanh, V.; Debut, L.; Chaumond, J.; Wolf, T. DistilBERT, a distilled version of BERT: Smaller, faster, cheaper and lighter. *arXiv* **2020**, arXiv:1910.01108.
39. Lutkevich, B. *BERT Language Model;* Technical report; TechTarget: Newton, MA, USA, 2020.
40. Tida, V.S.; Hsu, D.S.; Hei, D.X. Unified Fake News Detection using Transfer Learning of BERT Model. *IEEE* **2020**. Available online: https://d1wqtxts1xzle7.cloudfront.net/86079521/2202.01907v1-libre.pdf?1652817185=&response-content-disposition=inline%3B+filename%3DUnified_Fake_News_Detection_using_Transf.pdf&Expires=1723717032&Signature=SlJqui-38VOu3m7EAFYMcfZkoxq23tXKTFkq-wlwLHawKo0ibgs47MWTsCwm~7pRxvt4tl7LYN90t0QkZ7TNA8u30OuhD1JPpvNYhXoF4rYemFei0xLNEpYr4NkaPcsRshcrXcEuN0u1DTA5aR8TD1eZhJcU6x1~AZbl745yKnoIrztd032Gb2EVFS5VW~Gy3xxYIiAWD~HJ3zu5SFhTzdOcHChdGXexeXZ8Dls7N-UU-KGdGMWq4XnwnWXv9A20jpMYks6Dqcho9rutx~f3t3A0UyuCYilNghvcU-o0uGj4J4zGnEN1rhhCvtCUEAl1DMabCr-aCCW73t7Q9URcRg__&Key-Pair-Id=APKAJLOHF5GGSLRBV4ZA (accessed on 12 May 2024).
41. Lan, Z.; Chen, M.; Goodman, S.; Gimpel, K.; Sharma, P.; Soricut, R. Albert: A lite bert for self-supervised learning of language representations. *arXiv* **2019**, arXiv:1909.11942.
42. Luo, Y.; Shi, Y.; Li, S. Social media fake news detection algorithm based on multiple feature groups. In Proceedings of the 2023 IEEE 3rd International Conference on Information Technology, Big Data and Artificial Intelligence (ICIBA), Chongqing, China, 26–28 May 2023; Volume 3, pp. 91–95.
43. Bounaama, R.; Abderrahim, M.E.A. Classifying COVID-19 Related Tweets for Fake News Detection and Sentiment Analysis with BERT-based Models. *arXiv* **2023**, arXiv:2304.00636.
44. Essa, E.; Omar, K.; Alqahtani, A. Fake news detection based on a hybrid BERT and LightGBM models. *Complex Intell. Syst.* **2023**, *9*, 6581–6592. [CrossRef]
45. Shushkevich, E.; Cardiff, J.; Boldyreva, A. Detection of Truthful, Semi-Truthful, False and Other News with Arbitrary Topics Using BERT-Based Models. In Proceedings of the 2023 33rd Conference of Open Innovations Association (FRUCT), Zilina, Slovakia, 24–26 May 2023; pp. 250–256.
46. Sultana, R.; Nishino, T. Fake News Detection System: An implementation of BERT and Boosting Algorithm. In Proceedings of the 38th International Conference on Computers and Their Applications, Virtual, 20–22 March 2023;. Volume 91, pp. 124–137.
47. Alghamdi, J.; Lin, Y.; Luo, S. Towards COVID-19 fake news detection using transformer-based models. *Knowl.-Based Syst.* **2023**, *274*, 110642.
48. SATHVIK, M.; Mishra, M.K.; Padhy, S. Fake News Detection by Fine Tuning of Bidirectional Encoder Representations from Transformers. *IEEE Trans. Comput. Soc. Syst.* **2023**, *20*, 20.
49. Kitanovski, A.; Toshevska, M.; Mirceva, G. DistilBERT and RoBERTa Models for Identification of Fake News. In Proceedings of the 2023 46th MIPRO ICT and Electronics Convention (MIPRO), Opatija, Croatia, 22–26 May 2023; pp. 1102–1106.
50. Saini, K.; Jain, R. A Hybrid LSTM-BERT and Glove-based Deep Learning Approach for the Detection of Fake News. In Proceedings of the 2023 3rd International Conference on Smart Data Intelligence (ICSMDI), Trichy, India, 30–31 March 2023; pp. 400–406.
51. Fauzy, A.R.I.; Setiawan, E.B. Detecting Fake News on Social Media Combined with the CNN Methods. *J. Resti (Rekayasa Sist. Dan Teknol. Informasi)* **2023**, *7*, 271–277.
52. Nassif, A.B.; Elnagar, A.; Elgendy, O.; Afadar, Y. Arabic fake news detection based on deep contextualized embedding models. *Neural Comput. Appl.* **2022**, *34*, 16019–16032.

53. Ranjan, V.; Agrawal, P. Fake News Detection: GA-Transformer And IG-Transformer Based Approach. In Proceedings of the 2022 12th International Conference on Cloud Computing, Data Science & Engineering (Confluence), Virtual Conference, 27–28 January 2022; pp. 487–493.
54. Raza, S.; Ding, C. Fake news detection based on news content and social contexts: A transformer-based approach. *Int. J. Data Sci. Anal.* **2022**, *13*, 335–362.
55. Truică, C.O.; Apostol, E.S. MisRoBÆRTa: Transformers versus misinformation. *Mathematics* **2022**, *10*, 569.
56. Schütz, M.; Schindler, A.; Siegel, M.; Nazemi, K. Automatic fake news detection with pre-trained transformer models. In Proceedings of the Pattern Recognition. ICPR International Workshops and Challenges, Virtual Event, 10–15 January 2021; Springer: Berlin/Heidelberg, Germany, 2021; Part VII, pp. 627–641.
57. Huang, Y.; Gao, M.; Wang, J.; Shu, K. Dafd: Domain adaptation framework for fake news detection. In Proceedings of the Neural Information Processing: 28th International Conference, ICONIP 2021, Sanur, Bali, Indonesia, 8–12 December 2021; Springer: Berlin/Heidelberg, Germany, 2021; Part I 28, pp. 305–316.
58. Qazi, M.; Khan, M.U.; Ali, M. Detection of fake news using transformer model. In Proceedings of the 2020 3rd International Conference on Computing, Mathematics and Engineering Technologies (iCoMET), Sukkur, Pakistan, 29–30 January 2020; pp. 1–6.
59. Nirav Shah, M.; Ganatra, A. A systematic literature review and existing challenges toward fake news detection models. *Soc. Netw. Anal. Min.* **2022**, *12*, 168.
60. Kato, S.; Yang, L.; Ikeda, D. Domain Bias in Fake News Datasets Consisting of Fake and Real News Pairs. In Proceedings of the 2022 12th International Congress on Advanced Applied Informatics (IIAI-AAI), Kanazawa, Japan, 2–8 July 2022; pp. 101–106.
61. Hamed, S.K.; Ab Aziz, M.J.; Yaakub, M.R. A review of fake news detection approaches: A critical analysis of relevant studies and highlighting key challenges associated with the dataset, feature representation, and data fusion. *Heliyon* **2023**, *9*, e20382.
62. Ghosh, K.; Bellinger, C.; Corizzo, R.; Branco, P.; Krawczyk, B.; Japkowicz, N. The class imbalance problem in deep learning. *Mach. Learn.* **2024**, *113*, 4845–4901. https://doi.org/10.1007/s10994-022-06268-8.
63. Rastogi, S.; Bansal, D. A review on fake news detection 3T's: Typology, time of detection, taxonomies. *Int. J. Inf. Secur.* **2023**, *22*, 177–212.
64. Zhou, P.; Han, X.; Morariu, V.I.; Davis, L.S. Two-stream neural networks for tampered face detection. In Proceedings of the IEEE Conference on Computer Vision and Pattern Recognition Workshops, Honolulu, HI, USA, 21–26 July 2017; pp. 19–27.
65. Cardoso, E.F.; Silva, R.M.; Almeida, T.A. Towards automatic filtering of fake reviews. *Neurocomputing* **2018**, *309*, 106–116.
66. Castelo, S.; Almeida, T.; Elghafari, A.; Santos, A.; Pham, K.; Nakamura, E.; Freire, J. A topic-agnostic approach for identifying fake news pages. In Proceedings of the Companion Proceedings of the 2019 World Wide Web Conference, San Francisco, CA, USA, 13–17 May 2019; pp. 975–980.
67. Shu, K.; Mahudeswaran, D.; Wang, S.; Lee, D.; Liu, H. Fakenewsnet: A data repository with news content, social context, and spatiotemporal information for studying fake news on social media. *Big Data* **2020**, *8*, 171–188.
68. Ahmad, I.; Yousaf, M.; Yousaf, S.; Ahmad, M.O. Fake news detection using machine learning ensemble methods. *Complexity* **2020**, *2020*, 1–11.
69. Zubiaga, A.; Liakata, M.; Procter, R. Learning reporting dynamics during breaking news for rumour detection in social media. *arXiv* **2016**, arXiv:1610.07363.
70. Wang, W.Y. "Liar, liar pants on fire": A new benchmark dataset for fake news detection. *arXiv* **2017**, arXiv:1705.00648.
71. Diggelmann, T.; Boyd-Graber, J.; Bulian, J.; Ciaramita, M.; Leippold, M. CLIMATE-FEVER: A Dataset for Verification of Real-World Climate Claims. *arXiv* **2020**, arXiv:2012.00614.
72. Branco, P.; Torgo, L.; Ribeiro, R.P. A survey of predictive modeling on imbalanced domains. *ACM Comput. Surv. (CSUR)* **2016**, *49*, 1–50.
73. Agarwal, I.Y.; Rana, D.P. Fake News and Imbalanced Data Perspective. In *Data Preprocessing, Active Learning, and Cost Perceptive Approaches for Resolving Data Imbalance*; IGI Global: Hershey, PA, USA, 2021; pp. 195–210.
74. Salah, I.; Jouini, K.; Korbaa, O. On the use of text augmentation for stance and fake news detection. *J. Inf. Telecommun.* **2023**, *7*, 359–375.
75. Keya, A.J.; Wadud, M.A.H.; Mridha, M.; Alatiyyah, M.; Hamid, M.A. AugFake-BERT: Handling imbalance through augmentation of fake news using BERT to enhance the performance of fake news classification. *Appl. Sci.* **2022**, *12*, 8398.
76. Sastrawan, I.K.; Bayupati, I.; Arsa, D.M.S. Detection of fake news using deep learning CNN–RNN based methods. *ICT Express* **2022**, *8*, 396–408.
77. Mouratidis, D.; Nikiforos, M.N.; Kermanidis, K.L. Deep learning for fake news detection in a pairwise textual input schema. *Computation* **2021**, *9*, 20.
78. Al Obaid, A.; Khotanlou, H.; Mansoorizadeh, M.; Zabihzadeh, D. Multimodal fake-news recognition using ensemble of deep learners. *Entropy* **2022**, *24*, 1242.
79. Isa, S.M.; Nico, G.; Permana, M. Indobert for Indonesian fake news detection. *ICIC Express Lett.* **2022**, *16*, 289–297.
80. Szczepański, M.; Pawlicki, M.; Kozik, R.; Choraś, M. New explainability method for BERT-based model in fake news detection. *Sci. Rep.* **2021**, *11*, 23705.
81. Palani, B.; Elango, S.; Viswanathan K, V. CB-Fake: A multimodal deep learning framework for automatic fake news detection using capsule neural network and BERT. *Multimed. Tools Appl.* **2022**, *81*, 5587–5620.

82. Rai, N.; Kumar, D.; Kaushik, N.; Raj, C.; Ali, A. Fake News Classification using transformer based enhanced LSTM and BERT. *Int. J. Cogn. Comput. Eng.* **2022**, *3*, 98–105.
83. Gaudreault, J.G.; Branco, P.; Gama, J. An analysis of performance metrics for imbalanced classification. In Proceedings of the International Conference on Discovery Science. Springer, Virtual, 11–13 October 2021; pp. 67–77.

big data and
cognitive computing

MDPI

*Article*

# Does Social Media Enhance Job Performance? Examining Internal Communication and Teamwork as Mediating Mechanisms

**Satinder Kumar [1], Zohour Sohbaty [1], Ruchika Jain [2], Iqra Shafi [3] and Ramona Rupeika-Apoga [4,5,*]**

[1] School of Management Studies, Punjabi University, Patiala 147002, Punjab, India; kumarsatinder1981@gmail.com (S.K.)
[2] Gobindgarh Public College, Khanna 147301, Punjab, India; ruchikagarg82@gmail.com
[3] SKUAST-K, Kashmir University, Srinagar 191202, Jammu & Kashmir, India
[4] Faculty of Economics and Social Sciences, University of Latvia, LV-1586 Riga, Latvia
[5] Women Researchers Council (WRC), Azerbaijan State University of Economics (UNEC), Baku AZ1001, Azerbaijan
* Correspondence: rr@lu.lv

**Abstract:** This study investigates the impact of social media use on faculty job performance, exploring the mediating roles of internal communication and teamwork. Drawing on the Uses and Gratifications theory, we examine how faculty members utilize social media for three distinct purposes: social interaction (social use), enjoyment (hedonic use), and information seeking (cognitive use). We analyze how these three dimensions of social media use influence teachers' performance, encompassing both routine and innovative aspects. This analysis is based on data collected via an online survey completed by 456 faculty members at public state colleges in northern India in 2024. Structural Equation Modeling (SEM) was used to test the hypotheses. The findings reveal that social, hedonic, and cognitive use of social media positively affects faculty innovative and routine job performance, with teamwork and internal communication acting as partial mediators in this relationship. This research offers valuable insights for faculty development professionals, educational administrators, and policymakers.

**Keywords:** social media use (SMU); job performance (JP); internal communication (IC); teamwork (TW) and sequential mediation model

## 1. Introduction

The extensive spread of internet technology is fundamentally changing how academics interact, share information, and collaborate with colleagues and stakeholders [1]. Social media (SM) is a driving force in education, providing new avenues for academics to share their work and connect with a vast network of scholars and experts [2]. Similarly, new technologies empower organizations with capabilities they previously lacked. Social media, along with other technical advancements, has demonstrably improved organizations by enabling a wide range of applications and functionalities, from enhanced demand forecasting to the development of innovative marketing strategies and business models [3,4]. Unsurprisingly, SM has been described as a revolutionary force, transforming the ways people connect, communicate, consume information, and create valuable content [5].

Social media refers to a collection of online resources that facilitate the creation and exchange of user-generated content, building upon the foundations of Web 2.0, a concept that emphasizes user-generated content and collaboration [6]. Currently, social media empowers individuals to expand their social networks and maintain ongoing communication. Internal communication, on the other hand, refers to the communication channels organizations utilize to manage relationships with and among employees [7]. Tools like newsletters, intranets, and bulletin boards can streamline these interactions, fostering a more engaged workforce. Businesses leverage a variety of internal communication channels to enhance teamwork, communication between staff and stakeholders, and ultimately, job performance [8]. Social media's unprecedented power lies in its ability to create teams and improve teamwork across various departments within a company [9]. Social media also facilitates communication, team collaboration, and project management processes [10,11].

Similarly, social media use has permeated academia [12]. Because social media is becoming more and more ingrained in educators' personal and professional lives, research on how it affects work performance in the teaching profession is crucial [13–15]. These online social platforms serve a multitude of functions for academics, encompassing core academic tasks like research and teaching, as well as professional development, profile building, information sharing, and social and professional networking [16,17].

The aim of the study is to investigate the impact of social media use on faculty job performance, exploring the mediating roles of teamwork and internal communication. It offers several key contributions that differentiate it from previous research in this area. First, this study breaks new ground by identifying internal communication and teamwork as the mediating processes through which social media use influences faculty performance. While prior research has established a connection between social media and job performance [11,18], it often fails to explain the "how" behind this relationship. This study fills this gap by proposing a novel framework that reveals the crucial role of internal communication and teamwork in facilitating the positive impact of social media on faculty job performance.

Second, this study delves deeper by exploring the contingent effects of different social media usage dimensions (social, hedonic, cognitive) on job performance. Current research often treats social media use as a monolithic concept, limiting our understanding of its nuanced effects [10,19]. This study breaks away from this approach by examining how the type of social media use (focusing on social interaction, entertainment, or information gathering) influences the strength of the mediating effects (internal communication and teamwork). This study concentrated on employees who use Facebook and LinkedIn since these platforms offer three crucial components of usage: social, hedonic, and cognitive. Facebook is often used for social and hedonistic purposes, providing users with entertainment and personal connections, while LinkedIn is mostly utilized for cognitive and professional aims, such as networking, learning, and career development.

This reveals a more intricate picture, where the effectiveness of social media in enhancing faculty performance depends on the specific way it is used.

Third, this study addresses a critical gap in the literature by focusing on faculty performance, particularly examining its two key dimensions: innovative and routine tasks. While prior research explores social media's impact on general work performance (e.g., [20]), faculty members face unique demands. This study delves deeper by investigating how social media use influences both innovative tasks like research collaboration and knowledge dissemination, as well as routine tasks like student engagement. By examining these distinct performance dimensions, the study provides valuable insights for tailoring strategies and policies. Given the increasing prevalence of social media usage in both personal and professional contexts, it is critical to comprehend the effects these platforms have on teachers' efficiency, ability to manage their time, and job performance.

Fourth, by identifying internal communication and teamwork as mediating processes, this study offers valuable insights for faculty development and institutional policies. It informs strategies for promoting effective social media use within academic settings, fostering collaboration, and communication, and ultimately, enhancing faculty performance.

Additionally, this study specifically focuses on teachers, a unique workgroup whose job performance and effectiveness are distinct from other professions. The specificity of teachers' work is highlighted in this research, addressing the unique challenges and dynamics they face. This focus is crucial as it ensures that the findings are relevant and accurately reflect the academic context, which may not be directly applicable to other professions.

The paper is structured as follows. After the introduction, we present a research background that outlines the research model. This model depicts the relationships between social media use, job performance, internal communication, and teamwork, drawing upon established theoretical concepts. In light of these theories, we discuss the specific hypotheses of the study. Next, we conduct data analysis to test the hypotheses. Finally, the paper concludes by discussing the results and their corresponding theoretical and practical implications.

## 2. Theoretical Background and Research Hypothesis

### *2.1. Theoretical Background*

This research investigates the potential for social media to indirectly influence faculty job performance through the mediating factors of internal communication and teamwork. The model outlined in Figure 1 depicts this mediated relationship. This aligns with the concept of a mediated model in social science research, where an independent variable (social media use) can have an indirect effect on a dependent variable (job performance) through the influence of mediating variables (internal communication and teamwork) [21]. We develop hypotheses based on existing research to understand the specific relationships between these factors.

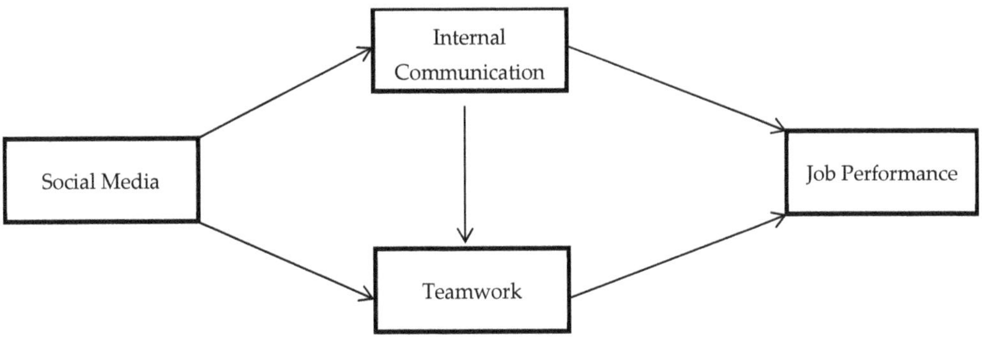


**Figure 1.** Research model.

Use of Social Media and Job Performance

This research investigates the motivations behind faculty members' engagement with social media and its potential influence on job performance. Social media use (SMU) serves as the independent variable, representing the faculty members' engagement with social media platforms. Job performance, on the other hand, is the dependent variable, reflecting the effectiveness of faculty members in their roles. The Uses and Gratifications (U&G) theory [22] provides a valuable framework for understanding the motivations behind SMU. U&G theory posits that individuals actively choose media to fulfill specific needs, categorized as social, hedonic, and cognitive [11,23]. Social media's unique characteristics make it a platform that can cater to all three need categories. Here is why we focus on these three dimensions:

- Social Use: Faculty members can leverage social media to maintain and strengthen relationships with colleagues [11].
- Hedonic Use: Social media can cater to hedonic needs by providing opportunities for entertainment and relaxation [11,23].
- Cognitive Use: Social media platforms are valuable tools for knowledge acquisition and information sharing [24,25].

Job performance refers to the effectiveness with which faculty members fulfill their job duties and responsibilities. This research focuses on two key dimensions of JP.

Innovative Job Performance. This dimension encompasses behaviors that go beyond the official job description and involve creativity [26]. It includes generating and implementing original ideas, tackling challenges with innovative solutions, and forming collaborations to bring new ideas to fruition [27–29].

Routine Job Performance. This refers to the consistent and dependable completion of essential tasks, obligations, and responsibilities [26,30]. Examples include meeting deadlines, delivering lectures effectively, and grading assignments accurately.

### 2.2. Direct Effect of Social Media Use on Job Performance (H1–H3)

The impact of social media use (SMU) on work performance has been a topic of growing interest [31,32]. This research delves deeper by examining the specific effects of SMU on two key dimensions of job performance: routine and innovative. Technology is a major driver of rapid change in the business world [33]. Social media has become a prominent tool for organizations, with some studies suggesting it can positively influence job performance [34]. Shang et al. [35] highlight the use of social media platforms by administrations for various functions and development purposes. Ali-Hassan et al. [11] found that social media use enhances workers' abilities to generate, share, and acquire information, potentially leading to increased productivity. However, Liu et al. [36] and Zahmat Doost and Zhang [37] suggest that while social media use in the office can improve communication quality, it can also lead to job interruptions. Therefore, effectively utilizing social media is crucial for organizations to improve job performance.

Social media platforms offer faculty members a unique set of tools for communication, collaboration, and knowledge acquisition, which can be particularly beneficial for fostering innovative job performance (IJP). Studies suggest that the social use of social media (SU) can facilitate the exchange of ideas and expertise among colleagues [11]. For example, faculty can leverage social media to connect with researchers in their field, participate in online discussions on emerging topics, and share their own findings. This exposure to diverse perspectives and the exchange of knowledge can stimulate creative thinking and problem-solving, potentially leading to the generation of innovative ideas and solutions. Additionally, SU can be used to build and maintain relationships with colleagues [11], fostering a sense of community that can lead to increased collaboration on innovative projects. Therefore, we propose the following hypothesis:

**H1a.** *Social use of social media (SU) is positively associated with innovative job performance (IJP).*

Studies suggest that social media can foster closer relationships and enhance communication within educational settings [15]. This aligns with the concept of SU, which encompasses activities like building connections, sharing information, and participating in online discussions with colleagues. Faculty can leverage these capabilities to facilitate communication and collaboration, potentially leading to streamlined completion of routine tasks. For example, social media platforms can be used to share course materials, updates on curriculum or departmental policies, and best practices for teaching. This can save time faculty would otherwise spend searching for information or reinventing the wheel. Additionally, faculty can utilize social media to collaborate on projects, troubleshoot challenges, and share resources, potentially leading to increased efficiency in completing routine tasks like lesson planning and grading.

Research highlights the potential of social media as a tool for communication and collaboration within organizations [9]. This directly connects to the potential benefits of SU for faculty members. By efficiently sharing resources and information with colleagues through social media platforms, faculty can save time searching for materials. For instance, a faculty member might use social media to inquire about specific software or request recommendations for textbooks, receiving quick and relevant responses from colleagues. This saved time can then be directed towards completing other routine job tasks more efficiently.

Studies also suggest that social media can contribute to a sense of community within a department or institution [18]. This sense of community fostered by SU can lead to increased morale, collaboration, and a more supportive work environment. When faculty feel connected and supported by colleagues through online interaction, it can improve motivation and potentially lead to increased efficiency in completing routine job duties. For example, social media can facilitate knowledge exchange and problem-solving among faculty members. A faculty member encountering a challenge can leverage social media to seek advice from colleagues with expertise in the specific area. This type of online support can lead to quicker resolution of issues and improved efficiency in completing routine tasks. Therefore, we hypothesize:

**H1b.** *Social use of social media (SU) is positively associated with routine job performance (RJP).*

The relationship between the hedonic use of social media (HU) and innovative job performance (IJP) appears complex and multifaceted. Some research suggests the potential benefits of social media platforms. They can help reduce stress and improve communication among colleagues [36,37]. This improved communication could foster collaboration and idea-sharing, potentially leading to more innovative outcomes. Additionally, social media can connect individuals with diverse perspectives and information sources, which can be valuable for creative problem-solving [11].

However, other studies highlight the potential drawbacks of HU. Social media's constant notifications and rapid updates can fragment attention and hinder the deep concentration needed for innovative thinking [38]. This fragmented attention could hinder the deep concentration needed for creative problem-solving and innovative thinking, potentially reducing IJP. Social media can significantly reduce the time dedicated to "deep work"—focused, uninterrupted effort required for innovation [39]. Mark et al. [40] found that even brief social media interruptions could significantly hinder task resumption and performance. This suggests HU could limit the time faculty have for deep work, potentially affecting their ability to generate innovative ideas. Taking into account the mixed nature of the existing literature, we hypothesize:

**H2a.** *There is a relationship between hedonic use of social media (HU) and innovative job performance (IJP).*

The relationship between hedonic social media use (HU) and routine job performance (RJP) appears multifaceted too. Research suggests that using technology for purely leisure purposes can lead to inefficiencies and negatively affect job performance [41]. This aligns with the concerns expressed by faculty members interviewed by Sobaih et al. [15] who viewed social media use for leisure during work hours as a waste of time that could be better spent on core job tasks. Studies suggest that engaging in hedonistic social media activities during work hours can be distracting, leading to decreased focus and ultimately impacting productivity on routine tasks [42]. The constant notifications and stimulation from social media can make it harder for employees to stay concentrated on routine tasks, leading to more task switching and potentially hindering performance. Studies have shown a negative correlation between excessive social media use and job satisfaction [43]. Dissatisfaction with work might lead to increased reliance on social media for enjoyment, creating a cycle that further hinders performance [43].

However, there are other studies pointing to potential benefits. Social media use can foster connections, improve communication, and even reduce stress, which could lead to positive effects on performance [6,44]. This mixed picture is further complicated by the potential influence of individual differences, work context, and specific types of social media use on the relationship between HU and RJP 08 [40]. Therefore, we hypothesize:

**H2b.** *There is a relationship between hedonic use of social media (HU) and routine job performance (RJP).*

The concept of cognitive social media use (CU) emphasizes utilizing social media platforms to create, share, and access information [45]. In the educational field, teachers leverage this cognitive dimension by generating and disseminating educational content [46]. This aligns with the findings of [47] who discovered a positive impact on innovative teaching practices when lecturers used social media for sharing and posting content. This makes sense, as sharing educational resources and collaborating online with colleagues through CU can foster innovative approaches to teaching. Further supporting this notion, [48] explore the potential of social networking sites (SNS) in academia. Their research highlights how researchers can utilize SNS for various purposes, including sharing research findings and collaborating with colleagues. These activities directly correspond with the core aspects of CU—content creation and knowledge sharing. In essence, these studies demonstrate the positive connection between CU and innovative practices within academic settings.

**H3a.** *Cognitive use of social media (CU) is positively associated with innovative job performance (IJP).*

Studies like the one by Ali-Hassan et al. [11] highlight the positive and indirect effect of social and cognitive technology use, including social media, on job performance. This effect likely stems from social capital—the benefits gained from social connections. Social media use, as highlighted in the [49] study, can help build social capital by creating information, maintaining social networks, and fostering trust among colleagues. Features of social media itself can contribute to better RJP. Jong et al. [42] suggest that aspects like easy access and information-sharing tools significantly influence work efficiency. This efficiency can translate to improved completion of routine tasks. The ability to share knowledge through social media is particularly valuable for routine tasks. Studies by [48,50] on faculty members demonstrate this. Sharing information, assigning tasks, and collaborating on projects through social media can streamline routine job processes for various professions, therefore:

**H3b.** *Cognitive use of social media (CU) is positively associated with routine job performance (RJP).*

### *2.3. Relationship between Social Media Use and Internal Communication (H4)*

Several scholars have explored the concept of internal communication. Ali and Anwar [51] and Anwar and Abdullah [52] define it as the sharing of ideas, data, and knowledge among multiple individuals with the goal of reaching a consensus. Within organizations, effective communication is considered a crucial management practice [53]. It is seen as a key way for employees at all levels to learn about their roles and responsibilities [54]. Abdullah and Rahman [53] further describe internal communication as the exchange of meanings and interactions that occur within an organization.

The rise of social media offers a transformative approach to internal communication (IC) within organizations. Traditionally, internal communication relied on top-down methods or limited channels for employee interaction. However, social media platforms hold the potential to revolutionize how employees connect and engage with one another [55]. Employee communication fosters the social and emotional benefits crucial for organizational success [56]. Social media platforms can facilitate a more informal and interactive communication style, fostering a sense of community and belonging. This can lead to improved employee morale, engagement, and overall well-being. Social media can promote both horizontal communication (between colleagues) and vertical communication (between leadership and employees) [57]. Features like groups, discussions, and employee recognition programs can foster stronger relationships between coworkers, leading to better collaboration and knowledge sharing.

Social media platforms can encourage a more transparent and open communication environment. Employees can readily access information, participate in discussions, and provide feedback through social media channels. This two-way communication can foster trust and a sense of shared purpose within the organization. Therefore, we hypothesize:

**H4a.** *Social use of social media (SU) is positively associated with internal communication (IC).*

The influence of hedonistic social media use (focusing on enjoyment and entertainment) on organizational communication has been a topic of debate, with research suggesting both positive and negative effects [44]. While concerns exist about reduced productivity due to distractions, social media platforms can also be valuable tools for enhancing internal communication. They can facilitate information sharing, announcements, and team collaboration, fostering a sense of community and employee engagement [58–60].

It is important to acknowledge that excessive use for personal enjoyment might lead to information overload and hinder communication effectiveness [61,62]. However, research by Sun and Chao [63] suggest that hedonic social media use can be a positive force for internal communication when used strategically and balanced with work-related communication practices. Consequently, we recommend the following hypothesis:

**H4b.** *There is a relationship between hedonic use of social media (HU) and internal communication (IC).*

Employers can leverage social media by promoting scouting behavior, which refers to the voluntary search, exchange, and transmission of information related to organizational interests [44]. Studies suggest that businesses use social media platforms to meet the individual communication needs of employees, ensure a smooth flow of information, and provide ongoing feedback on both personal and organizational matters [64]. However, social media can also affect the information shared, privacy settings, and communication styles among team members, potentially favoring certain groups over others [65]. Similarly, social networking platforms are used by some instructors and students in higher education for both personal and professional purposes [66]. Cognitive use of social media can enhance internal communication when aligned with organizational goals and established communication practices, therefore, we hypothesize:

**H4c.** *Cognitive use of social media (CU) is positively associated with internal communication (IC).*

### 2.4. Relationship between Social Media Use and Teamwork (H5)

The success of many endeavors hinges on the effectiveness of teamwork. Teams, composed of individuals with complementary skillsets, synergistically work towards achieving shared goals [67]. Collaboration often leads to superior outcomes compared to individual efforts [67]. Furthermore, effective teamwork fosters a robust and flourishing work environment [68]. The rise of social media (SM) presents a novel avenue for exploring communication and collaboration within teams.

Research suggests that social media can be a valuable tool for enhancing teamwork. Song et al. [69] highlight its role as a complementary resource that creates synergies to improve both individual and team performance. The increasing use of social media by businesses and individuals, for both professional and personal purposes [32], offers opportunities for improved collaboration. In particular, studies suggest that the combined use of social media for work and social interaction can be beneficial. Song et al. [69] found that social media-facilitated social interaction becomes a regular part of some employees' jobs and can even enhance and streamline work processes. Similarly, Lailiyah and Putra's [70] research on higher education students found positive perceptions of teamwork through social media use. Based on the potential benefits of social media for fostering interaction and collaboration, we propose the following hypothesis:

**H5a.** *Social use of social media (SU) is positively associated with teamwork (TM).*

The integration of leisure and work facilitated by social media features has been linked to changes in teamwork patterns [71]. For example, Bodhi et al. [72] found that restricting personal social media use in the office could lead to unintended consequences, such as reduced employee contributions. This suggests that some level of social media integration might be beneficial for teamwork. However, the impact of hedonistic use (focusing on entertainment and leisure) on teamwork appears more complex. While Dodokh [73] suggests that such use solely has negative impacts, like increased job burnout and turnover intention, other research points to potential benefits. For instance, social media can facilitate informal interactions and build rapport among team members, which could indirectly contribute to better teamwork [74]. Given this nuanced relationship, we propose the following hypothesis:

**H5b.** *There is a relation between hedonic use of social media (HU) and teamwork (TM).*

Organizations increasingly leverage social media due to its potential for knowledge creation and dissemination within and across teams [75]. This user-generated content can be a valuable resource for teams. Research suggests that knowledge sharing in teams is more effective when it is visible, persistent, and easily editable, characteristics facilitated by social media platforms [76]. Furthermore, Kadar et al. [77] found that participation in online groups through social media helps employees express their ideas and gain a deeper understanding of their work. This collaborative knowledge-building can be a significant asset for teamwork. Based on the potential for enhanced knowledge sharing and collaboration, we propose the following hypothesis:

**H5c.** *Cognitive use of social media (CU) is positively associated with teamwork (TM).*

### *2.5. Mediating Effects (H6–H7)*

#### 2.5.1. Mediating Effect of Internal Communication

Internal communication (IC) is a well-established factor influencing employee performance [78,79]. Experts like Meng and Berger [80] highlight its key functions: informing employees about organizational goals, fostering a sense of purpose, facilitating collaboration, and ultimately enhancing performance for all stakeholders. Research by Jacobs et al. [81] further emphasizes the critical role of IC in maximizing worker performance, optimizing network efficiency, and communicating strategic options, particularly during crises.

Qin and Men [82] highlight that effective internal communication (IC) fosters a range of positive outcomes within an organization. It strengthens collaboration, facilitates the transfer of processes and values, and ultimately empowers employees to achieve exceptional performance [83]. This link between IC and performance suggests that IC can be considered a core internal capability that contributes to organizational success [84].

The question this study explores is whether IC acts as a mediator in the relationship between social media use (SMU) and job performance. A mediator is a variable that explains how one variable (SMU in this case) influences another (job performance). In this context, we propose that social media can facilitate communication and collaboration, potentially improving IC, which in turn, could lead to enhanced job performance. Based on the potential mediating role of internal communication, we propose the following hypotheses:

**H6a1.** *Internal communication (IC) mediates the relationship between social use of social media (SU) and innovative job performance (IJP).*

**H6a2.** *Internal communication (IC) mediates the relationship between social use of social media (SU) and routine job performance (RJP).*

**H6b1.** *Internal communication (IC) mediates the relationship between hedonic use of social media (HU) and innovative job performance (IJP).*

**H6b2.** *Internal communication (IC) mediates the relationship between hedonic use of social media (HU) and routine job performance (RJP).*

**H6c1.** *Internal communication (IC) mediates the relationship between cognitive use of social media (CU) and innovative job performance (IJP).*

**H6c2.** *Internal communication (IC) mediates the relationship between cognitive use of social media (CU) and routine job performance (RJP).*

#### 2.5.2. Mediating Effect of Teamwork

Effective teamwork is essential for workplace success. Collaboration within teams fosters motivation, generates creative ideas, and leverages the strengths of diverse individuals to achieve common goals [85]. Teamwork success can be measured by achieving predetermined objectives [86]. Research by McEwan et al. [87] further highlights a positive correlation between teamwork and individual employee performance. This study examines whether teamwork acts as a mediator in the relationship between social media use (SMU) and job performance. Social media can potentially facilitate communication and collaboration within teams. If social media use enhances teamwork, this improved teamwork could then lead to better job performance. Based on the potential mediating role of teamwork, we propose the following hypotheses:

**H7a1.** *Teamwork (TM) mediates the relationship between social use of social media (SU) and innovative job performance (IJP).*

**H7a2.** *Teamwork (TM) mediates the relationship between social use of social media (SU) and routine job performance (RJP).*

**H7b1.** *Teamwork (TM) mediates the relationship between hedonic use of social media (HU) and innovative job performance (IJP).*

**H7b2.** *Teamwork (TM) mediates the relationship between hedonic use of social media (HU) and routine job performance (RJP).*

**H7c1.** *Teamwork (TM) mediates the relationship between cognitive use of social media (CU) and innovative job performance (IJP).*

**H7c2.** *Teamwork (TM) mediates the relationship between cognitive use of social media (CU) and routine job performance (RJP).*

### *2.6. Mediators and Job Performance (H8–H10)*

#### 2.6.1. Internal Communication and Job Performance

Internal communication is a cornerstone of successful job performance, influencing how employees understand expectations, collaborate with colleagues, and ultimately contribute to organizational goals [78]. Innovative job performance, on the other hand, reflects an employee's ability to develop and implement new ideas for products, services, processes, or business models, ultimately leading to a competitive advantage for the organization [88]. Research suggests that internal communication can enhance a company's innovative capabilities by fostering collaboration and knowledge sharing among employees [89]. Effective communication allows employees to share insights, learn from each other's expertise, and build upon existing knowledge, which can spark innovative ideas.

However, the impact of social media on this relationship is not entirely clear. While some studies (e.g., Soares et al. [90]) suggest that social media may not directly influence innovative behavior, others (e.g., Amalina and Pusparini, [91]) highlight a positive link between social media use and internal communication. Taking these mixed findings into account, we propose the following hypothesis:

**H8a.** *Internal communication (IC) is positively associated with innovative job performance (IJP).*

Effective internal communication fosters a supportive work environment and empowers employees to perform their regular tasks efficiently, transparency and open communication channels can lead to higher employee satisfaction and better performance [82,92].

Research also suggests that knowledge-sharing practices and clear communication contribute positively to routine job performance [93,94]. Based on the link between effective communication and routine job performance, we propose the following hypothesis:

**H8b.** *Internal communication (IC) is positively associated with routine job performance (RJP).*

#### 2.6.2. Teamwork and Job Performance

Diversity in knowledge and skills within a team is a well-established predictor of creativity [95]. However, simply having a diverse team is not enough. To harness the full potential of this diversity, teams need effective processes and collaboration skills. This is where teamwork comes in. Strong teamwork fosters a positive team climate, characterized by trust, psychological safety, and a shared focus on innovation [96,97]. In such an environment, team members feel comfortable sharing ideas, taking risks, and building upon each other's contributions, which ultimately leads to greater innovation [98,99]. Building on the link between teamwork and a positive climate for innovation, we propose the following hypothesis:

**H9a.** *Teamwork (TW) is positively associated with innovative job performance (IJP).*

Teamwork plays a significant role in enhancing routine job performance. Studies suggest that collaboration fosters information sharing and knowledge transfer among team members, allowing them to learn from each other and improve their efficiency on routine tasks [98,100]. Furthermore, strong teamwork involves well-established processes like communication, coordination, and cooperation [101]. These processes ensure tasks are clearly defined, responsibilities are shared effectively, and team members can support each other when needed, ultimately leading to smoother completion of routine tasks. Building on the evidence of how teamwork facilitates knowledge sharing, communication, and coordination, we propose the following hypothesis:

**H9b.** *Teamwork (TW) is positively associated with routine job performance (RJP).*

#### 2.6.3. Internal Communication and Teamwork

Internal communication and teamwork are intertwined concepts that contribute significantly to organizational success. Clear communication of goals and objectives is crucial for ensuring team members are aligned with each other and the broader organizational vision [102]. When employees understand the organization's goals, it fosters a sense of common purpose and motivates collaboration within teams [103]. Effective internal communication, characterized by open and transparent channels, facilitates the flow of information among team members [104]. This timely and relevant information allows teams to coordinate efforts, solve problems collaboratively, and ultimately work together more effectively. Research by Sari, Indrajaya, and Nurminingsih [105] further supports this notion, highlighting the positive influence of internal communication on teamwork. Based on the connection between clear communication and effective collaboration, we propose the following hypothesis:

**H10.** *Internal communication (IC) is positively associated with teamwork (TW).*

Considering the literature review above, Figure 2 illustrates the conceptual model, which shows the connections between the constructs and the proposed relationships.

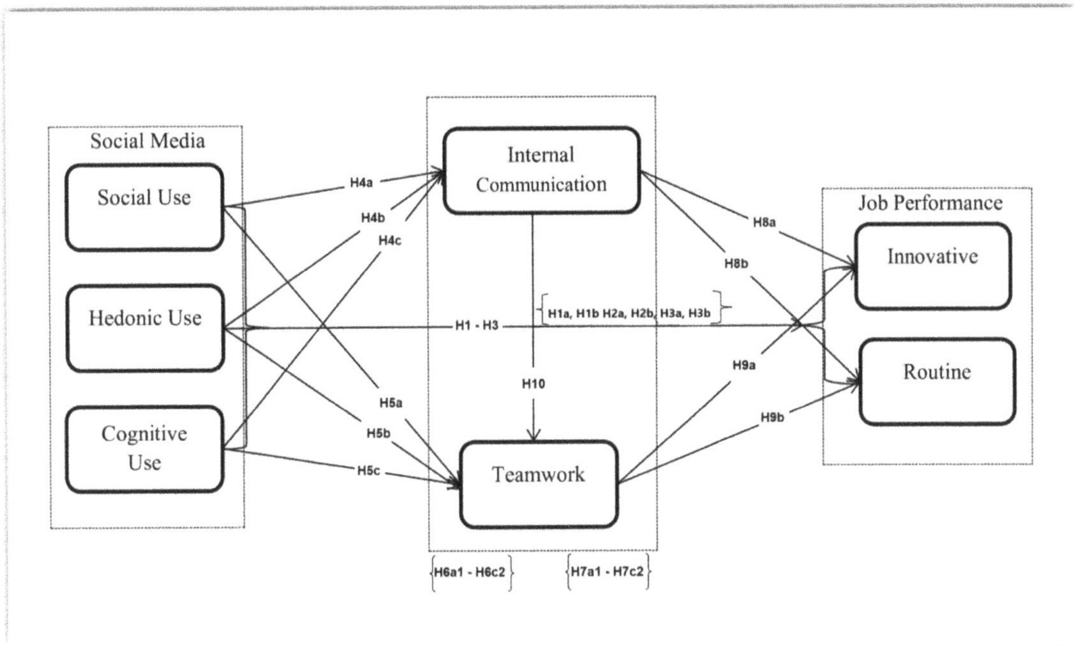


**Figure 2.** Conceptual model.

## 3. Research Methodology

### 3.1. Data Sampling and Collection

We collected data through a Google Form-based online survey distributed to teaching faculties of selected universities in northern India, using convenience and snowball sampling methods. We aimed to collect data from at least 450 participants, considering a minimum sample size of 384 determined through a power analysis. The power analysis, conducted using G*Power, aimed to detect a medium effect size of 0.5 (Cohen's d) with a desired power of 0.8 and a significance level of 0.05. A total of 550 surveys were distributed, resulting in a high response rate of 86% (478 completed surveys). However, during data refinement, 22 responses were excluded due to missing or contradictory data. Only surveys with complete responses for the final set of questions were included in the final analysis (456).

To address potential biases, such as self-selection bias, we employed several procedural measures. First, participants were assured of the survey's anonymity and encouraged to provide honest responses, emphasizing that the information would only be used for research purposes. Second, the cover page avoided mentioning any connections between the study variables to psychologically isolate the respondents. Finally, to further reduce participants' perception of connections, the survey asked about social media usage first, followed by questions on teamwork (TW), internal communication (IC), and job performance.

A self-reported questionnaire was developed to measure the key constructs of the study. The questionnaire employed a 5-point Likert scale ranging from 1 (Strongly Disagree) to 5 (Strongly Agree) for all items. The data collection for this study was conducted during Spring 2024.

To ensure the clarity and effectiveness of the questionnaire, we conducted a pilot study with a small group of 12 teaching faculty members before the main data collection phase. This pilot testing involved administering the survey to the pilot group and then analyzing the responses. The pilot testing helped us identify any issues with the survey instrument, such as unclear questions, ambiguous wording, or problematic flow. Based on the pilot study results, we refined the questionnaire by removing unclear questions or modifying the wording for better clarity. This iterative process ensured that the final survey instrument used for data collection was well-understood and easy for participants to complete.

### 3.2. Measurement

Most constructs' measurement scales were modified versions of previously used ones. Minor adjustments were made to the existing Janssen and Yperen [30] scales for innovative and routine job performance to fit the self-reported setup of this research. It is noteworthy that self-reporting of creative and/or standard job performance occurs frequently [41,106,107], especially when confidentiality concerns prevent access to objective performance measures. Researchers have not found the ideal metric for assessing an individual's performance, and self-reported data is no less accurate than other types of data [41].

The social use of the SM scale was borrowed from Ali-Hassan et al. [11], and the measurement scale for cognitive use was derived from social interaction ties found in Chiu et al., [108]. The knowledge-sharing instrument developed by Van den Hooff and Huysman [109] and the purpose to share implied knowledge scale developed by [110] informed the cognitive use component of SM. The hedonic use of SM items was inspired by Nevo and Nevo's [111] scale for using virtual environments and Agarwal and Karahanna's [112] scale for increased enjoyment of using the web. Measurement scales for measuring internal communication were adopted from Lee, Park and Lee [113], and for measuring teamwork, items were drawn from Jiang and Chen [61]. Table 1 presents the measurement items.

**Table 1.** Measurement items.

| Construct | Items | Source |
|---|---|---|
| Social Use of Social Media (SU) | Make new connections at work | [11] |
| | Get to know coworkers who have similar interests to me | |
| | Learn about professional opportunities from colleagues on social media | |
| | Connect with colleagues from different departments on social media | |
| | Build relationships with mentors or potential collaborators on social media | |
| Hedonic Use of Social Media (HU) | Use social media to appreciate my pause at work | [111,112] |
| | Take a break from my job by browsing social media for entertainment | |
| | Amuse myself on social media during my work breaks | |
| | Relax at my workstation by using social media for lighthearted interaction or browsing | |
| Cognitive Use of Social Media (CU) | Share documents, presentations, or other work-related content with colleagues on social media | [109,110] |
| | Use social media to stay updated on information and resources shared by colleagues | |
| | Collaborate with colleagues on creating content for work-related projects using social media | |
| | Post questions or requests for information on social media to leverage colleagues' expertise | |
| | Use social media to identify and learn from best practices shared by colleagues | |
| Internal Communication (IC) | I use social media to communicate with my coworkers | [113] |
| | Social media is an effective tool for communicating within our organization | |
| | Our organization's social media policies support effective internal communication | |
| | Social media has improved my ability to collaborate with colleagues | |
| | Social media has increased my awareness of organizational events and news | |
| Teamwork (TW) | Team members use social media to openly share information and knowledge with each other | [61] |
| | Participants of the team use SM to keep each other informed about progress | |
| | Members of the team work together to coordinate tasks and resources through social media | |
| | Members of the team respect each other's opinions and ideas on social media | |

**Table 1.** *Cont.*

| Construct | Items | Source |
|---|---|---|
| Innovative Job Performance (IJP) | Generate new and original ideas for improving work processes or products | [30] |
| | Champion creative ideas proposed by yourself or others | |
| | Seek out unconventional approaches to solve problems | |
| | Develop practical applications for creative ideas | |
| | Provide unique solutions to challenges faced at work | |
| Routine Job Performance (RJP) | I consistently finish the tasks outlined in my job description | |
| | I consistently fulfill all of my job's formal performance standards | |
| | I prioritize completing tasks outlined in my job description | |
| | I am able to meet all deadlines associated with my work responsibilities | |

To address potential common method bias arising from the self-reported survey data, we employed factor analysis using AMOS software. Specifically, Harman's one-factor test was conducted to assess the explained variance by a single common factor.

Confirmatory factor analysis (CFA) was performed using AMOS 24.0 to assess the measurement model fit. This analysis evaluated the reliability and validity of the constructs measured in our survey instrument. Various fit indices, including the Comparative Fit Index (CFI), Tucker–Lewis Index (TLI), Root Mean Square Error of Approximation (RMSEA), and Standardized Root Mean Square Residual (SRMR), were employed to assess the overall model fit and the fit of the measurement model. Thresholds for acceptable fit were established based on prior research [114].

Structural Equation Modeling (SEM) was then conducted using AMOS to test the proposed model. SEM allowed for the evaluation of the hypothesized relationships between the research variables. The analysis included assessing direct and indirect effects, mediation analysis, and testing for partial and serial mediation.

## 4. Results

### 4.1. Sample Demographics

Table 2 shows that male participants (55.5%) outnumbered female respondents (44.5%). This finding is fairly near to the sex distribution of Indian workers. Additionally, the majority of respondents (45.6%) were between the ages of 31 and 40. The majority of participants (50.7%) were assistant professors and worked as regular employees (67.5%). The respondents' demographic characteristics are presented in Table 2.

**Table 2.** Demographic profile.

| Measure | Item | Frequency | Percentage |
|---|---|---|---|
| Gender | Male | 253 | 55.5 |
| | Female | 203 | 44.5 |
| Marital Status | Single | 140 | 30.7 |
| | Married | 316 | 69.3 |
| Age | 21–30 | 68 | 14.9 |
| | 31–40 | 208 | 45.6 |
| | 41–50 | 137 | 30.0 |
| | 50 and above | 43 | 9.4 |
| Designation | Assistant Professor | 231 | 50.7 |
| | Associate Professor | 181 | 39.7 |
| | Professor | 44 | 9.6 |
| Experience | Less than 10 years | 242 | 53.07 |
| | 10–20 years | 176 | 38.5 |
| | More than 20 years | 38 | 8.3 |
| N (456) | | | |

### 4.2. Measurement Model Evaluation

Factor analysis was conducted to assess common method bias using Harman's one-factor test [115]. Only 12.825% of the total variance was explained by the first factor, well below the acceptable threshold of 40% [116]. This result suggests that common method bias was not a significant concern in our data.

While the chi-square test statistic ($\chi^2$) is often used in CFA, it can be sensitive to sample size, particularly in large samples like ours (456 participants). In our analysis, the chi-square value was 2.189 [117]. Therefore, we relied on alternative fit indices to provide a more robust assessment of model fit. The analysis yielded satisfactory results across all the alternative fit indices used. The Normed Fit Index (NFI) reached a value of 0.917, exceeding the recommended threshold of 0.90 [118]. The Comparative Fit Index (CFI) and the Tucker–Lewis Index (TLI) were both well above 0.90 (CFI = 0.953, TLI = 0.947) [118], further supporting a good model fit. Finally, the Root Mean Square Error of Approximation (RMSEA) fell below the recommended value of 0.08, with an actual value of 0.051 [118]. These results collectively suggest that the proposed model adequately represents the relationships between the study variables in our data.

To assess the internal consistency and reliability of our measures, we examined both alpha coefficients ($\alpha$) and composite reliability (C.R.) values. As shown in Table 3, all constructs achieved alpha coefficients and composite reliability estimates exceeding 0.7, which satisfy the established benchmarks for acceptable reliability [119,120].

**Table 3.** Outcomes of the measurement model.

| Variables | Items | Factor Loadings | CR | AVE | CA |
|---|---|---|---|---|---|
| Social Use of Social Media | SU1 | 0.827 | 0.902 | 0.649 | 0.897 |
| | SU2 | 0.778 | | | |
| | SU3 | 0.773 | | | |
| | SU4 | 0.726 | | | |
| | SU5 | 0.692 | | | |
| Hedonic Use of Social Media | HU1 | 0.826 | 0.910 | 0.717 | 0.908 |
| | HU2 | 0.816 | | | |
| | HU3 | 0.807 | | | |
| | HU4 | 0.726 | | | |
| Cognitive Use of Social Media | CU1 | 0.894 | 0.953 | 0.804 | 0.948 |
| | CU2 | 0.880 | | | |
| | CU3 | 0.864 | | | |
| | CU4 | 0.863 | | | |
| | CU5 | 0.809 | | | |
| Internal Communication | IC1 | 0.870 | 0.943 | 0.770 | 0.941 |
| | IC2 | 0.855 | | | |
| | IC3 | 0.821 | | | |
| | IC4 | 0.820 | | | |
| | IC5 | 0.808 | | | |
| Teamwork | TW1 | 0.816 | 0.910 | 0.717 | 0.905 |
| | TW2 | 0.811 | | | |
| | TW3 | 0.781 | | | |
| | TW4 | 0.715 | | | |
| Innovative Job Performance | IJP1 | 0.790 | 0.898 | 0.596 | 0.89 |
| | IJP2 | 0.787 | | | |
| | IJP3 | 0.670 | | | |
| | IJP4 | 0.658 | | | |
| | IJP5 | 0.594 | | | |
| Routine Job Performance | RJP1 | 0.786 | 0.842 | 0.572 | 0.838 |
| | RJP2 | 0.780 | | | |
| | RJP3 | 0.772 | | | |
| | RJP4 | 0.728 | | | |

Convergent validity refers to the extent to which a measure captures its intended construct. We evaluated convergent validity using average variance extracted (AVE), composite reliability (C.R.), factor loadings, and significance levels of factor loadings. [120]

suggest that acceptable convergent validity is indicated by a C.R. greater than 0.7, an AVE exceeding 0.5, statistically significant factor loadings ($p < 0.01$), and factor loadings with an absolute value greater than 0.7. The results presented in Table 3 demonstrate that all constructs meet these criteria for convergent validity.

In Table 4, the correlation analysis of seven constructs was conducted. The data suggest that the constructs are generally independent of one another because the square root of each AVE is bigger than its construct relationships.

**Table 4.** Square roots of AVEs.

| | IJP | SU | CU | HU | INC | TW | RJP |
|---|---|---|---|---|---|---|---|
| IJP | 0.772 | | | | | | |
| SU | 0.436 | 0.805 | | | | | |
| CU | 0.296 | 0.513 | 0.897 | | | | |
| HU | 0.520 | 0.573 | 0.347 | 0.847 | | | |
| INC | 0.535 | 0.407 | 0.293 | 0.485 | 0.877 | | |
| TW | 0.669 | 0.381 | 0.295 | 0.437 | 0.439 | 0.847 | |
| RJP | 0.301 | 0.471 | 0.528 | 0.365 | 0.372 | 0.373 | 0.756 |

The heterotrait–monotrait ratio of correlations (HTMT) matrix table values all fell below 0.85, thus supporting the discriminant validity of each construct. This result is presented in Table 5.

**Table 5.** Heterotrait–Monotrait ratio (HTMT).

| | SU | CU | HU | INC | TW | IJP | RJP |
|---|---|---|---|---|---|---|---|
| SU | | | | | | | |
| CU | 0.531 | | | | | | |
| HU | 0.589 | 0.354 | | | | | |
| INC | 0.412 | 0.303 | 0.497 | | | | |
| TW | 0.381 | 0.302 | 0.445 | 0.453 | | | |
| IJP | 0.463 | 0.304 | 0.531 | 0.536 | 0.689 | | |
| RJP | 0.478 | 0.534 | 0.369 | 0.382 | 0.370 | 0.315 | |

### 4.3. Validation of the Structural Model

Four statistical procedures were employed using CB-SEM with AMOS 24.0 to validate the hypothesized relationships in the structural model: mediation analysis, assessment of specific indirect effects, parallel mediation analysis to confirm individual mediator effects, and analysis of serial mediation to explore potential cascading effects between mediators.

The first step examined whether the independent variables (social, hedonic, and cognitive use of social media) exerted any direct effects on the dependent variable (innovative and routine job performance). The path coefficients in the model represent the total effects of social use, hedonic use, and cognitive use of social media on innovative and routine job performance, respectively (estimates: 0.198, 0.231, 0.418, 0.131, 0.087, and 0.416). All total effects were statistically significant ($p < 0.05$) except for the effects of social use ($p > 0.05$) and cognitive use ($p > 0.05$) on innovative job performance. Therefore, the results support all hypotheses H1–H3, but not H3a based on the total effects analysis.

In the second step, the hypothesized mediators (internal communication and teamwork) were incorporated into the path model alongside the direct effects. The results revealed that for the relationships between SU and IJP, HU and RJP, and CU and IJP, the introduction of the mediators explained a significant portion of the initial relationships. This is because the direct effects of the independent variables on the dependent variables became non-significant. This suggests that internal communication and teamwork likely play a mediating role in these specific relationships.

However, for the relationships between SU and RJP, HU and IJP, and CU and RJP, the introduction of the mediators did not fully explain the initial relationships. While the direct effects of the independent variables were somewhat reduced, they remained statistically significant. This suggests that internal communication and teamwork might partially explain these relationships, but there might be other factors at play.

The analysis revealed that social use (estimate: 0.177), hedonic use (estimate: 0.384), and cognitive use (estimate: 0.103) of social media directly influence internal communication, which in turn, impacts innovative (estimate: 0.217) and routine job performance (estimate: 0.130). Similarly, social use (estimate: 0.118), hedonic use (estimate: 0.228), and cognitive use (estimate: 0.100) of social media directly influence teamwork, which in turn, impacts innovative (estimate: 0.478) and routine job performance (estimate: 0.136). Additionally, the analysis found a significant direct effect of internal communication on teamwork (estimate: 0.264).

Most hypothesized relationships were statistically significant ($p < 0.05$), supporting hypotheses H4a through H5c (except H5a), H8a and H8b, and H9a through H10. The effect of social use of social media on teamwork (H5a) was not statistically significant ($p > 0.05$). This finding suggests no mediation for the relationship between social media use and teamwork (H5a), while the remaining significant relationships with internal communication (INC) as a mediator likely involve partial mediation (SU ---> INC ---> IJP, HU ---> INC ---> RJP, and CU ---> INC ---> IJP). The detailed results of the mediated effects analysis are presented in Table 6.

Figure 3 illustrates the strength and direction of the relationships between social media use (SU, HU, and CU), internal communication (INC), teamwork (TW), and job performance (IJP and RJP).

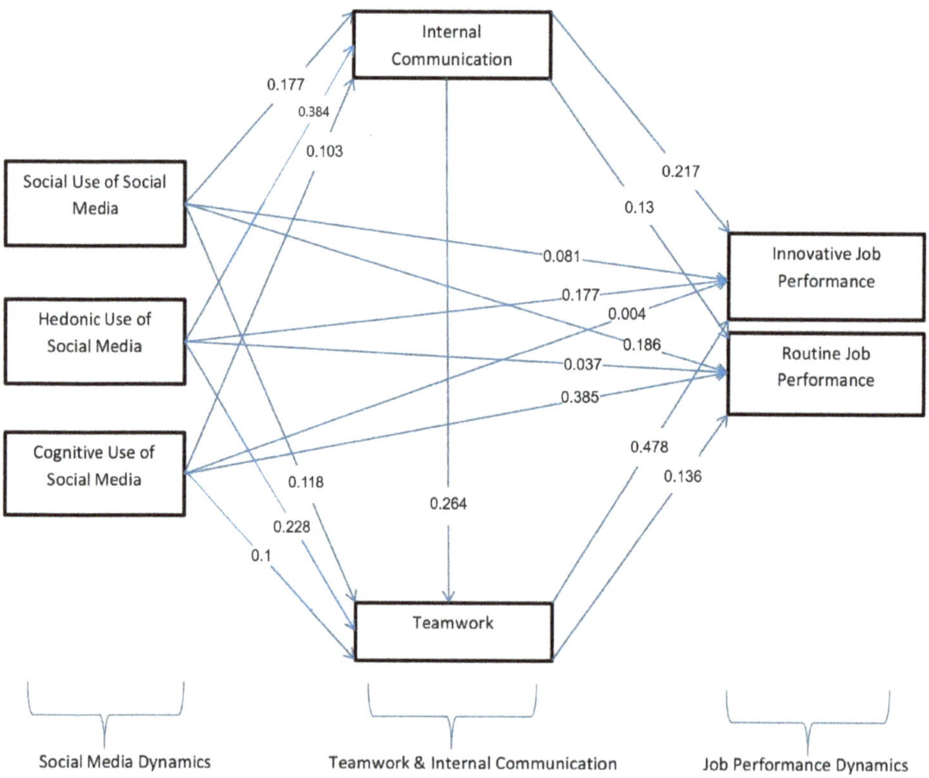


**Figure 3.** Path coefficients of the research model.

**Table 6.** Results of Parallel Mediation model.

| Path | Standardized Path Coefficients (β) | 95% Confidence Level (Lower Bound, Upper Bound) | Sig. Level | Hypothesis | Supported? |
|---|---|---|---|---|---|
| | | Total Effects | | | |
| SU ---> IJP | 0.198 | (0.056, 0.34) | 0.008 | H1a (total) | Yes |
| SU ---> RJP | 0.231 | (0.096, 0.352) | 0.002 | H1b (total) | Yes |
| HU ---> IJP | 0.418 | (0.293, 0.533) | 0.001 | H2a (total) | Yes |
| HU ---> RJP | 0.131 | (0.029, 0.25) | 0.014 | H2b (total) | Yes |
| CU ---> IJP | 0.087 | (−0.009, 0.182) | 0.074 | H3a (total) | No |
| CU ---> RJP | 0.416 | (0.296, 0.538) | 0.001 | H3b (total) | Yes |
| | | **Direct Effects** | | | |
| SU ---> IJP | 0.081 | (0.034, 0.198) | 0.191 | H1a (direct) | No |
| SU ---> RJP | 0.186 | (0.054, 0.307) | 0.004 | H1b (direct) | Yes |
| HU ---> IJP | 0.177 | (0.054, 0.3) | 0.006 | H2a (direct) | Yes |
| HU ---> RJP | 0.037 | (0.08, 0.17) | 0.547 | H2b (direct) | No |
| CU ---> IJP | 0.004 | (0.087, 0.093) | 0.934 | H3a (direct) | No |
| CU ---> RJP | 0.385 | (0.262, 0.509) | 0.001 | H3b (direct) | Yes |
| SU ---> INC | 0.177 | (0.058, 0.294) | 0.004 | H4a | Yes |
| HU ---> INC | 0.384 | (0.258, 0.497) | 0.001 | H4b | Yes |
| CU ---> INC | 0.103 | (−0.016, 0.194) | 0.02 | H4c | Yes |
| SU ---> TW | 0.118 | (0.003, 0.233) | 0.057 | H5a | No |
| HU ---> TW | 0.228 | (0.088, 0.359) | 0.002 | H5b | Yes |
| CU ---> TW | 0.1 | (0.012, 0.2) | 0.03 | H5c | Yes |
| INC ---> IJP | 0.217 | (0.116, 0.328) | 0.001 | H8a | Yes |
| INC ---> RJP | 0.13 | (0.024, 0.231) | 0.016 | H8b | Yes |
| TW ---> IJP | 0.478 | (0.353, 0.592) | 0.001 | H9a | Yes |
| TW ---> RJP | 0.136 | (0.022, 0.24) | 0.025 | H9b | Yes |
| INC ---> TW | 0.264 | (0.14, 0.364) | 0.002 | H10 | Yes |
| | | **Indirect Effects** | | | |
| SU ---> INC ---> IJP | 0.03 | (0.012, 0.064) | 0.002 | H6a1 | Yes (Full mediation) |
| SU ---> INC ---> RJP | 0.023 | (0.005, 0.055) | 0.008 | H6a2 | Yes (Partial Mediation) |
| HU ---> INC ---> IJP | 0.067 | (0.03, 0.114) | 0.001 | H6b1 | Yes (Partial Mediation) |
| HU ---> INC ---> RJP | 0.05 | (0.011, 0.1) | 0.01 | H6b2 | Yes (Full mediation) |
| CU ---> INC ---> IJP | 0.015 | (0.003, 0.033) | 0.014 | H6c1 | Yes (Full mediation) |
| CU ---> INC ---> RJP | 0.011 | (0.002, 0.03) | 0.014 | H6c2 | Yes (Partial Mediation) |
| SU ---> TW ---> IJP | 0.045 | (0.002, 0.096) | 0.043 | H7a1 | Yes (Partial Mediation) |
| SU ---> TW ---> RJP | 0.016 | (0.001. 0.046) | 0.032 | H7a2 | Yes (Partial Mediation) |
| HU ---> TW ---> IJP | 0.088 | (0.032, 0.166) | 0.001 | H7b1 | Yes (Partial Mediation) |
| HU ---> TW ---> RJP | 0.031 | (0.006, 0.075) | 0.014 | H7b2 | Yes (Partial Mediation) |
| CU ---> TW ---> IJP | 0.032 | (0.004, 0.067) | 0.029 | H7c1 | Yes (Partial Mediation) |
| CU ---> TW ---> RJP | 0.011 | (0.002, 0.035) | 0.022 | H7c2 | Yes (Partial Mediation) |

To examine the specific indirect effects of both mediators, a parallel mediation analysis was employed [121]. The results revealed significant indirect effects of social media use (social, hedonic, and cognitive) on innovative job performance through internal communication (b = 0.03, $p$ = 0.002; b = 0.067, $p$ = 0.001; b = 0.015, $p$ = 0.014, respectively). Similarly, significant indirect effects were found for social media use on innovative job performance through teamwork (b = 0.045, $p$ = 0.043; b = 0.088, $p$ = 0.001; b = 0.032, $p$ = 0.029, respectively).

Furthermore, the analysis revealed significant indirect effects of social media use on routine job performance, again mediated by internal communication (b = 0.023, $p$ = 0.008; b = 0.05, $p$ = 0.01; b = 0.011, $p$ = 0.014, respectively). Social media use also had a significant indirect impact on routine job performance through teamwork (b = 0.016, $p$ = 0.032; b = 0.031, $p$ = 0.014; b = 0.011, $p$ = 0.022, respectively). These findings suggest the presence of partial mediation in most relationships [122].

The significant direct effect of internal communication on teamwork (H10) prompted a follow-up hypothesis regarding a potential serial mediation process. This additional hypothesis was not initially considered because the focus was primarily on the direct influence of social media use on job performance, with internal communication seen as a single mediator. However, the unexpected finding of a strong relationship between internal communication and teamwork warranted further exploration of a multi-step mediation model. Serial mediation was examined to investigate the potential influence of internal communication on teamwork, following the alternative theory that emotions precede thoughts [123]. The results revealed significant effects for serial mediation, with detailed outcomes presented in Table 7.

**Table 7.** Result of serial mediation.

| Path | Standardized Path Coefficients (β) | 95% Confidence Level (Lower Bound, Upper Bound) | Sig. Level | Result |
|---|---|---|---|---|
| SU ---> INC ---> TW ---> IJP | 0.018 | (0.006, 0.037) | 0.002 | Significant |
| SU ---> INC ---> TW ---> RJP | 0.006 | (0.001, 0.017) | 0.01 | Significant |
| HU ---> INC ---> TW ---> IJP | 0.039 | (0.019, 0.066) | 0.001 | Significant |
| HU ---> INC ---> TW ---> RJP | 0.014 | (0.003, 0.033) | 0.014 | Significant |
| CU ---> INC ---> TW ---> IJP | 0.009 | (0.002, 0.019) | 0.014 | Significant |
| CU ---> INC ---> TW ---> RJP | 0.003 | (0, 0.008) | 0.018 | Significant |

## 5. Discussion

The study's findings offer a comprehensive understanding of how different dimensions of social media use influence job performance through the mediating roles of internal communication and teamwork.

Social Use (SU): The study found that social use of social media positively impacts routine job performance (RJP) both directly and indirectly through internal communication (INC) and teamwork (TW). Additionally, social use positively impacts innovative job performance (IJP) indirectly through INC and TW, although the direct effect is not significant. This finding highlights the importance of social interactions facilitated by social media in enhancing routine tasks. It suggests that social media can be a valuable tool for maintaining communication and collaboration, which are crucial for routine job performance. These results align with previous research indicating the positive effects of SU on workplace communication and collaboration [11,18].

Hedonic Use (HU): Hedonic use of social media, which involves using social media for pleasure and entertainment, was found to positively impact innovative job performance (IJP) both directly and indirectly through INC and TW. It also positively impacts routine job performance (RJP) indirectly through INC and TW, but the direct effect is not significant. This finding is particularly noteworthy because it contrasts with much of the existing literature, which often reports a negative relationship between hedonic use of social media and routine job performance. For instance, Ali-Hassan et al. [11] discuss how hedonic use generally has a negative impact on routine performance but can contribute positively to social ties and innovative performance. Similarly, Jong et al. [42] highlight the complex effects of social media use on work efficiency, noting that different types of use can have varying impacts. Kock and Moqbel [124] also suggest that positive emotions related to social media use can enhance job satisfaction and performance. Our study suggests that hedonic use can indeed have a beneficial effect, highlighting the complexity of social media's impact on work performance.

Cognitive Use (CU): The findings of this study highlight the significant role of cognitive use of social media in enhancing job performance. Specifically, cognitive engagement with social media positively impacts routine job performance (RJP) both directly and indirectly through improved internal communication (INC) and teamwork (TW). This aligns with previous research indicating that social media use can foster better communication and collaboration among employees, thereby enhancing their routine job performance [11,125]. Moreover, while cognitive use of social media positively influences innovative job performance (IJP) indirectly through INC and TW, the direct effect is not significant. This suggests that the benefits of cognitive engagement with social media on innovation are primarily mediated by enhanced communication and teamwork. These results are consistent with studies showing that social media can facilitate knowledge transfer and the formation of social capital, which are crucial for innovative performance [125]. Overall, the study extends the understanding of how cognitive engagement with social media can be leveraged to improve job performance. By using social media for information and knowledge sharing, employees can enhance their communication and teamwork, leading to better routine and innovative job performance. These insights are valuable for organizations aiming to optimize their social media strategies to boost employee performance.

The findings of this study underscore the significant mediating roles of Internal Communication (INC) and Teamwork (TW) in the relationship between social media use (SU, HU, and CU) and job performance (IJP and RJP). Both INC and TW are crucial in translating social media use into improved job performance. This aligns with the theory of parallel mediation, highlighting that effective communication and collaboration are key mechanisms through which social media use can enhance job performance [11,42].

The contrast comparison analysis revealed a stronger mediating effect of internal communication, particularly for the relationship between hedonic use (HU) and job performance. This finding extends existing knowledge by providing a more nuanced understanding of how social media use influences job performance. It suggests that internal communication is a critical factor in leveraging the benefits of hedonic social media use for job performance [126,127].

The study provides detailed insights into the pathways through which social media use influences job performance. Social use (SU) positively impacts both innovative job performance (IJP) and routine job performance (RJP) through improved internal communication and teamwork. Similarly, hedonic use (HU) enhances job performance through these mediators, with internal communication playing a particularly strong role. Cognitive use (CU) also positively influences job performance through the same pathways.

While the initial conceptual model focused on parallel mediation, the study also explored serial mediation. This analysis examined whether social media use influences job performance through a sequential process, where internal communication first affects teamwork, and then teamwork affects job performance. Interestingly, the results provided evidence supporting this unexpected serial mediation effect. This finding suggests that both internal communication and teamwork can play a role in influencing job performance but through different mediating pathways. This supports existing theories that emphasize the role of communication and collaboration in improving job performance through social media use [128,129].

Our findings align with those of Ali-Hassan et al. [11] and Jong et al. [42], who also identified internal communication and teamwork as key mediators in the relationship between social media use and job performance. These studies highlight the importance of effective communication and collaboration in leveraging social media for improved job outcomes. Furthermore, the stronger mediating effect of internal communication, particularly for hedonic use, is consistent with the findings of Moqbel et al. [126] and Fusi and Feeney [127], who emphasize that internal communication is crucial for translating the benefits of hedonic social media use into enhanced job performance.

Additionally, the unexpected serial mediation effect observed in our study, where internal communication influences teamwork, which in turn affects job performance, supports the theories proposed by [128,129]. These theories suggest that communication and collaboration are sequential processes that collectively enhance job performance through social media use.

## 6. Theoretical and Practical Implication

In this study, we explored whether and to what degree social media (SM) use influences routine and creative job performance. Our findings suggest that the context in which social media is used significantly influences job performance outcomes unique to the workplace. Specifically, using social media for knowledge exchange, entertainment, or socializing has different effects on work performance. We found that internal communication and teamwork are the mechanisms through which these changes occur. This discovery establishes a new practical connection between SM, internal communication, teamwork, and job performance.

Our results indicate that friendships among coworkers in a multi-person online relationship promote the development of strong bonds, which in turn open up opportunities for coworkers to communicate through comments. This can enhance teamwork and effective communication, leading to fewer misunderstandings. This finding is particularly relevant in today's increasingly digital work environments, where effective communication and collaboration are essential. Kamboj et al. [23] support this by highlighting the role of social media in fostering strong interpersonal relationships and reducing misunderstandings.

Additionally, our study contributes to the ongoing debate among management regarding the use of SM at work. While social media may negatively impact some jobs, such as routine ones, it can positively affect more innovative and creative roles. Our findings suggest that social media technologies when used appropriately, can improve employees' job performance. Ali-Hassan et al. [11] also found that social media use can enhance job performance, particularly in creative tasks. Employees must understand how social media contributes to internal communication and teamwork within their work environment culture, resulting in better job performance. This is supported by [130], who emphasizes the importance of understanding the cultural context in which social media is used.

In the context of universities, social media's advantages may include the ability to plan meetings, make appointments, exchange research papers, and share information about work activities with coworkers. University administration should clearly understand the kinds of work performance that are beneficial to them and plan their SM utilization according to their institutional culture. Çetinkaya and Sütçü [131] highlight the need for institutions to align social media use with their specific cultural and operational needs. By integrating concepts from the uses and gratification theory [22], our research provides a theoretical framework for understanding how people use SM at the workplace according to their institutional culture, as it differs from one institution to another.

Finally, we recommend legalizing the use of SM during work hours to plan work-related events and blur the distinction between business and social activities. These activities promote the teamwork and efficient internal communication needed to achieve high levels of job performance on SM platforms. Managers can benefit from teaching staff how to use social media to maximize their professional potential. Ghorbanzadeh et al. [18] support this by demonstrating the benefits of social media training for enhancing professional capabilities.

## 7. Limitation and Future Research Direction

This study differs from others in that it is primarily concerned with assessing the work performance of teachers. Teaching is distinct from other professions it is not exclusively concerned with output or productivity measurements since it entails not only presenting knowledge but also managing classroom dynamics, evaluating student achievement, and promoting social and emotional growth. Because of this, the research's findings and conclusions are specific to the teaching profession and might not apply to other professions whose performance standards and work circumstances are very different. The study's limitations should be taken into account when evaluating the findings.

Additionally, the findings may be limited to the specific cultural and institutional context of northern India. Cultural norms, institutional policies, and regional educational practices can significantly influence how social media is used and its impact on job performance. Therefore, the generalizability of the results to other regions or types of institutions may be limited.

A longer-term study is required to fully comprehend the effects of social media on teachers' job performance, even though this one offers insightful information in this area. The relationship between social media use and job performance at work may alter over time, and studies conducted over a short period can miss these dynamic shifts. Researchers could see patterns, trends, and possible behavioral changes over time with a longitudinal approach, leading to a more thorough understanding of how social media affects teachers' performance over the course of their careers. Such studies could provide more reliable results and guide future policy.

To provide a more thorough explanation of the phenomena, future research should examine how social media use affects job performance using a variety of theoretical frameworks. Applying additional theories, such as the Job Demands–Resources (JD-R) Model, Media Richness Theory (MRT), or the Technology Acceptance Model (TAM), could provide fresh perspectives even though this study may have concentrated on particular models. These theories may shed light on the ways in which social media affects engagement, stress, motivation, and teamwork. Researchers can reveal more intricate links and complexities by utilizing a variety of theoretical viewpoints, which will enhance their ability to analyze the effects of social media on job performance.

## 8. Conclusions

The findings of this study illuminate the multifaceted impact of social media use on job performance, mediated by internal communication and teamwork. This research underscores the potential benefits of social media in enhancing both routine and innovative job performance through improved communication and collaboration among employees.

Organizations that restrict or ban social media use may inadvertently forgo these advantages, missing out on opportunities to foster better teamwork, knowledge sharing, and problem-solving capabilities. By understanding the nuanced effects of different types of social media use—social, hedonic, and cognitive—managers can develop informed policies that leverage these tools to enhance employee performance while mitigating potential downsides.

This study contributes to the broader discourse on the role of social media in the workplace, offering valuable insights for organizations aiming to optimize their social media strategies. By embracing the positive aspects of social media use, educational institutes can create a more connected, innovative, and productive workforce, ultimately driving better organizational outcomes. University administrators must understand how social media contributes to internal communication and teamwork within their work environment culture, which results in better job performance. University administration should clearly understand the kinds of work performance that are beneficial to them and plan their SM utilization according to their institutional culture.

While this study provides valuable insights into the impact of social media use on faculty job performance, several limitations should be acknowledged. Firstly, the study was conducted among faculty members at public state colleges in northern India. Consequently, the findings may not be generalizable to faculty in other regions or countries with different cultural, educational, and technological contexts. Secondly, the data were collected through an online survey, which relies on self-reported measures. This method may introduce biases such as social desirability bias, where respondents might overreport positive behaviors or underreport negative ones. Thirdly, the study employs a cross-sectional design, which captures data at a single point in time. This limits the ability to infer causality between social media use and job performance.

To build on the findings of this study, future research could explore several areas. Conducting longitudinal studies would help establish causal relationships between social media use and job performance, providing a clearer picture of how these dynamics evolve over time. Expanding the research to include faculty from different regions, countries, and types of educational institutions would enhance the generalizability of the findings. Additionally, examining other professional groups could provide insights into how social media use impacts job performance across various occupations and industries. Investigating other dimensions of social media use, such as professional networking, marketing, and advocacy, could offer a more comprehensive understanding of how these platforms influence job performance. Exploring the role of organizational culture, leadership styles, and individual differences in technology proficiency could also provide a more nuanced view of the factors that mediate the relationship between social media use and job performance.

**Author Contributions:** Conceptualization, S.K., Z.S., R.J., I.S. and R.R.-A.; methodology S.K., Z.S., R.J., I.S. and R.R.-A.; validation, S.K., Z.S., R.J., I.S. and R.R.-A.; formal analysis, S.K., Z.S., R.J., I.S. and R.R.-A.; data curation, S.K., Z.S., R.J., I.S. and R.R.-A.; writing—original draft preparation, S.K., Z.S., R.J., I.S. and R.R.-A.; writing—review and editing, R.R.-A. All authors have read and agreed to the published version of the manuscript.

**Funding:** This research received no external funding.

**Informed Consent Statement:** Not applicable.

**Data Availability Statement:** Data are available from authors upon reasonable request.

**Acknowledgments:** This paper is part of the project COST CA19130 FinAI—Fintech and Artificial Intelligence in Finance—Towards a Transparent Financial Industry.

**Conflicts of Interest:** The authors declare no conflicts of interest.

## References

1. Chugh, R.; Ruhi, U. Social media for tertiary education. In *Encyclopedia of Education and Information Technologies*; Tatnall, A., Ed.; Springer Nature: Cham, Switzerland, 2019. [CrossRef]
2. Carrigan, M. *Social Media for Academics*; Sage Publications Limited: Thousand Oaks, CA, USA, 2019. [CrossRef]
3. Cao, X.; Yu, L. Exploring the influence of excessive social media use at work: A three-dimension usage perspective. *Int. J. Inf. Manag.* **2019**, *46*, 83–92. [CrossRef]
4. Jain, R.; Kumar, S.; Sood, K.; Grima, S.; Rupeika-Apoga, R. A Systematic Literature Review of the Risk Landscape in Fintech. *Risks* **2023**, *11*, 36. [CrossRef]
5. Chang, T.S.; Hsiao, W.H. Time spent on social networking sites: Understanding user behavior and social capital. *Syst. Res. Behav. Sci.* **2014**, *31*, 102–114. [CrossRef]
6. Kaplan, A.M.; Haenlein, M. Users of the world, unite! The challenges and opportunities of Social Media. *Bus. Horiz.* **2010**, *53*, 59–68. [CrossRef]
7. Welch, M.; Jackson, P.R. Rethinking internal communication: A stakeholder approach. *Corp. Commun. Int. J.* **2007**, *12*, 177–198. [CrossRef]
8. Etter, M. Broadcasting, reacting, engaging–three strategies for CSR communication in Twitter. *J. Commun. Manag.* **2014**, *18*, 322–342. [CrossRef]
9. Leonardi, P.M.; Vaast, E. Social media and their affordances for organizing: A review and agenda for research. *Acad. Manag. Ann.* **2017**, *11*, 150–188. [CrossRef]
10. O'Leary, M.B.; Mortensen, M.; Woolley, A.W. Multiple team membership: A theoretical model of its effects on productivity and learning for individuals and teams. *Acad. Manag. Rev.* **2011**, *36*, 461–478. [CrossRef]
11. Ali-Hassan, H.; Nevo, D.; Wade, M. Linking dimensions of social media use to job performance: The role of social capital. *J. Strateg. Inf. Syst.* **2015**, *24*, 65–89. [CrossRef]
12. Vandeyar, T. The academic turn: Social media in higher education. *Educ. Inf. Technol.* **2020**, *25*, 5617–5635. [CrossRef]
13. Manca, S.; Ranieri, M. Yes for sharing, no for teaching!": Social media in academic practices. *Internet High. Educ.* **2016**, *29*, 63–74. [CrossRef]

14. Mou, Y. Presenting professorship on social media: From content and strategy to evaluation. *Chin. J. Commun.* **2014**, *7*, 389–408. [CrossRef]
15. Sobaih, A.E.E.; Moustafa, M.A.; Ghandforoush, P.; Khan, M. To use or not to use? Social media in higher education in developing countries. *Comput. Hum. Behav.* **2016**, *58*, 296–305. [CrossRef]
16. Knight, C.G.; Kaye, L.K. 'To tweet or not to tweet? a comparison of academics' and students' us-age of twitter in academic contexts. *Innov. Educ. Teach. Int.* **2016**, *53*, 145–155. [CrossRef]
17. Murphy, J. We Asked Teachers about Their Social Media Use. Some of Their Answers Surprised Us. 2018. Available online: https://mdreducation.com/2019/01/17/teachers-social-media-use/ (accessed on 25 January 2020).
18. Ghorbanzadeh, D.; Khoruzhy, V.I.; Safonova, I.V.; Morozov, I.V. Relationships between social media usage, social capital and job performance: The case of hotel employees in Iran. *Inf. Dev.* **2021**, *39*, 6–18. [CrossRef]
19. Ellison, N.B.; Gibbs, J.L.; Weber, M.S. The use of enterprise social network sites for knowledge sharing in distributed organizations. *Am. Behav. Sci.* **2015**, *59*, 103–123. [CrossRef]
20. Soni, V.D. Importance and strategic planning of team management. *Int. J. Innov. Res. Technol.* **2020**, *7*, 47–50.
21. Preacher, K.J.; Hayes, A.F. Asymptotic and Resampling Strategies for Assessing and Comparing Indirect Effects in Multiple Mediator Models. *Behav. Res. Methods* **2008**, *40*, 879–891. [CrossRef]
22. Katz, E.; Blumler, J.G.; Gurevitch, M. Uses and gratifications research. *Public Opin. Q.* **1973**, *37*, 509–523. [CrossRef]
23. Kamboj, S.; Kumar, V.; Rahman, Z. Social media usage and firm performance: The mediating role of social capital. *Soc. Netw. Anal. Min.* **2017**, *7*, 35–51. [CrossRef]
24. Quan-Haase, A.; Young, A.L. Uses and gratifications of social media: A comparison of Facebook and instant messaging. *Bull. Sci. Technol. Soc.* **2010**, *30*, 350–361. [CrossRef]
25. Whiting, A.; Williams, D. Why people use social media: A uses and gratifications approach. *Qual. Mark. Res. Int. J.* **2013**, *16*, 362–369. [CrossRef]
26. Sparrowe, R.T.; Liden, R.C.; Wayne, S.J.; Kraimer, M.L. Social networks and the performance of individuals and groups. *Acad. Manag. J.* **2001**, *44*, 316–325. [CrossRef]
27. Amabile, T.M.; Conti, R.; Coon, H.; Lazenby, J.; Herron, M. Assessing the work environment for creativity. *Acad. Manag. J.* **1996**, *39*, 1154–1184. [CrossRef]
28. Scott, S.G.; Bruce, R.A. Determinants of innovative behavior: A path model of individual innova-tion in the workplace. *Acad. Manag. J.* **1994**, *37*, 580–607. [CrossRef]
29. Kanter, R.M. Three Tiers for Innovation Research. *Commun. Res.* **1988**, *15*, 509–523. [CrossRef]
30. Janssen, O.; Van Yperen, N.W. Employees' goal orientations, the quality of leader-member ex-change, and the outcomes of job performance and job satisfaction. *Acad. Manag. J.* **2004**, *47*, 368–384. [CrossRef]
31. Suryanto, A.; Fitriati, R.; Natalia, S.I.; Oktariani, A.; Munawaroh, M.; Nurdin, N.; Ahn, Y.H. Study of working from home: The impact of ICT anxiety and smartphone addiction on lecturers at NIPA School of Administration on job performance. *Heliyon* **2022**, *8*, e11980. [CrossRef]
32. Ali, A.; Wang, H.; Khan, A.N. Mechanism to enhance team creative performance through social media: A transactive memory system approach. *Comput. Hum. Behav.* **2019**, *91*, 115–126. [CrossRef]
33. Rupeika-Apoga, R.; Wendt, S. FinTech Development and Regulatory Scrutiny: A Contradiction? The Case of Latvia. *Risks* **2022**, *10*, 167. [CrossRef]
34. Chauhan, R. Impact of social media usage on job performance and employee retention: Role of knowledge sharing and organizational commitment. *Glob. Bus. Organ. Excell.* **2023**, *43*, 19–34. [CrossRef]
35. Shang, Y.; Pan, Y.; Richards, M. Facilitating or inhibiting? The role of enterprise social media use in job performance. *Inf. Technol. People* **2023**, *36*, 2338–2360. [CrossRef]
36. Liu, X.; Zheng, B.; Liu, H. Understanding the social media interactivity paradox: The effects of social media interactivity on communication quality, work interruptions and job performance. *Inf. Technol. People* **2022**, *35*, 1805–1828. [CrossRef]
37. Zahmat Doost, E.; Zhang, W. The Effect of Social Media Use on Job Performance with Moderating Effects of Cyberloafing and Job Complexity. *Inf. Technol. People* **2024**, *37*, 1775–1801. [CrossRef]
38. Czerwinski, M.; Horvitz, E.; Wilhite, S. A Diary Study of Task Switching and Interruptions. In Proceedings of the SIGCHI Conference on Human Factors in Computing Systems, Vienna, Austria, 24–29 April 2004; ACM: New York, NY, USA, 2004; pp. 175–182. [CrossRef]
39. Newport, C. *Deep Work: Rules for Focused Success in a Distracted World*, 1st ed.; Grand Central Publishing: New York, NY, USA; Boston, MA, USA, 2016.
40. Mark, G.; Gudith, D.; Klocke, U. The Cost of Interrupted Work: More Speed and Stress. In Proceedings of the SIGCHI Conference on Human Factors in Computing Systems, Florence, Italy, 5–10 April 2008; ACM: New York, NY, USA, 2008; pp. 107–110. [CrossRef]
41. Teigland, R.; Wasko, M. Knowledge transfer in MNCs: Examining how intrinsic motivations and knowledge sourcing impact individual centrality and performance. *J. Int. Manag.* **2009**, *15*, 15–31. [CrossRef]
42. Jong, D.; Chen, S.-C.; Ruangkanjanases, A.; Chang, Y.-H. The Impact of Social Media Usage on Work Efficiency: The Perspectives of Media Synchronicity and Gratifications. *Front. Psychol.* **2021**, *12*, 693183. [CrossRef]

43. Scarpi, D. Work and Fun on the Internet: The Effects of Utilitarianism and Hedonism Online. *J. Interact. Mark.* **2012**, *26*, 53–67. [CrossRef]
44. Kim, J.N.; Rhee, Y. Strategic thinking about employee communication behavior (ECB) in public relations: Testing the models of megaphoning and scouting effects in Korea. *J. Public Relat. Res.* **2011**, *23*, 243–268. [CrossRef]
45. Raacke, J.; Bonds-Raacke, J. MySpace and Facebook: Applying the Uses and Gratifications Theory to Exploring Friend-Networking Sites. *CyberPsychol. Behav.* **2008**, *11*, 169–174. [CrossRef]
46. Sobaih, A.E.E.; Hasanein, A.M.; Abu Elnasr, A.E. Responses to COVID-19 in higher education: Social media usage for sustaining formal academic communication in developing countries. *Sustainability* **2020**, *12*, 6520. [CrossRef]
47. Murire, O.T.; Cilliers, L. Social media adoption among lecturers at a traditional university in East-ern Cape Province of South Africa. *S. Afr. J. Inf. Manag.* **2017**, *19*, 1–6. [CrossRef]
48. Nentwich, M.; König, R. Academia goes Facebook? The potential of social network sites in the scholarly realm. In *Opening Science: The Evolving Guide on How the Internet Is Changing Research, Collaboration and Scholarly Publishing*; Springer: Berlin/Heidelberg, Germany, 2014; pp. 107–124. [CrossRef]
49. Ndung'U, J.; Vertinsky, I.; Onyango, J. The Relationship between Social Media Use, Social Media Types, and Job Performance amongst Faculty in Kenya Private Universities. *Heliyon* **2023**, *9*, e22946. [CrossRef]
50. Chawinga, W.D. Taking social media to a university classroom: Teaching and learning using Twitter and blogs. *Int. J. Educ. Technol. High. Educ.* **2017**, *14*, 3. [CrossRef]
51. Ali, B.J.; Anwar, G. A Study of Knowledge Management Alignment with Production Management: A Study of Carpet Manufacture in Kurdistan Region of Iraq. *Int. J. Engl. Lit. Soc. Sci.* **2021**, *6*, 346–360. [CrossRef]
52. Anwar, G.; Abdullah, N.N. Inspiring future entrepreneurs: The effect of experiential learning on the entrepreneurial intention at higher education. *Int. J. Engl. Lit. Soc. Sci.* **2021**, *6*, 183–194. [CrossRef]
53. Abdullah, N.N.; Rahman, M.F.A. The Use of Deliberative Democracy in Public Policy Making Process. *Public Policy Adm. Res.* **2015**, *5*, 221–229. [CrossRef]
54. Ali, B.J.; Saleh, P.F.; Akoi, S.; Abdulrahman, A.A.; Muhamed, A.S.; Noori, H.N.; Anwar, G. Impact of Service Quality on the Customer Satisfaction: Case Study at Online Meeting Platforms. *Int. J. Eng. Bus. Manag.* **2021**, *5*, 65–77. [CrossRef]
55. Madsen, V.T. Internal Social Media and Internal Communication. In *Current Trends and Issues in Internal Communication: Theory and Practice*; Palgrave Macmillan: Cham, Switzerland, 2021; pp. 57–74. [CrossRef]
56. Fang, R.; McAllister, D.J.; Duffy, M.K. Down but not out: Newcomers can compensate for low vertical access with strong horizontal ties and favorable core self-evaluations. *Pers. Psychol.* **2017**, *70*, 517–555. [CrossRef]
57. Namhata, R.; Patnaik, P. The 'Verticals', 'Horizontals', and 'Diagonals' in Organisational Communication: Developing Models to Mitigate Communication Barriers Through Social Media Applications. In *Digital Business*; Patnaik, S., Yang, X.-S., Tavana, M., Popentiu-Vlădicescu, F., Qiao, F., Eds.; Lecture Notes on Data Engineering and Communications Technologies; Springer International Publishing: Cham, Switzerland, 2019; Volume 21, pp. 343–373. [CrossRef]
58. Ernst, C.-P.H. Hedonic and Utilitarian Motivations of Social Network Site Usage. In *Factors Driving Social Network Site Usage*; Springer Fachmedien Wiesbaden: Wiesbaden, Germany, 2015; pp. 11–28. [CrossRef]
59. Tamborini, R.; Grizzard, M.; Bowman, N.D.; Reinecke, L.; Lewis, R.J.; Eden, A. Media Enjoyment as Need Satisfaction: The Contribution of Hedonic and Nonhedonic Needs. *J. Commun.* **2011**, *61*, 1025–1042. [CrossRef]
60. Rupeika-Apoga, R.; Solovjova, I. Profiles of SMEs as Borrowers: Case of Latvia. *Contemp. Stud. Econ. Financ. Anal.* **2016**, *98*, 63–76. [CrossRef]
61. Jiang, Y.; Chen, C.C. Integrating Knowledge Activities for Team Innovation: Effects of Transfor-mational Leadership. *J. Manag.* **2018**, *44*, 1819–1847. [CrossRef]
62. Falco, A.; Girardi, D.; Marcuzzo, G.; De Carlo, A.; Bartolucci, G.B. Work Stress and Negative Affectivity: A Multi-Method Study. *Occup. Med.* **2013**, *63*, 341–347. [CrossRef] [PubMed]
63. Sun, W.; Chao, M. Exploring the Influence of Excessive Social Media Use on Academic Performance through Media Multitasking and Attention Problems: A Three-Dimension Usage Perspective. *Educ. Inf. Technol.* **2024**. [CrossRef]
64. Chau, H.T.H.; Bui, H.P.; Dinh, Q.T.H. Impacts of online collaborative learning on students' intercultural communication apprehension and intercultural communicative competence. *Educ. Inf. Technol.* **2023**, *29*, 7447–7464. [CrossRef]
65. Hajli, N.; Lin, X. Exploring the Security of Information Sharing on Social Networking Sites: The Role of Perceived Control of Information. *J. Bus. Ethics* **2016**, *133*, 111–123. [CrossRef]
66. Sabah, N.M. The impact of social media-based collaborative learning environments on students' use outcomes in higher education. *Int. J. Hum.–Comput. Interact.* **2023**, *39*, 667–689. [CrossRef]
67. Khawam, A.M.; DiDona, T.; Hernández, B.S. Effectiveness of teamwork in the workplace. *Int. J. Sci. Basic Appl. Res. (IJSBAR)* **2017**, *32*, 267–286. Available online: https://www.gssrr.org/index.php/JournalOfBasicAndApplied/article/view/7134 (accessed on 12 July 2024).
68. Schmutz, J.B.; Meier, L.L.; Manser, T. How Effective Is Teamwork Really? The Relationship between Teamwork and Performance in Healthcare Teams: A Systematic Review and Meta-Analysis. *BMJ Open* **2019**, *9*, e028280. [CrossRef]
69. Song, Q.; Wang, Y.; Chen, Y.; Benitez, J.; Hu, J. Impact of the usage of social media in the workplace on the team and employee performance. *Inf. Manag.* **2019**, *56*, 103160. [CrossRef]

70. Lailiyah, M.; Putra, S.P. Integrating the use of social media for group collaboration in ESP classroom. *Engl. Teach. J. J. Engl. Lit. Lang. Educ.* **2022**, *10*, 60–66. [CrossRef]
71. Thom, J.; Millen, D.; DiMicco, J. Removing gamification from an enterprise SNS. In Proceedings of the ACM 2012 Conference on Computer Supported Cooperative Work, Washington, DC, USA, 11–15 February 2012; pp. 1067–1070. [CrossRef]
72. Bodhi, R.; Luqman, A.; Hina, M.; Papa, A. Work-Related Social Media Use and Employee-Related Outcomes: A Moderated Mediation Model. *Int. J. Emerg. Mark.* **2023**, *18*, 4948–4967. [CrossRef]
73. Dodokh, A.M.I. Examining the Impact of Personal Social Media Use at Work on Workplace Outcomes. Ph.D. Thesis, University of Plymouth, Plymouth, UK, 2022. [CrossRef]
74. ÅžAHÄ, A.; Firat, A. The Effect of Social Media on Hedonic Consumption of Individuals. *Turk. J. Mark.* **2018**, *3*, 1–16.
75. Weber, M.S.; Shi, W. Enterprise social media. In *The International Encyclopedia of Organizational Communication*; Scott, C., Lewis, L., Eds.; Wiley Blackwell: Hoboken, NJ, USA, 2016; Volume 4, pp. 1–9. [CrossRef]
76. Dhar, S.; Bose, I. Corporate users' attachment to social networking sites: Examining the role of so-cial capital and perceived benefits. *Inf. Syst. Front.* **2023**, *25*, 1197–1217. [CrossRef]
77. Kadar, N.S.A.; Yadri, W.S.W.; Faris, N.D.; Johari, M.D.M.; Nurgeldiyevna, K.N.; Rahmat, N.H. Exploring Online Group Work Through The Social Cognitive Theory. *Int. J. Acad. Res. Bus. Soc. Sci.* **2023**, *13*, 1684–1698. [CrossRef]
78. Tourish, D.; Hargie, C. Internal communication: Key steps in evaluating and improving perfor-mance. *Corp. Commun. Int. J.* **1996**, *1*, 11–16. [CrossRef]
79. Rupeika-Apoga, R.; Saksonova, S. SMEs' Alternative Financing: The Case of Latvia. *Eur. Res. Stud. J. XXI* **2018**, 43–52. [CrossRef]
80. Meng, J.; Berger, B.K. Measuring return on investment (ROI) of organizations' internal commu-nication efforts. *J. Commun. Manag.* **2012**, *16*, 332–354. [CrossRef]
81. Jacobs, M.A.; Yu, W.; Chavez, R. The effect of internal communication and employee satisfaction on supply chain integration. *Int. J. Prod. Econ.* **2016**, *171*, 60–70. [CrossRef]
82. Qin, Y.S.; Men, L.R. Exploring the Impact of Internal Communication on Employee Psychological Well-Being During the COVID-19 Pandemic: The Mediating Role of Employee Organizational Trust. *Int. J. Bus. Commun.* **2023**, *60*, 1197–1219. [CrossRef]
83. Men, L.R.; Yue, C.A. Creating a Positive Emotional Culture: Effect of Internal Communication and Impact on Employee Supportive Behaviors. *Public Relat. Rev.* **2019**, *45*, 101764. [CrossRef]
84. Danso, A.; Poku, K.; Agyapong, A. Mediating Role of Internal Communications in Market Orientation and Performance of Mobile Telecom Firms: Evidence from Ghana. *Cogent Bus. Manag.* **2017**, *4*, 1403713. [CrossRef]
85. Octavia, D.H.; Budiono, B. Pengaruh Teamwork terhadap Kinerja Karyawan melalui Job Satisfac-tion. *J. Ilmu Manaj.* **2021**, *9*, 954–965. [CrossRef]
86. Awalia, A.R.; Fania, D.; Setyaningrum, D.U. Pengaruh Teamwork Terhadap Kinerja Karyawan (Study Kasus Pada Pt. XYZ Jatinangor). *E-J. Equilib. Manaj.* **2020**, *6*, 12–19.
87. McEwan, D.; Ruissen, G.R.; Eys, M.A.; Zumbo, B.D.; Beauchamp, M.R. Beauchamp. The Effectiveness of Teamwork Training on Teamwork Behaviors and Team Performance: A Systematic Review and Meta-Analysis of Controlled Interventions. *PLoS ONE* **2017**, *12*, e0169604. [CrossRef]
88. Shin, S.J.; Yuan, F.; Zhou, J. When Perceived Innovation Job Requirement Increases Employee Innovative Behavior: A Sensemaking Perspective. *J. Organ. Behav.* **2017**, *38*, 68–86. [CrossRef]
89. Söderlund, M.; Öhman, P. Internal communication and innovation capability. *J. Bus. Commun.* **2016**, *53*, 207–224.
90. Soares, M.E.; Mosquera, P.; Cid, M. Antecedents of innovative behaviour: Knowledge sharing, open innovation climate and internal communication. *Int. J. Innov. Learn.* **2021**, *30*, 241–257. [CrossRef]
91. Amalina, H.; Pusparini, E.S. Influence of Social Media Communication on Employee Innovative Work Behavior: Mediating Role of Work Engagement. In *7th Global Conference on Business, Management, and Entrepreneurship (GCBME 2022)*; Atlantis Press: Amsterdam, The Netherlands, 2023; pp. 1518–1526. [CrossRef]
92. Zaidi, S.H.; Rupeika-Apoga, R. Liquidity Synchronization, Its Determinants and Outcomes under Economic Growth Volatility: Evidence from Emerging Asian Economies. *Risks* **2021**, *9*, 43. [CrossRef]
93. Kuo, H.; Lee, T.; Fang, C. Free Trade and Economic Growth. *Aust. Econ. Pap.* **2014**, *53*, 69–76. [CrossRef]
94. Baralou, E.; Dionysiou, D.D. Routine dynamics in virtual teams: The role of technological artifacts. *Inf. Technol. People* **2022**, *35*, 1980–2001. [CrossRef]
95. West, M.A. Sparkling fountains or stagnant ponds: An integrative model of creativity and innova-tion implementation in work groups. *Appl. Psychol.* **2002**, *51*, 355–387. [CrossRef]
96. Anderson, N.R.; West, M.A. Measuring climate for work group innovation: Development and validation of the team climate inventory. *J. Organ. Behav.* **1998**, *19*, 235–258. [CrossRef]
97. Johari, A.; Abdul Wahat, N.W.; Zaremohzzabieh, Z. Innovative work behavior among teachers in Malaysia: The effects of teamwork, principal support, and humor. *Asian J. Univ. Educ. (AJUE)* **2021**, *7*, 72–84. [CrossRef]
98. Ariefahnoor, D.; Nugroho, R. The effect of innovation, organizational learning, and teamwork on managerial and organizational performance of rural banking in the province of south kalimantan. *J. Mod. Proj. Manag.* **2022**, *10*, 134–143. [CrossRef]
99. Grima, S.; Rupeika-Apoga, R.; Kizilkaya, M.; Romānova, I.; Gonzi, R.D.; Jakovljevic, M. A Proactive Approach to Identify the Exposure Risk to COVID-19: Validation of the Pandemic Risk Exposure Measurement (PREM) Model Using Real-World Data. *Risk Manag. Healthc. Policy* **2021**, *14*, 4775–4787. [CrossRef]

100. Hansen, M.T. Knowledge networks: Explaining effective knowledge sharing in multiunit compa-nies. *Organ. Sci.* **2002**, *13*, 232–248. [CrossRef]
101. Phulpoto, N.H. Teamwork and its impact on employee performance mediated by job satisfaction: A comprehensive study of services sector of Pakistan. *J. Innov. Sustain. RISUS* **2023**, *14*, 21–31. [CrossRef]
102. Katzenbach, J.R.; Smith, D.K. *The Wisdom of Teams: Creating the High-Performance Organization*; Harvard Business Review Press: Cambridge, MA, USA, 1993.
103. Miller, K.; Barbour, J. *Organizational Communication: Approaches and Processes*; Cengage Learning: Belmont, CA, USA, 2019.
104. Mercader, V.; Galván-Vela, E.; Ravina-Ripoll, R.; Popescu, C.R.G. A focus on ethical value under the vision of leadership, teamwork, effective communication and productivity. *J. Risk Financ. Manag.* **2021**, *14*, 522. [CrossRef]
105. Sari, M.; Indrajaya, T.; Nurminingsih. The influence of internal communication and teamwork on employee performance in the microwave dismantle project division pt. panca karsa sejahtera bekasi city. *J. Humanit. Soc. Sci. Bus. (JHSSB)* **2023**, *3*, 1–13. [CrossRef]
106. Teigland, R.; Wasko, M.M. Integrating knowledge through information trading: Examining the re-lationship between boundary spanning communication and individual performance. *Decis. Sci.* **2003**, *34*, 261–286. [CrossRef]
107. Bakker, A.B.; Van Emmerik, H.; Van Riet, P. How job demands, resources, and burnout predict objective performance: A constructive replication. *Anxiety Stress Coping* **2008**, *21*, 309–324. [CrossRef] [PubMed]
108. Chiu, C.M.; Hsu, M.H.; Wang, E.T. Understanding knowledge sharing in virtual communities: An integration of social capital and social cognitive theories. *Decis. Support Syst.* **2006**, *42*, 1872–1888. [CrossRef]
109. Van den Hooff, B.; Huysman, M. Managing knowledge sharing: Emergent and engineering ap-proaches. *Inf. Manag.* **2009**, *46*, 1–8. [CrossRef]
110. Bock, G.W.; Zmud, R.W.; Kim, Y.G.; Lee, J.N. Behavioral intention formation in knowledge shar-ing: Examining the roles of extrinsic motivators, social-psychological forces, and organizational climate. *MIS Q.* **2005**, 87–111. Available online: https://www.jstor.org/stable/25148669 (accessed on 9 May 2024). [CrossRef]
111. Nevo, S.; Nevo, D. Re-invention of applicable innovations: The case of virtual worlds. In Proceedings of the 44th Hawaii International Conference on System Sciences, Kauai, HI, USA, 4–7 January 2011. [CrossRef]
112. Agarwal, R.; Karahanna, E. Time flies when you're having fun: Cognitive absorption and beliefs about information technology usage. *MIS Q.* **2000**, *24*, 665–694. [CrossRef]
113. Lee, S.; Park, J.G.; Lee, J. Explaining knowledge sharing with social capital theory in information systems development projects. *Ind. Manag. Data Syst.* **2015**, *115*, 883–900. [CrossRef]
114. Hu, L.T.; Bentler, P.M. Cutoff criteria for fit indexes in covariance structure analysis: Convention-al criteria versus new alternatives. *Struct. Equ. Model.* **1999**, *6*, 1–55. [CrossRef]
115. Podsakoff, P.M.; Organ, D.W. Self-reports in organizational research: Problems and prospects. *J. Manag.* **1986**, *12*, 531–544. [CrossRef]
116. Podsakoff, P.M.; MacKenzie, S.B.; Lee, J.Y.; Podsakoff, N.P. Common method biases in behavior-al research: A critical review of the literature and recommended remedies. *J. Appl. Psychol.* **2003**, *88*, 879. [CrossRef]
117. Bentler, P.M. *EQS Structural Equations Program Manual*; Multivariate Software: Los Angeles, CA, USA, 1995.
118. Hair, J.F. *Multivariate Data Analysis with Readings*, 4th ed.; Prentice Hall: Englewood Cliffs, NJ, USA, 1995.
119. Cronbach, L.J. Coefficient alpha and the internal structure of tests. *Psychometrika* **1951**, *16*, 297–334. [CrossRef]
120. Fornell, C.; David, F.L. Structural Equation Models with Unobservable Variables and Measurement Error: Algebra and Statistics. *J. Mark. Res.* **1981**, *18*, 382–388. [CrossRef]
121. Hayes, A.F. Process: A Versatile Computational Tool for Observed Variable Mediation, Moderation, and Conditional Process Modeling [White Paper]. 2012. Available online: http://www.afhayes.com/public/process2012.pdf (accessed on 11 August 2024).
122. Collier, J. *Applied Structural Equation Modeling Using AMOS: Basic to Advanced Techniques*; Routledge: London, UK, 2020. [CrossRef]
123. Seaver, F.A.; Cummings, T.G.; Molloy, E.S. *Improving Productivity and the Quality of Work Life*; Praeger: Oxford, UK, 1977.
124. Kock, N.; Moqbel, M. Social Networking Site Use, Positive Emotions, and Job Performance. *J. Comput. Inf. Syst.* **2021**, *61*, 163–173. [CrossRef]
125. Cao, X.; Guo, X.; Vogel, D.; Zhang, X. Exploring the Influence of Social Media on Employee Work Performance. Edited by Professor Pan Wang, Professor Sohail Chaudhry, Professor Ling Li. *Internet Res.* **2016**, *26*, 529–545. [CrossRef]
126. Moqbel, M.; Nevo, S.; Kock, N. Organizational Members' Use of Social Networking Sites and Job Performance: An Exploratory Study. *Inf. Technol. People* **2013**, *26*, 240–264. [CrossRef]
127. Fusi, F.; Feeney, M.K. Social Media in the Workplace: Information Exchange, Productivity, or Waste? *Am. Rev. Public Adm.* **2018**, *48*, 395–412. [CrossRef]
128. Leonardi, P.M.; Huysman, M.; Steinfield, C. Enterprise Social Media: Definition, History, and Prospects for the Study of Social Technologies in Organizations. *J. Comput.-Mediat. Commun.* **2013**, *19*, 1–19. [CrossRef]
129. Treem, J.W.; Leonardi, P.M. Social Media Use in Organizations: Exploring the Affordances of Visibility, Editability, Persistence, and Association. *SSRN Electron. J.* **2012**. [CrossRef]

130. Verheyden, M. Social Media and the Promise of Excellence in Internal Communication. *J. Organ. Ethnogr.* **2017**, *6*, 11–25. [CrossRef]
131. Çetinkaya, L.; Sütçü, S.S. The Effects of Facebook and WhatsApp on Success in English Vocabulary Instruction. *J. Comput. Assist. Learn.* **2018**, *34*, 504–514. [CrossRef]

MDPI AG
Grosspeteranlage 5
4052 Basel
Switzerland
Tel.: +41 61 683 77 34

*Big Data and Cognitive Computing* Editorial Office
E-mail: bdcc@mdpi.com
www.mdpi.com/journal/bdcc

www.ingramcontent.com/pod-product-compliance
Lightning Source LLC
Chambersburg PA
CBHW041018270926
51850CB00024B/1471
*9783725847693*